Lexicon of Pulse Crops

Lexicon of Pulse Crops

Aleksandar Mikić

with original color drawings by the author

CRC Press is an imprint of the
Taylor & Francis Group, an **informa** business

CRC Press
Taylor & Francis Group
6000 Broken Sound Parkway NW, Suite 300
Boca Raton, FL 33487-2742

Printed on acid-free paper

International Standard Book Number-13: 978-1-138-08943-3 (Paperback)
International Standard Book Number-13: 978-1-138-08951-8 (Hardback)

Library of Congress Cataloging-in-Publication Data

Names: Mikić, Aleksandar, author.
Title: Lexicon of pulse crops / author: Aleksandar Mikić.
Description: Boca Raton, FL : CRC Press, Taylor & Francis Group, 2018. | Includes bibliographical references.
Identifiers: LCCN 2018006068 | ISBN 9781138089433 (pbk.)
Subjects: LCSH: Legumes. | Legumes--Nomenclature.
Classification: LCC QK495.L52 M55 2018 | DDC 583/.63--dc23
LC record available at https://lccn.loc.gov/2018006068

Visit the Taylor & Francis Web site at
http://www.taylorandfrancis.com

and the CRC Press Web site at
http://www.crcpress.com

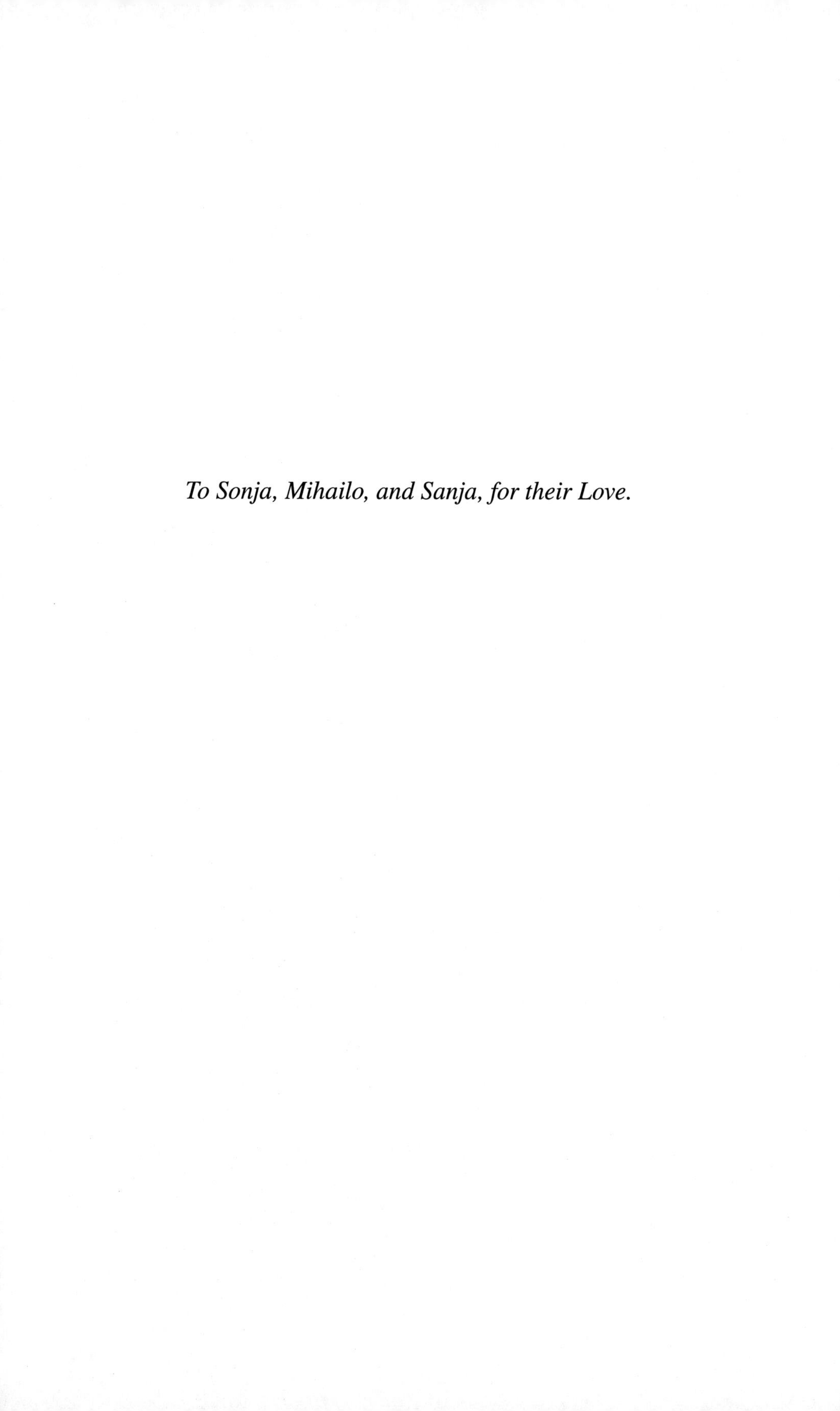

To Sonja, Mihailo, and Sanja, for their Love.

Contents

Foreword

Prove thy servants, I beseech thee, ten days; and let them give us pulse to eat, and water to drink.
(Daniel 1:12)

Therefore is the name of it called Babel; because the Lord did there confound the language of all the earth: and from thence did the Lord scatter them abroad upon the face of all the earth.
(Genesis 11:9)

These two paragraphs may describe best what this book is about and what is its goal. It is about the cultivated plants, commonly known as pulses or grain legumes, about the origin and diversity of their popular names in the languages of the world, and about the beauty of both. Being a plant scientist, specializing in grain legume genetics, breeding, and genetic resources, I have been impressed for decades by the striking similarities among common names denoting pea, faba bean, or lentil in geographically distant languages. These impressions had merely been deposited in some side corridors of my mind until, almost 11 years ago, I decided to venture into first attempts of gathering, systematizing, and writing down something on this topic. I was extremely cautious, almost scared, because although I felt quite confident to deal with the issues that link plant-related disciplines, I was aware that I absolutely had no qualifications in linguistics. Moreover, I was challenged not only by lexicology, but also by etymology, meaning that I could easily produce and publish completely wrong and misleading results I would have been ashamed of and, worst of all, ridiculed for by the linguistic community. Thank God, it has not been so, save for few times. Both negative and positive critics I received were very helpful for my further work, and the last decade brought forth a number of results, published in both plant science and linguistic international journals. In light of all that has been said, this book should be considered as a database of the vernacular names relating to pulse crops in various languages and a study of their attested and possible origin, development, and mutual impacts.

The first two chapters aim to provide both agronomic and linguistic perspective: the first chapter is designed to present pulse crops to the readers dealing with languages, while the second one intended to give an account on ethnolinguistcs families to those belonging to the plant sciences community. At the same time, both chapters may address the experts in their own topics and, after all, inform any reader interested in these two subjects without necessary scientific background. The remaining fourteen chapters have identical structure. In each first section, the species and subtaxa are listed after their scientific names and according to several most widely accepted classifications, such as *The Genetic Resources Information Network* (GRIN), *The International Legume Database & Information Service* (ILDIS), or *The Plant List*, together with their synonyms, and followed by the compiled common names in every language I found available. The second section of Chapters 3 through 14 represents an analysis of the common names from an etymological viewpoint, delivering a review of the widely accepted explanations and assumptions for those that have not been sufficiently clarified yet. The lexicon contains more than 9500 popular names in more than 900 living and extinct languages, dialects, and speeches of all the ethnolinguistic world families for about 1500 species and their subtaxa of 12 main grain legume genera.

At the beginning I had a somewhat obscure idea of what number of vernacular names for one grain legume taxon is adequate for listing them in the form of tables, but, gradually, while systematizing the data, the number 12 seemed to be most suitable: the reader will find more than 80 tables, with the common names shown after their alphabetically listed languages. Unfortunately, it was not feasible to include the synonyms for all languages, dialects, and speeches and I was forced to choose those that seemed to be most in use. When the source did not specify the variety of some language, then the name of the language is given generally, such as, for instance, *Chinese*, when it was not stated if *Cantonese* or *Mandarin Chinese*. There are also 14 drawings that symbolically depict the evolution of some proto-word into its mediating derivatives and contemporary descendants. The references for all the chapters are unified, mostly because the sources for lexicological data were, more or less, identical for all.

I would like to thank the following people for providing me with hardly accessible information about the names in various languages, dialects and speeches, most of which are gravely endangered or on an inevitable way to perish forever: the esteemed elders of the Miriwoong people with Ms. Maryann of the Yawuru people, Amanda Lissarague, Clarrie Kemarr Long, Frances Kofod, Knut J. Olawsky, and Jane Simpson for the Australian Aboriginal languages, Kenny Coeck for the Dutch dialects of Belgium and the Netherlands, Anicet Gbaguidi for the languages of Benin, Ol'ga Borisovna Kovan with Hidetoshi Shiraishi and Miki Mizushima for Nivkh, Alija Kurtiši for Gora Serbian, Ismo Porna with Kirsti Aapala, Timo Kunnari, Tanja Kyrö, Irma Lahti, Tauno Ljetoff and Bengt Pohjanen for Finnic and Sami, Dragica Radovanović for Dalmatia Serbian, Roman Rausch for the Tolkienian Elvish, Lars Steensland for Elfdalian and Claus Wenicker for Colognian.

An immense gratitude is owed to many colleagues and friends of mine, who have been encouraging me constantly to carry on this sort of research over the years, especially to John Bengtson, Aleksandra and Branko Ćupina, Antonio De Ron, Gérad Duc, Frank Dugan, Noel Ellis and Julie Hofer, Howard Huws, Brigitte Maass, Aleksandar Medović, Andrey Sinjushin, Richard Thompson, Astrid van Nahl, Margarita Vishnyakova, Tom Warkentin and Bojan Zlatković, as well as to the three pivotal persons in my life that have essentially helped me to articulate and develop my scientific and artistic interests, namely Alfredo Castelli, Vladimir R. Đurić (1947–2010) and Aleksandar B. Popović (1973–1999). I also owe a deep respect to the reviewers of the book manuscript and the Taylor & Francis team led by most kind Ms. Randy Brehm for their assistance, efficiency and patience.

The research leading to these results has received funding from the European Community's Seventh Framework Programme (FP7/2007–2013) under the grant agreement n°FP7-613551, LEGATO project. This book is also a result of the project

TR-31024 of the Ministry of Education, Science and Technological Development of the Republic of Serbia. It is also a tribute to Sergei A. Starostin (1953–2005), the founder, and George S. Starostin, the keeper of *The Tower of Babel, Evolution of Human Language Project*, together with their fellow colleagues, all genial linguistic minds and genuine polymaths.

I sincerely wish that the reader will enjoy browsing and reading this lexicon of pulses, the crops so essential and sufficient for the nutrition and health of the speakers of all the world's languages, like they were for Daniel with his comrades and despite the geographical dispersal of the human race.

Aleksandar Mikić
Novi Sad, Serbia

Author

Aleksandar Mikić, PhD, born in Pančevo, Serbia, in 1974, is Research Associate and annual legume breeder at the Institute of Field and Vegetable Crops in Novi Sad, Serbia. His fields of interest encompass conventional and molecular genetics, genetic resources, breeding, agronomy, agroecology, and crop history of annual legumes. Dr. Mikić co-authored more than 350 journal papers, about 120 international conference papers, over 10 book chapters, and more than 30 legume cultivars. He is also one of the founding members of the *International Legume Society* (2012), among the creators of the *UN FAO 10-Year Research Strategy For Pulse Crops* (2016) and is a member of the Editorial Board of the journal *Genetic Resources and Crop Evolution* (since 2018).

1 World's Pulses

Where the global pulse beats mightiest was the title of the seventh issue of the journal *Legume Perspectives*, devoted to the major grain legume scientific event in 2014, held in Saskatoon, Canada, one of the pivotal places where the research of various pulse crops advances in great moves to the common welfare. This title was, of course, a wordplay with two meanings of the word *pulse*, the one in an agronomic sense, where it denotes a grain legume crop used for human and animal nutrition, and the other from a medical viewpoint, referring to a normally regular beat caused by the pumping action of the heart. A similar wordplay was used to entitle the *carte blanche* of this journal's issue, *A meeting with pulse beating* (Warkentin 2014). Both titles, as well as the front cover artwork, showing a stripe made of pulse grains running across our planet in the form of a normal electrocardiography line, referred not only to the contribution made by pulses to the global health, but also attempted to point out how these two meanings of this word, or, more precisely, these two homonyms, rhythmically pulse in human metabolism and human diets. The modern English term *pulse*, in its botanical sense, either came together with the Norman conquest of England in the eleventh century, having evolved from the Old French *pols*, *pouls*, or directly from the Latin *puls*, denoting meal or porridge; in its turn, the latter is a borrowing of the Ancient Greek *póltos*, relating to porridge, and, ultimately, originates from the Proto-Indo-European **pel*, **pelə*, **plē-*, meaning *dust* or *flour* (Pokorny 1959, Nikolayev 2012). The pulse crops are, as already said, legumes that belong to the immensely abundant plant family of *Fabaceae* Lindl. (syn. *Leguminosae* Juss., *Papilionaceae* Giseke), with between 700 and 800 genera and around 19,000 systematized species (Christenhusz and Byng 2016). Among the economically most important pulse crops in the world and throughout the known human history are the taxa of the genera *Arachis* L., *Cajanus* Adans, *Cicer* L., *Glycine* Willd., *Lablab* Adans, *Lathyrus* L., *Lens* Mill., *Lupinus* L., *Phaseolus* L., *Pisum* L., *Vicia* L., and *Vigna* Savi and the species *Vicia ervilia* (L.) Willd., and *Vicia faba* L. (Figure 1.1). As a rule, the pulses are considered food legumes, in the form of an immature or mature grain and, sometimes, an immature pod, which is the reason why the terms *pulse* and *grain legume* are regarded as synonyms (Turner et al. 2001). Their additional forms of use often comprise mature grain in animal feeding, fresh and field-dried forage, forage meal, straw as both feed and biofuel and, in wild flora, grazing and browsing by livestock and other animals (Voisin et al. 2014). The cool season pulses exist in spring- and autumn-sown forms (Mikić et al. 2011), while all pulses are frequently cultivated together, or, in more specialized words, *intercropped*, most often with cereals (Bedoussac et al. 2015), but also with crucifers (Marjanović-Jeromela et al. 2017) or with each other (Antanasović et al. 2011, Zorić et al. 2015). The pulse grain and other plant parts are precious sources not only for food and feed, but also in various industries (Vaz Patto et al. 2015), medicine, and pharmacy (Lin et al. 2001).

FIGURE 1.1 **(See color insert.)** Some of the most significant pulse crops, today and in the past: (from left to right and from above to below) *Arachis hypogaea* L., *Cajanus cajan* (L.) Huth, *Cicer arietinum* L., *Vicia ervilia* (L.) Willd., *Vicia faba* L., *Glycine max* (L.) Merr., *Lablab purpureus* (L.) Sweet, *Lathyrus odoratus* L., *Lens culinaris* Medik., *Lupinus albus* L., *Lupinus texensis* Hook., *Phaseolus lunatus* L., *Pisum sativum* L., *Vicia villosa* Roth, *Vigna angularis* (Willd.) Ohwi & H. Ohashi, *Vigna subterranea* (L.) Verdc.

Having defined the meaning of the word *pulse*, we may proceed with attempting to answer the question, what exactly is a crop? Summarizing numerous uses of this word, primarily from the agronomic viewpoint and merging it with its meaning in a broader, economic or industrial sense, we may say that a crop is a plant species that is purposely grown for a specific product or utilization in a process that requests a full attention of the human factor—ranging from the very beginnings, in the form of sowing, planting, or propagating—over various measures during the plant growth and development until the end, such as cutting, harvesting, or gathering. Strictly, crops refer to the plants cultivated for food or feed in the field and in small-scale land and water resources, such as gardens. Broadly, crops encompass the fruit trees, medicinal, ornamental and biofuel plants, fungi, and, extremely rarely, certain animals or microorganisms. The status of any plant as crop is not definite and may vary during the time, being conditioned by various factors and this is fully valid for the pulses and all legumes. There are a large number of those that are regarded as a kind of *always-have-been-and-always-will-be* crops, such as chickpea (*Cicer arietinum* L.), lentil (*Lens culinaris* L.), or pea (*Pisum sativum* L.). Some may be considered as both crops and as a part of local wild flora, such as hairy vetch (*Vicia villosa* Roth). Although they may be used for the same purpose, either as a crop or a part of wild flora, these plants must be sown to be considered crops (Mikić and Mihailović 2014a). We also distinguish plant species that used to be crops and today are, almost or completely, not, such as bitter vetch (*Vicia ervilia* [L.] Willd.), as well as those which manage to make a kind of revival, such as French serradella (*Ornithopus sativus* L.) (Mikić 2015b). There are species that could have been crops, but which attempts to be domesticated were abandoned, ending with their return to wild flora, such as in the case of *Vicia peregrina* L. (Melamed et al. 2008) and, on the other hand, those currently wild, but with a potential for becoming a crop, like *Vicia noeana* Reut. ex Boiss. (Mikić et al. 2016). After all, there is a countless wealth of crop wild relatives (Maxted et al. 2006), many of which have various beneficial characteristics, which may be introgressed into their cultivated cousins. Such are undomesticated species, like red-yellow pea (*Pisum fulvum* Sm.) (Mikić et al. 2013b), or semi-domesticates, which often shift from wild to agricultural flora and *vice versa*, like *P. sativum* L. subsp. *elatius* (Steven ex M. Bieb.) Asch. & Graebn. (Zlatković et al. 2010, Ćupina et al. 2011, Mikić and Mihailović 2014a). In addition, at any point during the long history of agriculture, we always may easily find examples of a plant that is a more or less important crop in a certain environment, while, concurrently and only a few hundred miles away, it is completely unknown or is considered wild or weed. As may be seen, the story of the pulse crops is, similarly to the abovementioned homonym, fluctuating and dynamic, pulsing at its own pace through space and time.

For the purpose of this chapter, the data on production, area, and yield of selected pulse crops, provided by the Food and Agriculture Organization (FAO) of the United Nations (FAOSTAT 2017), are presented on the basis of these three criteria and are given for the first five leading countries and the world's average and for the last available year, that is, 2014 (Table 1.1).

Without any need to perform some rather thorough analysis of the shown data, it would be quite enough to note several facts. Among the top five producers of monitored pulse crops, there are countries from all the continents and contrasting climates. Opposing this diversity, China is present in almost all cases, confirming its role as the largest global producer (and consumer) of the vast majority of pulse crops (Table 1.1). Also, there are many obvious differences in all three parameters among individual crops, such as tens of millions of tons in soybean (*Glycine max* [L.] Merr.) in contrast to tens of thousands of tons in bambara groundnut (*Vigna subterranea* [L.] Verdc.) or yield, being mostly around one ton per hectare of dry grain and with more than five times higher yield in soybean in comparison to the one in cowpea (*Vigna unguiculata* [L.] Walp.). All this, in brief, confirm that the yield in grain legumes is generally lower than, for instance, in cereals, leading to most often a greater interest among the farmers in the latter, because of higher

TABLE 1.1
The Data on Production, Harvested Area, and Yield of Pulse Crops in the World in 2014, Provided by the Food and Agriculture Organization (FAO) of the United Nations (FAOSTAT 2017); for Each Crop, the Five Greatest Producers Are Given, Listed Alphabetically, as Well as the World's Average

Country	Production (t)	Area (ha)	Yield (kg ha^{-1})
	***Arachis hypogaea* L. (Groundnuts, with Shell)**		
China	16,481,700	4,603,850	3,580
India	6,557,000	4,685,000	1,400
Nigeria	3,413,100	2,770,100	1,232
Sudan	1,767,000	2,104,000	840
United States	2,353,540	535,200	4,397
World	43,915,365	26,541,660	1,655
	***Cajanus cajan* L. (Huth) (Pigeon Peas)**		
India	3,290,000	5,602,000	587
Kenya	274,523	276,124	994
Malawi	335,165	81,753	4,100
Myanmar	575,100	611,600	940
Tanzania	248,000	250,509	990
World	4,890,099	7,033,049	695
	***Cicer arietinum* L. (Chickpeas)**		
Australia	629,400	507,800	1,239
Ethiopia	458,682	239,755	1,913
India	9,880,000	9,927,000	995
Myanmar	562,163	384,217	1,463
Turkey	450,000	388,169	1,159
World	13,730,998	13,981,218	982
	***Glycine max* (L.) Merr. (Soybeans)**		
Argentina	53,397,715	19,252,552	2,774
Brazil	86,760,520	30,273,763	2,866
China	12,154,000	6,799,900	1,787
India	10,528,000	10,908,000	965
United States	106,877,870	33,423,750	3,198
World	306,519,256	117,549,053	2,608
	***Lens culinaris* Medik. (Lentils)**		
Australia	238,120	162,400	1,466
Canada	1,987,000	1,217,100	1,633
India	1,100,000	1,800,000	611
Nepal	226,830	205,939	1,101

(Continued)

TABLE 1.1 (*Continued*)
The Data on Production, Harvested Area, and Yield of Pulse Crops in the World in 2014, Provided by the Food and Agriculture Organization (FAO) of the United Nations (FAOSTAT 2017); for Each Crop, the Five Greatest Producers Are Given, Listed Alphabetically, as Well as the World's Average

Country	Production (t)	Area (ha)	Yield (kg ha^{-1})
Turkey	345,000	243,370	1,418
World	4,827,122	4,524,043	1,067
	***Lupinus* spp. (Lupins)**		
Australia	625,600	387,400	1,615
Belarus	34,137	13,448	2,538
Germany	40,800	21,400	1,907
Poland	139,802	80,022	1,747
Russia	75,690	50,355	1,503
World	1,014,022	622,427	1,629
	***Phaseolus vulgaris* L. (Beans, Dry)**		
Brazil	3,294,586	3,185,745	1,034
India	4,110,000	1,000,000	4,110
Mexico	1,273,957	1,680,897	758
Myanmar	4,651,094	3,017,250	1,542
United States	1,311,340	667,170	1,966
World	26,529,580	30,612,842	867
	***Phaseolus vulgaris* L. (Beans, Green)**		
China	17,017,405	7,890	26,877
India	636,103	225,727	2,818
Indonesia	855,958	113,233	7,559
Thailand	305,002	170,791	1,786
Turkey	638,469	74,000	8,628
World	21,720,588	1,527,613	14,219
	***Pisum sativum* L. (Peas, Dry)**		
Canada	3,444,800	1,467,000	2,348
China	1,350,000	950,000	1,421
India	600,000	730,000	822
Russia	1,502,845	896,923	1,676
United States	778,140	364,020	2,138
World	11,186,123	6,931,941	1,614
	***Pisum sativum* L. (Peas, Green)**		
China	10,711,208	1,338,469	8,003
Egypt	184,018	18,471	9,963
France	185,692	24,255	7,656

(*Continued*)

TABLE 1.1 (*Continued*)
The Data on Production, Harvested Area, and Yield of Pulse Crops in the World in 2014, Provided by the Food and Agriculture Organization (FAO) of the United Nations (FAOSTAT 2017); for Each Crop, the Five Greatest Producers Are Given, Listed Alphabetically, as Well as the World's Average

Country	Production (t)	Area (ha)	Yield (kg ha^{-1})
India	3,868,630	433,560	8,923
United States	329,180	75,920	4,336
World	17,426,421	2,356,340	7,396
	***Vicia faba* L. (Broad Beans, Horse Beans, Dry)**		
Australia	327,700	152,100	2,155
China	1,428,700	701,600	2,036
Ethiopia	838,944	443,107	1,893
France	278,545	74,884	3,720
Morocco	166,680	190,966	873
World	4,139,972	2,150,905	1,925
	***Vicia* spp. Vetches**		
Belarus	86,797	30,273	2,867
Ethiopia	251,439	136,884	1,837
Mexico	116,684	97,050	1,202
Russia	127,003	76,495	1,660
Spain	107,000	122,000	877
World	905,002	541,699	1,671
	***Vigna subterranea* (L.) Verdc. (Bambara Beans)**		
Burkina Faso	56,264	50,428	1,116
Cameroon	38,075	43,516	875
DR Congo	10,741	25,235	426
Mali	22,930	37,702	608
Niger	32,383	70,505	459
World	160,378	227,386	705
	***Vigna unguiculata* (L.) Walp. (Cowpeas, Dry)**		
Burkina Faso	573,048	1,205,162	475
Cameroon	174,251	209,019	834
Niger	1,593,166	5,325,168	299
Nigeria	2,137,900	3,701,500	578
Tanzania	190,500	197,323	965
World	5,589,216	12,610,956	443

profit (Welch and Graham 1999). For this reason, many pulses are still widely underutilized and neglected, being accompanied with a rapidly decreased use of inexhaustible existing biodiversity in breeding programs (Doyle and Luckow 2003). Finally, the attested great genetic potential of pulses for high-quality and stable yield remains untapped, mainly due to a number of irregularities in applying adequate production technology, despite their remarkable adapting ability to sustainable agriculture and diverse farming systems (Rubiales and Mikić 2015).

From a paleontological point of view, pulse crops are adapted to a remarkably wide range of climates (Table 1.2) and are found in nearly all centers of diversity or

TABLE 1.2
Centers of Diversity of Some of the Economically Most Important Pulse Crops in the World and Throughout the History, according to One of the Most Traditional Classifications

Species and Their Subtaxa	Center of Origin
Arachis hypogaea L.	South American′, African″
Cajanus cajan (L.) Huth.	African′, Hindustani″
Cicer arietinum L.	Near Eastern′, Central Asian″, Hindustani″, Mediterranean″
Vicia ervilia (L.) Willd.	Mediterranean′, Near Eastern″
Vicia faba L.	Central Asian′, Mediterranean″
Vicia faba L. var. *equina* St.-Amans	Central Asian′, Mediterranean″
Vicia faba L. var. *faba*	Central Asian′, Mediterranean″
Vicia faba L. var. *minuta* (hort. ex Alef.) Mansf.	Central Asian′, Mediterranean″
Glycine max (L.) Merr.	Chinese-Japanese′, Near Eastern″
Glycine soja Siebold & Zucc.	Chinese-Japanese
Lablab purpureus (L.) Sweet	African
Lathyrus annuus L.	Mediterranean
Lathyrus cicera L.	Mediterranean
Lathyrus clymenum L.	Mediterranean
Lathyrus hirsutus L.	Mediterranean
Lathyrus ochrus (L.) DC.	Mediterranean
Lathyrus odoratus L.	Mediterranean
Lathyrus sativus L.	Mediterranean′, Central Asian″
Lathyrus sylvestris L.	European Siberian
Lathyrus tingitanus L.	Mediterranean
Lathyrus tuberosus L.	European Siberian
Lens culinaris Medik. subsp. *culinaris*	Near Eastern

(*Continued*)

TABLE 1.2 (*Continued*)
Centers of Diversity of Some of the Economically Most Important Pulse Crops in the World and Throughout the History, according to One of the Most Traditional Classifications

Species and Their Subtaxa	Center of Origin
Lupinus albus L.	Mediterranean′, Near Eastern″
Lupinus angustifolius L.	Mediterranean
Lupinus luteus L.	Mediterranean
Lupinus mutabilis Sweet	South American
Lupinus nootkatensis Donn ex Sims	North American
Lupinus perennis L.	North American
Lupinus polyphyllus Lindl.	North American
Phaseolus acutifolius A. Gray	Central American and Mexican
Phaseolus coccineus L.	Central American and Mexican
Phaseolus lunatus L.	Central American and Mexican′, South American′
Phaseolus vulgaris L.	Central American and Mexican′, South American″
Pisum sativum L.	Near Eastern′, Mediterranean″
Pisum abyssinicum A. Braun	African
Vicia articulata Hornem.	Mediterranean
Vicia benghalensis L.	Mediterranean
Vicia cracca L.	European Siberian
Vicia hirsuta (L.) Gray	European Siberian
Vicia narbonensis L.	Near Eastern′, Mediterranean″
Vicia pannonica Crantz	Near Eastern′, European Siberian″
Vicia sativa L. subsp. *sativa*	Near Eastern
Vicia villosa Roth	Near Eastern
Vigna angularis (Willd.) Ohwi & H. Ohashi var. *angularis*	Chinese-Japanese
Vigna lanceolata Benth.	Australian
Vigna mungo (L.) Hepper var. *mungo*	Hindustani
Vigna radiata (L.) R. Wilczek var. *radiata*	Indochinese-Indonesian
Vigna subterranea (L.) Verdc.	African
Vigna umbellata (Thunb.) Ohwi & H. Ohashi	Hindustani′, Indochinese-Indonesian″
Vigna unguiculata (L.) Walp.	African′, Hindustani″

Source: Zeven, A.C. and Zhukovsky, P.M., *Dictionary of Cultivated Plants and their Centres of Diversity*, Centre for Agricultural Publishing and Documentation, Wageningen, the Netherlands, 1975.

′ Primary center.

″ Secondary center.

centers of origin (Zeven and Zhukovsky 1975). This is confirmed by the works of the famous Russian and Soviet geneticist and botanist Nikolay I. Vavilov (1887–1943), who was the first to conceive the idea of the centers of origin, and by numerous and mutually different classifications of these centers (Corinto 2014). The most important facts presented in this compiled list comprise the existence of primary and secondary centers of diversity in many of the grain legume species, which are the subject of this book, equally in more moderate environments, such as between the Near Eastern and Mediterranean centers, and in equatorial regions, including Central American and Mexican, African, South American, Hindustani, or Indochinese-Indonesian centers.

Recently, there began to appear the first attested evidences of grain legumes being a part of the diets of the Neanderthal man, along with cereals (Henry et al. 2011). There are also numerous archaeological findings from various stages of Palaeolithic, ranging from the Mediterranean coastal regions of the Iberian Peninsula, with the remains of the *Lathyrus cicera* L. or *Lathyrus sativus* L. (Aura et al. 2005), to the famous Franchthi cave in Greece, with lentil (Sonnante et al. 2009), and the Busmpra Cave in Ghana, with cowpea (Oas et al. 2015), all dated as back as more than 10,000 years BP (before present). The archaeobotanical data on the presence of domesticated pulse crops is rather rich, constantly updating the timescale of the course of grain legume domestication. The dates assessed for diverse pulses are 8500 BP for peanut (Dillehay et al. 2007), at least 1500 BC for pigeon pea (*Cajanus cajan* [L.] Huth) (Fuller and Harvey 2006), 9300 BP for chickpea (Tanno and Willcox 2008), around 11,000 BC for bitter vetch (Fuller et al. 2012), more than 10,000 BP for faba bean (*Vicia faba* L.), *Lathyrus* spp., lentil, pea, and Narbonne vetch (*Vicia narbonensis* L.) (Caracuta et al. 2017), up to 9000 BP for soybean (Sedivy et al. 2017), around 2800 BP for hyacinth bean (*Lablab purpureus* [L.] Sweet) (Fuller et al. 2004); 7800 BP for South American *Lupinus* species (Jantz and Behling 2012), 9000 BP for the cultivated *Phaseolus* species (Piperno and Dillehay 2008), more than 6500 BC for some *Vicia* species other than bitter vetch and faba bean (Fairbairn et al. 2007) and at least 4500 BP for the South Asian *Vigna* species (García-Granero et al. 2017). All this confirms a rather solid status of pulses and legumes in general as one of the primary domesticated plant families in the world with a persisting and remarkable wealth of genetic resources (Hammer et al. 2015).

2 World's Languages

For the basic purpose of this book and considering its targeted readership, perhaps it would be quite sufficient to say that language is, in the narrowest sense, usually defined as a privileged human ability to communicate in a spoken and written form, while the science studying this phenomenon is called linguistics (Halliday 2003). Among the most continuing debates among linguists are those whether language followed the human evolution (Anderson 2012) or appeared suddenly, caused by some unidentified external stimulus (Chomsky 2000), and if all the human languages descended from one single language (Gell-Mann and Ruhlen 2011) or if that language evolved independently in mutually sundered regions of the world (Campbell and Poser 2008). It is worth mentioning that, relatively recently, it was definitely assessed that Neanderthal man was also able to speak (D'Anastasio et al. 2013).

It is estimated that there is between 7100 and 7200 living languages (Ethnologue 2017), with an unknown number of the already extinct ones and an alarmingly increase of those that are inevitably perishing forever, mainly due to globalization. The languages are generally grouped into families, which number significantly varies depending on individual viewpoints, with a certain number of so-called language isolates (Figure 2.1). Each family or isolate is considered a development of what is referred to as a protolanguage (Koerner 1999). In the next several paragraphs, we shall try to give a rather rough and provisional overview of the major global ethnolinguistic families and thus make this topic more familiar to the plant scientific community.

2.1 LANGUAGES OF SUB-SAHARAN AFRICA

Although comprising 13 endangered members and less than half a million speakers, such as Khoikhoi and San (syn. Bushmen), the Khoisan languages are remarkably distinctive for having a large number of different click consonants. Preliminary genetic analyses propose that the Khoisan homeland was in East Africa, from where, not knowing when, these peoples migrated far to the south (Hammer et al. 2001).

Encompassing more than 1500 languages, with Fula, Igbo, Shona, Swahili, Yoruba, and Zulu as the most widespread ones, the Niger-Congo is the third richest language family in the world. Its 400 million speakers live in almost the whole territory of Sub-Saharan Africa. The homeland of the Proto-Niger-Congo language, formed at least by 3000 BC, was in western or central Africa (Diamond 1997).

The Nilo-Saharan ethnolinguistic family is supposed to consist of between 50 and 60 million speakers belonging to Nilotic peoples, mostly in Kenya, South Sudan, and Uganda. It has more than 10 subdivisions and about 200 languages. One of the most common viewpoints is that the hypothetical Proto-Nilo-Saharan language existed in eastern Sudan earlier than 10,000 BC (Campbell and Tishkoff 2010).

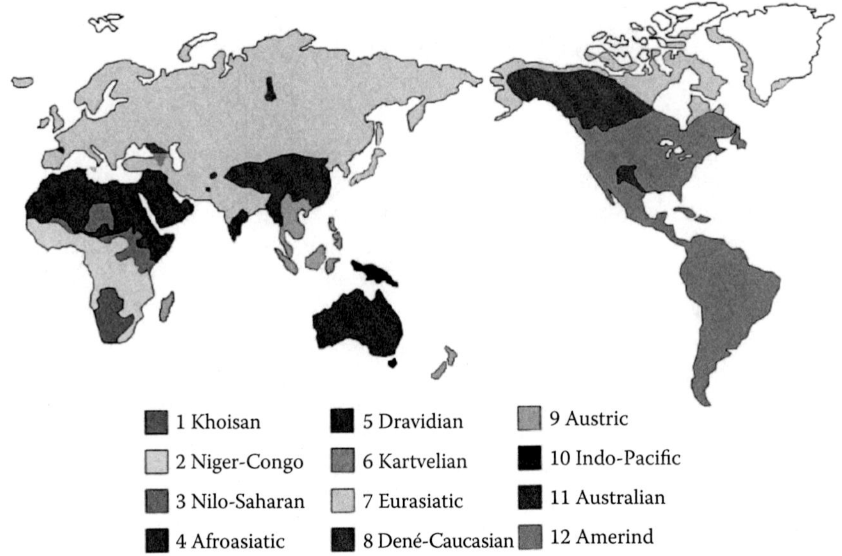

FIGURE 2.1 **(See color insert.)** A simplified map of the major world ethnolinguistic families; *Eurasiatic* encompasses Indo-European, Altaic, Uralic, and Paleosiberian languages, *Dené-Caucasian* comprises Basque, Caucasian, Burushaski, Yenissenian, Sino-Tibetan, and Na-Dené languages, *Austric* denotes Austroasiatic and Austronesian languages and *Indo-Pacific* designates Andamanese, Trans-New Guinea, and Tasmanian languages. (Modified from Starostin, G., *The Tower of Babel, Evolution of Human Language Project*, http://starling.rinet.ru. With permission.)

2.2 LANGUAGES OF AUSTRALIA AND NEW GUINEA

The Australian Aboriginal and the Trans-New Guinea languages comprise about 350 and more than 500 languages, respectively, including Tiwi and Warlpiri in the first group and Enga and Makasae in the second group. An exceptional internal diversity in both islands may be explained by the fact that they were inhabited by humans about 40,000 years ago, mutually splitting very early and with almost no external contacts (Dixon 2002). So far, it has not been possible to assess the exact time when the proposed Proto-Australian language was spoken, that is, earlier or later than 30,000 years ago (Dixon and Dixon 2011), and the relationships among the Australian Aboriginal languages have not been sufficiently clarified, which is the reason they are currently and merely for practical reasons divided into the Pama-Nyungan and *non-Pama-Nyungan* groups (Bowern and Atkinson 2012). Merging Kusunda, a language isolate in Nepal, Andamanese, such as Aka-Jeru, non-Austronesian Melanesian, Papuan Aboriginal, and now extinct Tasmanian languages into the hypothetical Indo-Pacific superfamily has been rejected by mainstream linguistics, although it was acknowledged for partially contributing to the establishment of the currently accepted Trans-New Guinea family (Wurm 1982). However, its apologists argue that its extraordinary oldness of up to 55,000 years and the subsequent divisions to the mutually remote places are the main cases of the obscurity of their common origin, which, nevertheless, may be demonstrated on very fundamental levels

(Greenberg 1971, Ruhlen 1994, Whitehouse et al. 2004). The debate *pro et contra* the existence of the Indo-Pacific family and its structure is still going on, addressing not only linguists, but also anthropologists and archaeologists (Clark et al. 2017).

2.3 LANGUAGES OF THE AMERICAS

The hypothetical Amerind superfamily is at least 10,000 years old and comprises about 600 indigenous languages of the North, Central and South Americas, except those belonging to Na-Dené and Eskimo-Aleut appearing later, with about 18 million remaining native speakers (Greenberg 1987, Ruhlen 1991, Greenberg 1996). Assessing the position of the Proto-Amerind people and their language remains extremely difficult and is currently not in favor of the mainstream linguistics, which sees nearly each segment of the proposed Amerind superfamily as a language family of its own and with barely sufficient or completely absent evidence to associate it with the others (Campbell 2000).

2.4 LANGUAGES OF ASIA AND PACIFIC

Comprising nearly 170 languages, including Khmer and Vietnamese and spoken by almost 100 million persons, the Austroasiatic family is concentrated in Southeast Asia and additionally ranging from India to southern China. One of the many classifications assumes that, by 6300 BC and in the middle Mekong, the Proto-Austroasiatic language was divided into the Munda-Khmer and Nicobarese branches (Peiros 2004, Sidwell and Blench 2011).

The Austronesian ethnolinguistic family consists of more than 1200 languages, including Fijian, Hawaiian, Indonesian, Javanese, Malagasy, Malay, and Filipino (Tagalog), and nearly 400 million speakers from Madagascar, over Maritime Southeast Asia, to the Pacific Ocean. The main branches of Austronesian are Nuclear Austronesian, Puyuma, Rukai, and Tsomu (Ross 2009). The Proto-Austronesian developed in Taiwan, about 6000 BC (Bellwood 1997).

One of the less known East Asian ethnolinguistic families is called Hmong-Mien, nowadays scattered in a large number of communities in China and the countries of Southeast Asia, with a basic division into Hmongic, also known as Miao, and Mienic, or Yao, branches, and with around 8 million speakers (Ethnologue 2017). Its ancestor, the Proto-Hmong-Mien language, is estimated to exist at least 4500 years ago, with a possibility that this distance may be extended for another two millennia (Ratliff 2010).

Judging from the attested linguistic diversity, the Tai-Khadai languages are supposed to originate in the southern Chinese provinces of Guizhou and Hainan, where from the more abundant Tai branch migrated southwards and produced Lao and Thai, the most widely spoken members of this family. Altogether, there are nearly 100 Tai-Kadai languages, with around 100 million speakers around the world (Diller et al. 2008). The internal classification of this ethnolinguistic group is still not assessed to a sufficient extent, with more recent suggestions of its restructuring and renaming into Kra-Daic (Srithawong et al. 2015).

2.5 LANGUAGES OF EURASIA

The Basque language is distributed on both sides of the western Pyrenees, with more than 700,000 speakers. Basque and extinct Aquitanian could form the Vasconic family, with a very vague attestation of their protolanguage (Trask 1997). The Basque people may have descended from the Ice Age European hunter-gatherers, who retreated into the mountains after the first farmers had arrived (Ruhlen 2001).

One of the most intriguing languages is Burushaski, spoken by almost 90,000 persons belonging to the Burusho people, concentrated in the isolated valleys of Hunza, Nagar, and Yasin in northernmost Pakistan (Holst 2014). The relationship of these languages to their neighboring ethnolniguistic families has been unresolved, except for the attested cases of certain word exchanges, as well as the course of their origin and development. Among numerous theories, there is one linking them with the Caucasian languages (Bengtson 1997), considering both a kind of language islands of the Palaeolithic Eurasian population in the sea of the Neolithic peoples subsequently surrounding them.

On the northern slopes of the Caucasus, there exist two groups of languages, commonly named Northwest Caucasian and Northeast Caucasian. The first one comprises five languages, such as Abkhaz, Adyghe, and Kabardian, with up to 2 million speakers in total, while the second one consists of nearly 30 languages, such as Avar, Chechen, Ingush or Lezgian, spoken by more than 4 million people. Although the mainstream linguistics considers their mutual relationship still insufficiently explored, there are views that they form a single ethnolinguistic family, known as simply Caucasian (Bengtson 1998). A possible Proto-Caucasian homeland may be the Near East (Wuethrich 2000).

The most renowned representatives of the Na-Dené ethnolinguistic family are the languages of the Dené people, inhabiting northern Canada, and of the Apache and Navajo peoples, living in the southwest regions of the United States. It is supposed that they share the common homeland with the Yenisseian languages somewhere in Beringia (Sicoli and Holton 2014).

With about 1.4 billion speakers, Sino-Tibetan is the second most widely spoken ethnolinguistic family in the world, following Indo-European. It is primarily spoken in East Asia, South Asia, and Southeast Asia. Myanmar (Burmese), Chinese, and Tibetan are the main groups among the more than 400 languages of Sino-Tibetan. During the first half of the twentieth century, the first classifications of Sino-Tibetan were proposed and have been continually debated over because of the large number of languages and still unsatisfactorily determined borders between single clusters (Handel 2008). It is widely accepted that the homeland of Proto-Sino-Tibetan is close to the upper flows of Brahmaputra, Mekong, Salween, and Yangtze (van Driem 1993).

The Yenisseian family, placed around the upper and middle flow of the river Yenissei in central Siberia, comprise only one living language, Ket, with 200 speakers, and several extinct languages. Despite these verily sad facts, these languages are still an objective of numerous linguistic and philological studies. According to one of the supposed timelines, the Proto-Yenisseian was formed as a distinct language at least a few thousand years ago but definitely began to split by 500 BC into

Northern Yenisseian, comprising the Ket language, and Southern Yennisseian. One of the viewpoints regarding their origin and relationship to other families links the Yenisseian languages to Burushaski (van Driem 2001).

2.6 LANGUAGES OF EURASIA AND NORTH AFRICA

Afroasiatic, also spelled as Afro-Asiatic and known as Afrasian and Hamito-Semitic, is one of the four African language families. Although there is no consensus regarding its exact division, it basically comprises six main branches, namely Berber, Chadic, Cushitic, Egyptian, Omotic, and Semitic, with Arabic, Amharic, Aramaic, Coptic, Hausa, Hebrew, Oromo, and Somali as some of its members (Diakonoff 1988). Afroasiatic languages are spoken by more than 350 million speakers in North Africa, a part of the Sahel, the Horn of Africa, and the Near East. It is estimated that the Proto-Afroasiatic, the supposed ancestor of all extinct and living Afroasiatic languages (Ehret 1995), existed as early as 10,000 BC or even by 16,000 BC, which makes it older than the majority of the other protolanguages. One of its possible homelands is Levant (Blench 2006), with Natufian culture, and North Africa, with the Halfan culture (Midant-Reynes 1999). Among the extinct Afroasiatic languages, the most renowned are Akkadian, Egyptian, Old Aramaic, and Old Hebrew.

The Altaic ethnolinguistic family is still considered by many as hypothetic and controversial (Georg et al. 1999). It comprises Japonic, Koreanic, Mongolic, Tungusic, Turkic, and perhaps Ainu languages (Blažek 2006), with a total number of about 70 and about 400 million speakers from East Europe to northeast Asia. Our knowledge of the prehistory of the Altaic peoples is still very limited, and it is hard to suggest where exactly their homeland was (Miller 1996). One of the possible locations is southeast Mongolia, southern Manchuria, and North Korea (Janhunen 2010). There are estimates that Proto-Altaic could have been spoken as early as 6000 BC (Kuz'mina 2007).

The Dravidian peoples mainly inhabit the Indian subcontinent, with more than 200 million native speakers of Kannada, Malayalam, Tamil, Telugu, and other between 80 and 90 living languages (Ethnologue 2017). Many archaeologists consider Dravidian much more widely spread before the arrival of the Indo-Aryan tribes, as well as the bearers of the Indus Valley Civilization, lasting roughly from 3300 BC to 1300 BC (Wright 2009). It is estimated that the Proto-Dravidian, conceived in either central or northeast regions of India, was actively spoken during 4th millennium BC and that it began to split about 1000 years later into its Central, Eastern, and Northern branches (Andronov 2003).

Today, the Indo-European ethnoliniguistic family has numerous subdivisions (Anthony 2007), such as Balto-Slavic, Germanic, Hellenic, Indo-Iranian, or Italic. Among its about 450 living languages, spoken by almost 3 billion speakers, are English, German, Greek, Hindi, Persian, Russian, and Spanish. This group encompasses almost all the languages of Europe, as well as a large number of those in West, Central, and South Asia. According to one of the most widely accepted opinions, the Kurgan hypothesis, the Proto-Indo-European, conceived in the Pontic-Caspian steppe, began to split between 5000 BC and 4000 BC (Gimbutas et al. 1997). The Indo-European linguistics exists for more than two centuries and, initially based upon the similarities among Archaic Greek, Latin, and Sanskrit, has produced a considerable number of etymological dictionaries and databases (Mallory 1989). Some extinct Indo-European languages, such as Hittite (Beckman 2011), Old Prussian (Mažiulis 2004), and Tocharian (Winter 1998), have a remarkable significance for comparative linguistics.

Represented with four languages, spoken by between 5 million and 6 million people, mostly in Georgia, the Kartvelian ethnolinguistic family holds a genuine and remarkable position in both historical and modern linguistics (Boeder 2005). Although Kartvelian, also known as Iberian or South Caucasus languages, are geographically approximate to other various distinctive families, no firm attestation with any of them has been found so far. Its homeland is most likely to be identical with its current distribution area, while the Proto-Kartvelian language is supposed to be split into its two main branches, Proto-Georgian-Zan and Laz, by the end of 3rd millennium BC (Klimov 1998).

The Paleosiberian languages, today spoken by somewhat more than 20,000 persons, antedate all other language groups spoken in northeast Siberia and the Russian Far East. Their long-term mutual isolation may explain why they are often viewed not as a single ethnolinguistic family in its strict sense. They are generally divided into Chukotko-Kamchatkan and Eskimo-Aleut families, with few isolates, such as Nivkh (Fortescue 2005).

About 25 million people speak nearly 40 languages attributed to the Uralic family, extending mainly in northern Eurasia and with Estonian, Finnish, and Hungarian as the most numerous (Michalove 2002). It is usually assumed that the homeland of the Uralic peoples is westernmost Siberia or, more precisely, the eastern slopes of the Ural Mountains (Hajdú 1969). There, the Proto-Uralic language was spoken as a single language at least by 5000 BC and then, about 4500 BC, began to primarily divide between Finno-Ugric and Samoyedic (Janhunen 2009). The latter group is critically endangered, with a decreasing number of native speakers and several languages already extinct (Janhunen 1998).

2.7 OTHER LANGUAGES

Apart from all the listed language families described earlier, it should be mentioned that there are also numerous kinds of mixed languages (Meakins 2013), usually a consequence of bilingualism, such as creole or pidgin languages, and constructed languages, designed as a tool of international communication (Eco 1995) with Esperanto as its most famous representative.

3 *Arachis* L.

3.1 LIST OF TAXA SCIENTIFIC AND POPULAR NAMES

This section brings an overview of the most widely accepted species of the genus *Arachis* L. and their subtaxa, along with their synonyms in various botanical classifications and vernacular names, listed alphabetically and according to the official or most used language designations (ISTA 1982, Krapovickas and Gregory 1994, Rehm 1994, Gledhill 2008, Porcher 2008, The Plant List 2013, Ecocrop 2017, EPPO 2017, Ethnologue 2017, IBIS 2017, ILDIS 2017, Logos 2017, NPGS 2017, Wikipedia 2017, Wiktionary 2017).

Our knowledge on the extinct and modern words for peanut in the Native American languages is rather limited, since their official status is recognized mainly as vulnerable or critically endangered (Zepeda and Hill 1991, Moseley 2010). The number of active speakers of nearly all native languages of the Americas, especially in the north, is rapidly decreasing every day (Crystal 2000). However, we are able to find out certain and academically accurate knowledge on the lexicology relating to *Arachis* from the available resources, especially early lexicons and dictionaries, conversation books, folk tales and other forms of popular literature and etymological studies (Powell 1891, Boas 1911, Shapiro 1987, Fabre 2005, 2016).

- *Arachis appressipila* Krapov. & W. C. Greg.
English: flat-haired peanut
Portuguese (*Brazil*): amendoim-bravo
- *Arachis archeri* Krapov. & W. C. Greg.
Synonyms: *Arachis diogoi* sensu auct.
English: Archer's pea
Portuguese (*Brazil*): amendoim-do-campo-limpo
- *Arachis batizocoi* Krapov. & W. C. Greg.
English: Batizoco's peanut; forest peanut
Spanish: manduví; maní silvestre
- *Arachis duranensis* Krapov. & W. C. Greg.
Synonyms: *Arachis argentinensis* Speg.; *Arachis spegazzinii* M. Greg. & W. C. Greg.
English: wild peanut; yellow peanut
Spanish: sacha maní
- *Arachis glabrata* Benth. var. glabrata
Chinese: duōniánshēng huāshēng
English: cocos; creeping forage peanut; golden glory; ornamental peanut grass; perennial forage peanut; perennial peanut; rhizoma peanut; rhizoma perennial peanut

Portuguese: amendoim-bravo; amendoim-forrageiro; mendoim-do-campo-baixo
Spanish: maní perenne
Vietnamese: cỏ lạc; lạc tiên; lạc trường niên hay còn gọi
- *Arachis hypogaea* L. (Table 3.1)
Synonyms: *Arachis africana* Lour.; *Arachis americana* Ten.; *Arachis asiatica* Lour.; *Arachis hypogaea* L. subsp. *oleifera* A. Chev.; *Arachis nambyquarae* Hoehne; *Arachis rasteiro* A. Chev.; *Arachidna hypogaea* (L.) Moench
- *Arachis hypogaea* L. subsp. *fastigiata* Waldron var. *aequatoriana* Krapov. & W. C. Greg.
English: equatorial peanut
Spanish: huasquillo
Spanish (*Ecuador*): zaruma
- *Arachis hypogaea* L. subsp. *fastigiata* Waldron var. *fastigiata* (Waldron) Krapov. & W. C. Greg.
English: Valencia peanut
- *Arachis hypogaea* L. subsp. *hypogaea* var. *hypogaea*
English: Virginia peanut
- *Arachis hypogaea* L. subsp. *fastigiata* Waldron var. *vulgaris* Harz
English: Spanish peanut
- *Arachis kretschmeri* Krapov. & W. C. Greg.
English: Pantanal peanut
- *Arachis kuhlmannii* Krapov. & W. C. Greg.
English: Kuhlmann's peanut
Portuguese (*Brazil*): Amendoim-bravo
- *Arachis macedoi* Krapov. & W. C. Greg.
English: cold peanut; Macedo's peanut
Portuguese (*Brazil*): amendoim do resfriado
- *Arachis major* Krapov. & W. C. Greg.
Synonyms: *Arachis diogoi* Hoehne subsp. *major* Hoehne
English: big peanut
Spanish: amendoim de Aquidauana
- *Arachis pintoi* Krapov. & W. C. Greg.
English: forage peanut; pinto peanut
Portuguese (*Brazil*): amendoim-forrageiro
Spanish: maní forrajero perenne; maní perenne
- *Arachis pusilla* Benth.
English: petty peanut
Portuguese (*Brazil*): amendoim-de-caracará
- *Arachis repens* Handro
English: crawling peanut; creeping peanut
Portuguese (*Brazil*): amendoim-rasteiro
Spanish (*Colombia*): tepe colombiano
- *Arachis sylvestris* (A. Chev.) A. Chev.
Synonyms: *Arachis hypogaea* L. subsp. *sylvestris* A. Chev.
English: forest peanut; pig peanut

Portuguese (*Brazil*): amendoim do porco; mandubi do porco; mundubi; mundubi do joazeiro; mundubim bravo
- *Arachis tuberosa* Benth.
English: tuberous peanut
Portuguese (*Brazil*): amendoim do tubéras
- *Arachis veigae* S. H. Santana & Valls
English: Veiga's peanut
Portuguese: mundubi; mundubi-do-Joazeiro
- *Arachis villosulicarpa* Hoehne
English: hairy-podded peanut
Portuguese (*Brazil*): amendoim-bravo; wi-ki-rin-gui
- *Arachis williamsii* Krapov. & W. C. Greg.
English: Williams' peanut
Spanish: manicillo

3.2 ORIGIN OF SCIENTIFIC AND POPULAR TAXA NAMES

The scientific name of the genus *Arachis* L. (Linnaeus 1753, 1758) is based upon the Ancient Greek word *ӑrakos*, with its variations *árako-s* and *árak-s* (Pokorny 1959). It is considered denoting a weedy grain legume species growing in the lentil (*Lens culinaris* Medik.) crop, most likely annual vetchling (*Lathyrus annuus* L.) or, alternatively, some other semi-domesticated and occasionally cultivated leguminous plant with similar growth habit (Nikolayev 2012). According to the historical linguistic analysis, this Ancient Greek word has its ultimate origin in the attested Proto-Indo-European root **arenko-*, **arn k*(')-, meaning both a kind of cereal (Pokorny 1959) and a leguminous plant in general (Nikolayev 2012). This root also gave the Latin *arinca*, denoting spelt (*Triticum aestivum* L. subsp. *spelta* [L.] Thell.) or *Gallarium propria*, an undefined plant species mentioned by Pliny the Elder (Pokorny 1959). In modern Indo-European languages, the Ancient Greek *ӑrakos* survived in the contemporary Greek *arakás*, meaning *pea* (*Pisum sativum* L.), and in several Indo-Aryan languages, at least in the Lakhimpur Awadhi *arrhī*, the Hindi *arhaṛ* and, still speculatively, the Sanskrit *āḍhakī'* and the Suśruta Prakrit *āḍhaī* (Turner 1962–1966, Southworth 2004), in all of which it means *pigeon pea* (*Cajanus cajan* [L.] Huth) and with no attested mediating Proto-Indo-Aryan root yet. At any rate, the scientific name for the genus, *Arachis peanut*, has entered into many languages of diverse ethnolinguistic families, such as Belarusian, Cebuano, or Turkmen, as well as constructed languages, such as Esperanto, Interlingua, and Volapük (Table 3.1).

Among the attested roots of the hypothetical Proto-Amerind language, there is one that designs seed in general, namely **ica* (Greenberg and Ruhlen 2007). This Proto-Amerind root gave the words with the same meaning in many extinct and living languages and dialects in North, Central, and South Americas, such as the Blackfoot *kiníínoko*, the Mohawk *enhnekeri*, the Cheyenne *ugata*, the Nahuatl *inach*, the Q'anjob'al *inat*, the Xavante *'ï-jë*, the Akwáva *a'yni*, or the Kawésqar (*ye*)*c'oy*. The root **ica* could have also produced the words denoting not only seed, but specifically peanut, especially in the languages that are geographically close

TABLE 3.1
Popular Names Denoting *Arachis hypogaea* L. in Some World Languages and Dialects

Language/Dialect	Name
Abenaki	skibô+k
Afrikaans	apeneutjie; grondboontjie
Aka-Jeru	uta
Albanian	badiava; kikirik
Amharic	ocholonī
Antillean Creole	pistach
Arabic	fawall sudạni
Aragonese	cacagüet; calcagüet; cascagüet
Armenian	getnanush
Asturian	cacagüesa; cacagüeses
Atikamekw	pakan
Aymara	chuqupa
Azerbaijani	yerfındığı
Badînî	fstaq j'abid
Basque	kakahuete
Belarusian	arachis
Bengali	cīnābādāma; māṭa-kalāi
Berber	akawkaw
Bislama	pinat
Bosniak	kikiriki
Breton	kakaouetenn; kraoñenn-varmouz
Bulgarian	fŭstŭk
Catalan	cacao; cacauet
Cebuano	arachis
Cherokee	tuya aniladisgi
Chinese (Cantonese)	huāshēng
Chinese (Hakka)	fân-theu
Chinese (Mandarin)	chang sheng guo; huāshēng; luo hua sheng
Choctaw	bahpo
Cree	pâkân
Croatian	kikiriki
Czech	burský oříšek; podzemnice olejná
Danish	jordnød
Dari	badam zameeni
Dené	dlíenı
Dhao	kabui ae lèu
Dutch	apennoot; grondnoot; olienoot; pinda aardnoot
Dyula	tigba
Enga	kalípu
English	earth-nut; goober; goober bean; goober pea; gouber pea; ground-pea; groundnut; monkeynut; peanut; pindar nut

(*Continued*)

TABLE 3.1 (*Continued*)
Popular Names Denoting *Arachis hypogaea* L. in Some World Languages and Dialects

Language/Dialect	Name
Esperanto	arakido; ternukso
Estonian	arahhis; harilik maapähkel; hiina pähkel
Faroese	iarðnøt
Ferraresi	bagìga
Fijian	pinati
Filipino	mani
Finnish	maapähkinä
French	arachide; cacahuèt; cacahuète; pistache de terre
French (Canada)	pinotte
Frisian	apenút
Friulian	bagjigji; pistaç di tiere
Galician	cacahuete
Genoese	pistàccio
Georgian	mitsis t'khili
German	aschanti; Arachis; Erdmandel; Erdnuss; Kamerunnuss
Greek	arahída; arápiko fystíki
Greek (Cyprus)	foustoukoúdi
Guarani	manduvi
Gullah	guba
Hausa	gyaɗa
Hawaiian	pineki
Hebrew	'gvz 'dmh
Hindi	cīnā-badāma; moongaphalee; mosaṃbī caṇā; mumphali; mungphali
Hmong	txiv lws suav
Hungarian	amerikaimogyoró; földimogyoró
Icelandic	jarðhneta
Igbo	ahuekere; asiboko; opupa
Ilocano	maní
Indonesian	kacang tanah
Interlingua	arachide
Irish	phis talún
Italian	arachide; aracide; bagigi; caccaetti; cecini; nocciolina; noccioline americane; scachetti; spagnolette
Japanese	nankin-mame; piinatsu; rakkasei
Javanese	kacang brol
Kalaallisut	jordnøddi
Kannada	kaḍalēkāyi
Kapampangan	mani
Kazakh	arahis; jañğağı; jañğağı qıtai
Khmer	santek dei
Kimbundu	nguba

(*Continued*)

TABLE 3.1 (*Continued*)
Popular Names Denoting *Arachis hypogaea* L. in Some World Languages and Dialects

Language/Dialect	Name
Kongo	mpinda; nguba
Korean	ttangkong
Kriol	pinat
Kupang Malay	kacang
Kurdish (Central)	paqlx'i suudani
Kurdish (Northern)	fisteqê erdê; zirfisteq
Kyrgyz	jer jaŋgak
Lao	thouadin
Latvian	arahiss; zemesrieksts
Ligurian	pistàccio
Lithuanian	arachis; arachisu; valgomasis arachis; žemės riešutu
Luxembourgish	kakuett
Macedonian	kikiritka
Makasae	uta
Malagasy	kapika; voanjo
Malay	kacang tanah
Malayalam	nilakkaṭala
Malecite-Passamaquoddy	ktahkitom
Maltese	karawetta
Mam	ch'i'lilj
Manx	cro thallooin
Māori	pīnati
Marathi	śēṅgadāṇā
Maung	jangkurri
Min (Eastern)	huă-sĕng
Min (Southern)	thô˙-tāu
Mirandese	cascaboi
Moldovan	alune americane; alune de pământ; arahidă; arahide
Mongolian	gazryn samryn; khuasan
Montagnais	pakan
Myanmar	mway pell
Nahuatl (Classical)	tlalcacahualotl
Nahuatl (Huasteca)	tlālcacahuatl
Nepali	badāma
Norwegian (Bokmål)	jordnøtt; peanøtt
Norwegian (Nynorsk)	jordnøtt; peanøtt
Ojibwe	bagaanag
Otomi	maní
Palenquero	ngubá
Pashto	mmpliu

(*Continued*)

TABLE 3.1 (*Continued*)
Popular Names Denoting *Arachis hypogaea* L. in Some World Languages and Dialects

Language/Dialect	Name
Persian	badam zmini
Polish	fistaszki; orzacha podziemna; orzech arachidowy; orzech ziemny
Portuguese	alcagoita; amendoim; caranga; largo; macarra; mandobi; mendobi
Portuguese (Brazil)	amendoí; amendoís; mandobi; mandubi; mendubi; menduí; mindubi; minuim
Punjabi (Eastern)	mūgaphalī
Punjabi (Western)	moong phali
Quechua (Bolivia)	inchik; chuqupi
Quechua (Peru)	inchis
Réunion Creole	pistache de terre
Romani	fastătaja
Romanian	alun de pământ; alune americane; arahide
Russian	arakhis kul'turnyi; arakhis podzemnyi; kitaiskie oreshki; zemlianoi orekh
Samoan	pīnati
Sango	karakō
Sanskrit	kalāyaḥ
Scottish Gaelic	cnò-thalmhainn
Serbian	arahis; kikiriki
Shipibo	táma
Shona	nzungu
Sinhalese	raṭakaju
Slovak	arachida; arašid; arašida; podzemnica olejná
Slovenian	arašidovo; kikiriki
Somali	loos; lows
Sorbian (Lower)	zemski wórješk
Sorbian (Upper)	zemski worjech
Spanglish	pínut
Spanish	avellana americana; cacahuate; cacahuete; cocos; maní; pistache
Spanish (Puerto Rico)	pindels
Sundanese	suuk
Swahili	karanga; mjugu nyasa; mnjugu; mnjugu nyasa
Swedish	jordnöt
Tagalog	mani
Tajik	arakhis
Tamil	nilakkaṭalai
Tatar	jir çikqlävege
Telugu	vēruśanaga
Tetum	forai
Thai	thạ̀w

(*Continued*)

TABLE 3.1 (*Continued*)
Popular Names Denoting *Arachis hypogaea* L. in Some World Languages and Dialects

Language/Dialect	Name
Tigrinya	fuli
Tiwi	pinati; wurranyini
Tongan	pīnati
Turkish	yer fıstığı
Turkmen	arahis
Ukrainian	arakhis kul'turnyi; arakhis pidzemnyi; kitaiskyi bib; zemlianyi gorikh
Urdu	moong phali
Uyghur	khasaang
Uzbek	yeryong'oq
Valencian	cacau
Venetian	bagigio
Vietnamese	đậu phộng; đậu phụng; lạc
Volapük	raagid
Waray	mani
Welsh	cnau mwnci; cneuen y ddaear
Xhosa	indongomane
Yiddish	fistashke; st'ashk'\r\n
Yoruba	epa
Yucatec	cacahuete; maní
Zulu	ikinati; indlubu; intongomane

to its center of origin. Among such examples are Aymara, with *chuqupa*, both Bolivian and Peruvian Quechua, with *chuqupi*, *inchik* and *inchis*, and Mam, with *ch'i'lilj*, as well as the sole assessed case among the North Amerind languages, Malecite-Passamaquoddy, with *ktahkitom* (Figure 3.1). It is curious that the word denoting peanut in a language from the other, eastern, side of the Atlantic Ocean, namely *ikinati* in Zulu, has a strong morphological resemblance to those used in the Americas—the question if it is a pure coincidence or a consequence of the conquests by the West European colonialists remains a currently unsolved linguistic issue.

There is another attested Proto-Amerind root, namely **man* ~, **min*, originally meaning *tree* and giving several mediating words relating to diverse plants in its descending languages (Greenberg and Ruhlen 2007). Hypothetically, this could give the word *mani* in the language of the Arawak people called Taíno, now unhappily extinct and once inhabiting the Greater Antilles and the Bahamas and responsible for transferring the word *mahiz*, denoting maize (*Zea mays* L.) into Spanish and subsequently into numerous other Old World languages. The same Proto-Amerind root may be responsible for the words denoting peanut in some other American languages, such as the Otomi *maní*, the Yucatec *maní* and the Guarani *manduvi*.

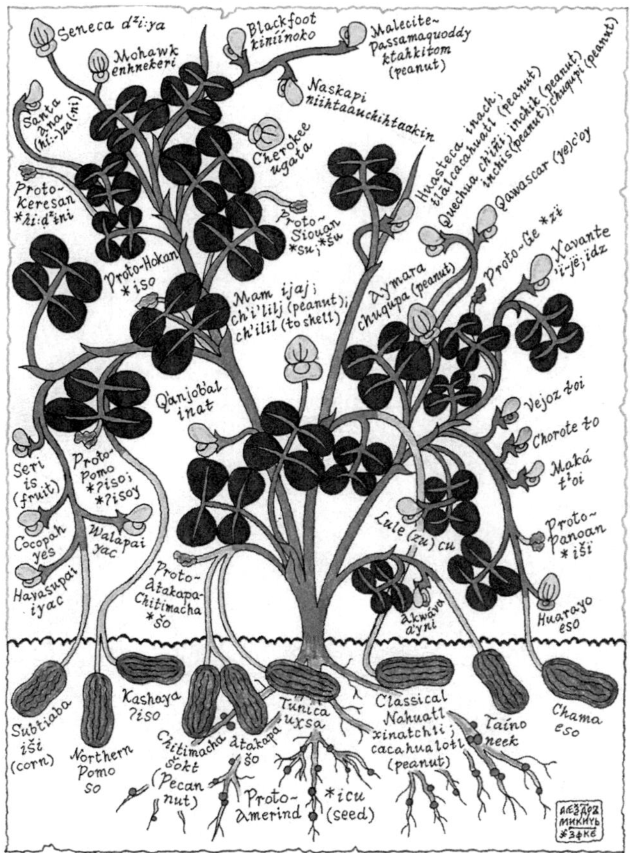

FIGURE 3.1 **(See color insert.)** One of the possible evolutions of the root **icu* of the proposed Proto-Amerind language, denoting seed (Greenberg and Ruhlen 2007), into its extinct and modern descendants in North, Central, and South America; in all cases where the initial meaning was changed, the actual one is given within the brackets; the mediating protolanguages are shown as withered flowers, while the extinct languages are depicted as pods.

Any of these words, regardless of their initial meaning and quite possibly having descriptively become associated with the peanut plant, were borrowed by Spanish and Portuguese. Following the invasions of these two then naval powers across the world, both crop and word reached as far as the Philippines with its many languages, as may be seen from the examples in Ilocano, Kapampangan, Filipino/Tagalog, and Waray (Table 3.1).

It is a remarkably curious how peanut was named in mutually distant and unrelated languages in the almost same fashion, that is, by reminding their speakers mainly about various kinds of familiar local grain legumes and fruits. Among pulses to which the names denoting peanut are based upon, the most frequent are pea and beans, whether common (*Phaseolus vulgaris* L.) or faba bean (*Vicia faba* L.), while the most referred fruits are nuts, in a wider botanical sense (Alasalvar and Shahidi 2008), such as hickories (*Carya* spp.) hazelnut (*Corylus avellana* L.), walnut

(*Juglans regia* L.), pistachio (*Pistacia vera* L.) and almond (*Prunus dulcis* [Mill.] D. A. Webb). The names designating peanut also quite often stress that peanut is a crop essentially connected to earth, ground, or soil and sometimes refer to its quality properties, animal species, human body organs, and some specific distant country, imagined as its homeland or the place where it came from.

One of the most noticeable examples for these various associations among the languages of Americas is the Huasteca Nahuatl word denoting peanut, *tlālcacahuatl*, where *tlāl* stands for *soil* and *cacahuatl* for *cocoanut* (*Theobroma cacao* L.), the latter being derived from the reconstructed Proto Mije-Sokean root with the same meaning, **kakaw~*, **kakawa* (Kaufman and Justeson 2006). The Classical Nahuatl word denoting peanut, *tlalcacahualotl*, was borrowed in its shortened form by the conquistadors in early sixteenth century and began to denote this crop in many European and creole languages and dialects, such as Aragonese, Asturian, Basque, Breton, Catalan, French, Galician, Italian, Luxembourgish, Maltese, Mirandese, Portuguese, Sango, Spanish, and Valencian (Table 3.1). In several other native North American languages, such as Atikamekw, Cree, and Montagnais, the names describing peanut are based upon those denoting equally the fruits of pecan (*Carya illinoinensis* [Wangenh.] K. Koch) with other hickory species and walnut.

The very name *peanut* is certainly one of the most widely known among those that denote the species *Arachis hypogaea* and its numerous varieties, forms, and cultivars outside the global English-speaking community. It is a complex word that equally associates peanut to the pea crop, because of the shape of its seeds, and the nut fruits, due to the morphological characteristics of its pods. Through the English language, the word peanut entered diverse languages across the world, such as Bislama, Fijian, French in Canada, Hawaiian, Japanese, Māori, Norwegian, Samoan, Spanglish, Tiwi, and Tongan (Table 3.1). The alternatively used term *groundnut* has identical meaning as *peanut* only in everyday speech; otherwise, the former has a broader sense and denotes several other plants than *Arachis hypogaea*.

Linking the peanut crop with pea is also present in Bengali, Irish, Japanese, or Myanmar, while the reference to beans may be found in Afrikaans, Albanian, Arabic, Central Kurdish, Dutch, Hindi, Indonesian, Khmer, Malay, Punjabi, Tigrinya, Ukrainian, Urdu, or Vietnamese (Table 3.1). The words denoting peanut in certain Southeast European languages, such as Albanian, Macedonian, and Serbian, remain without deciphered etymology: there is a possibility that in their core is present the Proto-Indo-European root **kek-*, **kik-*, originally designating pea (Nikolayev 2012), although the way it survived solely to denote peanut, a non-native crop to the region, still seeks a thorough linguistic explanation. Certain alternative words for peanut in Italian are derived from the names denoting chickpea (*Cicer arietinum* L.), perhaps equally because of the shape of their seeds and the way of their use as a roasted snack. Whereas several languages and dialects of the Italian Peninsula have the words denoting peanut based upon the vernacular terms relating to pod, such as in Ferraresi, Friulian, standard Italian, or Venetian. Somewhat similar circumstances are found in numerous Dravidian languages, with Kannada, Malayalam, or Tamil, and the neighboring Indo-Aryan languages, such as Sinhalese, where the names denoting peanut contain the morphemes relating to pigeon pea.

In a large number of the languages belonging to most of the world's ethnolinguistic families, the words denoting peanut are connected with either hazelnut or walnut. Among the Indo-European ones, such are Celtic, with Breton or Welsh; Germanic, with Danish and its borrowing into the Greenlandic Eskimo-Aleut language of Kalaallisut, Dutch, English, Faroese, German, Icelandic, Norwegian, or Swedish; Romance, with Italian or Spanish; and Slavic, with Czech, Polish, Russian, or Lower and Upper Sorbian (Table 3.1). The names denoting peanut In the Uralic ethnolinguistic family, with Estonian, Finnish, or Hungarian, are primarily in relation to hazelnut. Pistachio, with an ultimate etymological origin in the Old Persian *pistak*, with the same meaning (Nişanyan 2017), is present in the words designating peanut in Antillean Creole, Badînî, Bulgarian, French, Genoese, German, Cypriot Greek, Ligurian, Polish, Portuguese, Réunion Creole, Romani, or Yiddish, while almond is a part of the names for peanut mostly in the Indo-Aryan and Iranian branches of the Indo-European languages, such as Bengali, Dari, Hindi, Nepali, or Persian.

Since cultivating the peanut crop is substantially associated with its maturing below the surface, it is no wonder that the terms linked to soil, earth, or ground are an inevitable segment of the names denoting this plant in numerous languages of diverse families across the globe (Table 3.1). In that way, we encounter the words designating peanut in the said Amerind language, Huasteca Nahuatl; the Indo-European languages, comprising Baltic, with Latvian or Lithuanian; all Celtic, all Germanic, Iranian, with Northern Kurdish or Persian; Romance, with Friulian, French, or Romanian; and all West and East Slavic languages; nearly all Turkic languages of the speculative Altaic superfamily, such as Azerbaijani, Kazakh, Kyrgyz, Tatar, Turkish, or Uzbek; the Uralic languages, with Hungarian; the Kartvelian languages, with Georgian; and the Austronesian languages, with Indonesian and Malay.

One of the names in Czech connects peanut to *peasants*, without any pejorative background and clearly meaning that it is a nut, not of a tree, but from the field (Table 3.1). It is quite interesting how individual peoples tend to explain where the peanut crop came from to their country: it is America in Hungarian or Moldovan, Arabia in Greek, Cameroon in German, China in Hindi, Japanese or Kazakh, Spain in Italian, and Sudan in Arabic. By some reason, peanut is considered as a monkey food: therefore, the alternative names in Afrikaans, Dutch, English, Frisian, or Welsh. Few languages, such as Czech, Dutch, or Slovak, point out that the peanut crop is rich in oil.

The peanut crop was introduced not only to Europe but farther to Asia. Still in the days of their dominance on the world's oceans, Spanish and Portuguese traders brought it to Central Africa and adjacent regions. The peanut was embraced by the local population and, in many cases, either complemented or replaced by bambara groundnut (*Vigna subterranea* [L.] Verdc.), a native African crop with similar morphology and cultivation, especially in terms of ripening pods under the soil surface. That is how peanut received its first African names, belonging to the languages of the vast Niger-Congo ethnolinguistic family, many of which originally denoted bambara groundnut. Among many examples, there may be mentioned the mutually close words in Igbo and Yoruba, as well as the Kongo *mpinda*, that was adopted and modified by Dutch, English, and Puerto Rican Spanish (Table 3.1). The name *nguba* in both Kimbundu and Kongo means *kidney-like* and is related to the words in other

Niger-Congo languages, such as Dioula, Shona, Swahili, Xhosa, and Zulu. A few centuries later and after the change on the throne of the global seafarers, the countless ships were heading for the Americas, carrying unfortunate slaves with both the peanut crop and their native names denoting it. In that way, peanut finished its kind of the *Tour d'Atlantique* and enriched the languages of Central and North Americas with new words designating it, such as *ngubá* in the creole language Palenquero (Domonoske 2014) and *goober pea* in U.S. English. During the American Civil War, peanut played the same role of staple crop among the Confederate soldiers as it had done before among the black people in their African homelands, serving as their only food and saving them from starvation, as immortalized by the famous traditional folk song *Goober Peas* with its moving ending lines: 'I wish the war was over, so free from rags and fleas / We'd kiss our wives and sweethearts, and gobble goober peas' (Dunaway 1913).

4 *Cajanus* Adans.

Synonyms: *Atylosia* Wight & Arn.; *Cajanus* DC., *Cantharospermum* Wight & Arn.; *Endomallus* Gagnep.; *Peekelia* Harms

4.1 LIST OF TAXA SCIENTIFIC AND POPULAR NAMES

The following paragraphs give a list of the most widely accepted species of the genus *Cajanus* Adans. and their subtaxa, along with their synonyms in various botanical classifications and vernacular names, listed alphabetically and according to the official or most used language designations (ISTA 1982, van der Maesen et al. 1986, van der Maesen 1990, 1998, 2003, Rehm 1994, Gledhill 2008, Porcher 2008, Lim 2012, Quattrocchi 2012, The Plant List 2013, Ecocrop 2017, EPPO 2017, Ethnologue 2017, IBIS 2017, ILDIS 2017, Logos 2017, NPGS 2017, Wikipedia 2017, Wiktionary 2017).

- *Cajanus acutifolius* (F. Muell.) Maesen

Synonyms: *Atylosia acutifolia* F. Muell.; *Rhynchosia acutifolia* (F. Muell.) F. Muell.

English: sharp-leafed cajanus

- *Cajanus albicans* (Wight & Arn.) Maesen

Synonyms: *Atylosia albicans* (Wight & Arn.) Benth.; *Cajanus wightii* Wight & Arn.; *Cantharospermum albicans* Wight & Arn.

English: whitening cajanus

- *Cajanus aromaticus* Maesen

English: aromatic cajanus

- *Cajanus cajan* (L.) Huth (Table 4.1)

Synonyms: *Cajanus cajan* (L.) Millsp.; *Cajanus cajan* (L.) Millsp. var. *cajan*; *Cajanus indorum* Medik.; *Cajanus bicolor* DC.; *Cajanus flavus* DC.; *Cajanus indicus* Spreng.; *Cajanus indicus* var. *bicolor* (DC.) Kuntze; *Cajanus indicus* var. *flavus* (DC.) Kuntze; *Cajanus indicus* var. *maculatus* Kuntze; *Cajanus inodorus* Medik.; *Cajanus luteus* Bello; *Cajanus obcordifolius* V. Singh; *Cajanus pseudo-cajan* (Jacq.) Schinz & Guillaumin; *Cajanus striatus* Bojer; *Cytisus cajan* L.; *Cytisus guineensis* Schum. & Thonn.; *Cytisus pseudocajan* Jacq.; *Phaseolus balicus* L.

- *Cajanus cajan* (L.) Millsp. var. *bicolor* (DC.) Purseglove

Chinese: chi dou; chi xiao dou

English: Congo pea; spotted pigeon pea; red gram

- *Cajanus cajan* (L.) Millsp. var. *flavus* (DC.) Purseglove

Chinese: huang dou shu

English: green pigeon pea; no-eye-pea; yellow dhal; yellow-flowered pigeon pea; yellow-seeded pigeon pea

TABLE 4.1
Popular Names Denoting *Cajanus cajan* (L.) Huth. in Some World Languages and Dialects

Language/Dialect	Name
Acholi	Lapena
Adilabadi	tūri
Afrikaans	duiwe-ertjie; Kongoboontjie
Amharic	yewof ater
Arabic	bisillah Hindîyah; lûbyâ Sûdânî
Arabic (Sudan)	lubia addassy
Assamese	rahar dal; rōhōra
Asturian	bimbu; frijol de palu; guandú; guandul; quinchoncho
Azerbaijani	göyərçin noxudu
Balinese	kekace; undis
Batak Karo	ritik lias
Bengali	arhar; aṛahara
Bikol	tabois
Bontoc	kidis
Buginese	kance
Catalan	pèsol verd tropical
Cebuano	tabois
Chamorro	lenteha franchesa; lenteja francesa
Chewa	nandalo
Chinese (Mandarin)	chieh tu; chieh tu tzu; huang dou shu; mù dòu; shan tou ken
Chinese (Taiwan)	shu dou
Comorian	mtsongi
Czech	kajan; kajan indický; mestelice
Danish	ærtebønne; Angolaært
Dutch	katjang geode; struikerwt
English	Congo-pea; gandul; guandul; gunga pea; gungo pea; no-eye pea; pigeon pea; pigeon-pea; pigeonpea; Puerto Rico pea; red gram; spotted pigeon pea; tur; yellow dhal
English (Cayman Islands)	gungo pea
English (Jamaica)	Congo pea; gungo pea
English (Seychelles)	red gram
Esperanto	kajano; kajan-pizo; kajano-pizo
Estonian	harilik tuvihernes
Fijian	nggiringgiri; pi; rahar
Filipino	kadios
Finnish	kyyhkynherne
French	ambrévade; pois cajan; pois d'Angole
French (Congo)	pois cajan
French (Haiti)	pois Congo
French (West Africa)	pois d'Angola
Galician	frijol de arbol

(*Continued*)

TABLE 4.1 (*Continued*)
Popular Names Denoting *Cajanus cajan* (L.) Huth. in Some World Languages and Dialects

Language/Dialect	Name
Georgian	mtredi ts'erts'va
German	Strauchbohne; Straucherbse; Taubenerbse
Gondi	tōri
Gujarati	tuvara; tuvēra; tuvēranī dāḷa
Haitian Creole	pwa kongo
Halmahera Sea	acang iris; kacang turis; lebui; legui; puwe jai
Hausa	waaken santanbul; waken turawa
Hawaiian	pī nūnū; pī pokoliko
Hindi	arahar daal; arahara; arhar; dahl; dhal; ihora; laher; oroha; oror; toor; tor; tuar; tur; tuur; tuvar; tuver; tuver dahl
Hungarian	galambborsó
Ibanag	kardis
Icelandic	dúfnabaun
Ifugao	kusia
Igorot	kaldis; kardis; kusia
Ilocano	kaldis; kardis; kidis
Indonesian	gude; kacang bali; kacang gude; kacang kayo
Italian	caiano; pisello d'Angola; pisello del tropico
Japanese	ki-mame; pijonpii
Javanese	gude; kacang gude; kacang kayu
Kannada	athaki; baele; byale; dhaal; kari uddu; thogari bele; thurukara, thogari; togari; togari kāḷu
Khmer	sândaèk dai; sândaèk kléng; sândaèk klöng; sândaèk kroap sâ'; sândaèk kroëb sâ
Kinwatkar	togar
Kinyarwanda	itenderwa; umukunde
Kirundi	agacaruzo; inkunde; urucaruzo
Kodava	tōri bēḷe
Kolami	togar
Konda	keŋ
Lango	apena; lapena
Lao	thwàx h'ê
Madiya	tōra
Maithili	arahar daal
Makassarese	binatung
Malagasy	ambarivatry; ambatribe; ambaty; amberivatry; ambote; ambraty; ambrevade; ambrevate; antsotry
Malay	dhal; kacang bali; kacang dal; kacang hiris
Malayalam	thora-paerou; thuvarappayar; toovara paruppu; tuvara; tuvarapparippu
Maldivian	toḷi mugu

(*Continued*)

TABLE 4.1 (*Continued*)
Popular Names Denoting *Cajanus cajan* (L.) Huth. in Some World Languages and Dialects

Language/Dialect	Name
Mangyan	kadios
Marathi	thoora; thoori; thoovar; toor toovar; tūra; tuur
Meitei	mairongbi
Myanmar	pellhcainnngone
Naiki	togari
Nepali	rahara
Nkore	enkuuku
Norwegian (Bokmål)	ertebønne
Norwegian (Nynorsk)	ertebønne
Nyoro	enkuuku
Odia	arhar; harara
Panay	kadios
Papiamento	wandu
Persian	nxud kftri
Polish	nikla indyjska
Portuguese	andu; anduzeiro; ervilha de Angola; ervilha-de-pombo; ervilha do Congo; feijão boer; feijão-guandu; guisante-de-Angola; guandeiro; guando; guandu; jinsonge
Portuguese (Brazil)	andu; anduzeiro; ervilha-de-angola; ervilha-de-árvore; ervilha-de-congo; ervilha-de-sete-anos; feijão-de-árvore; feijão-guando; feijão-guandu; guandeiro; guando; guandu; tantaraga
Punjabi	toor; toovar
Rotenese	tulis
Russian	golubinyi gorokh; kaian
Sanskrit	adhaki; tuvarī
Serbian	kajan
Sinhalese	wal-kollu
Spanish	bimbu; cachito; chícharo de árbol; frejol de palo; fríjol de árbol; frijol de la India; frijol de palu; frijol del monte; frijol guandul; frijol quinchancho; guisante-de-Angola; guando; guandú; guandul; guisante de paloma; guisante gunga; guisante gungo; quinchonchillo; quinchocho; vaina del guandú
Spanish (Argentina)	planta de guandú
Spanish (Colombia)	frijol quinchancho
Spanish (Cuba)	gandul
Spanish (Peru)	frijol de palo
Spanish (Puerto Rico)	gandul; gandule; gandures
Spanish (Venezuela)	quinchonchillo; quinchoncho
Sundanese	gude; kacang bali; kacang gude; kacang hiris
Swahili	apena; empinamuti; enkolimbo; mbaazi; ntondigwa mbaazi
Swedish	duvärt

(*Continued*)

TABLE 4.1 (*Continued*)
Popular Names Denoting *Cajanus cajan* (L.) Huth. in Some World Languages and Dialects

Language/Dialect	Name
Tagalog	gablos; kadios; kagyos; kagyus; kalios; tabios
Tamil	amakam; amam; atacai; ataki; atakam; curattam; duvarai; impurupali; irumpali; iruppappuli; iruppuli; iruppulikam; iruppulikamaram; iyavu; iyavucceti; kacci; kaccikacceti; kaccikam; kalvayam; kalvayamaram; kanti; karaviram; karkai; karkaicceti; kattu-thovarai; kaycci; kecapukacceti; kecapukam; kuvalam; kuvalamaram; malaittuvarai; malikaittuvarai; malur; miruttalakam; miruttanam; miruttanam yarai; muluttuvarai; naiciravam; naiciravamaram; nattuttuvarai; pataippeyan; pataippeyanmaram; toovaram paruppu; torai; thuvarai; tuvarai; tuvaraippayaru; tuvarankay; tuvari; tuvarikam
Telugu	aadhaki; errakandulu; kandi; kandul; kandulu; kondakandi; peddakandi; peddakondakandi; polukandi; potujandalu; sinnakandi
Ternate	fouhate
Teso	ekilimite
Tetum	tunis
Thai	ma hae; thua maetaai; thua rae
Tibetan	tubari; tuparip
Tidore	fouhate
Timorese	tunis
Tongan	pī kula; piisi kula
Tooro	enkuuku
Tulu	togarè; togari
Vietnamese	cay dau chieu; đậu cọc rào; đậu săng; đậu thong; đậu triều tên khác
Wakatobi (Tomia)	koloure
Waray	cajanus
Yavatmal	tūriŋ
Yoruba	otili; otinli

- *Cajanus cajanifolius* (Haines) Maesen
Synonyms: *Atylosia cajanifolia* Haines; *Cantharospermum cajanifolium* (Haines) Raizada
English: cajan-leafed cajanus
Telugu: banokandulo
- *Cajanus cinereus* (F. Muell.) F. Muell.
Synonyms: *Atylosia cinerea* F. Muell.
English: ashen cajanus
- *Cajanus confertiflorus* F. Muell.
Synonyms: *Atylosia pluriflora* Benth.
English: densely-flowered cajanus
- *Cajanus crassicaulis* Maesen
English: thick-stemmed cajanus

- *Cajanus crassus* (Prain ex King) Maesen
Synonyms: *Atylosia crassa* Prain ex King; *Atylosia mollis* Benth.; *Atylosia volubilis* (Blanco) Gamble; *Cantharospermum volubile* (Blanco) Merr.; *Cantharospermum volubilis* (Blanco) Merr.
Bengali: mewape
English: dense cajanus; thick cajanus
Hindi: jangal baler
- *Cajanus crassus* (Prain ex King) Maesen var. *burmanicus* (Collett & Hemsl.) Maesen
English: Burmese cajanus
- *Cajanus elongatus* (Benth.) Maesen
Synonyms: *Atylosia elongata* Benth.; *Cantharospermum elongatum* (Benth.) Raizada
English: elongated cajanus
- *Cajanus geminatus* Pedley ex Maesen
- *English*: paired cajanus; twin cajanus
- *Cajanus goensis* Dalzell
Synonyms: *Atylosia barbata* (Benth.) Baker; *Atylosia calycina* (Miq.) Kurz; *Atylosia goensis* (Dalzell) Dalzell; *Atylosia siamensis* Craib; *Cantharospermum barbatum* (Benth.) Koord.; *Dolichos ornatus* Wall., nom. nud.; *Dunbaria barbata* Benth.; *Dunbaria calycina* Miq.; *Dunbaria stipulata* Thuan; *Dunbaria thorelii* Gagnep.; *Endomallus pellitus* Gagnep.; *Endomallus spirei* Gagnep.
English: Goan cajanus
- *Cajanus grandiflorus* (Benth. ex Baker) Maesen
Synonyms: *Atylosia grandiflora* Benth. ex Baker; *Dunbaria pulchra* Baker; *Pueraria seguinii* H. Lev.
English: large-flowered cajanus
- *Cajanus heynei* (Wight & Arn.) Maesen
Synonyms: *Atylosia kulnensis* (Dalzell) Dalzell; *Cajanus kulnensis* Dalzell; *Dunbaria heynei* Wight & Arn.; *Dunbaria oblonga* Arn.
English: Heyn's cajanus
- *Cajanus hirtopilosus* Maesen
English: hairy-shaggy cajanus
- *Cajanus kerstingii* Harms
English: Kertsing's cajanus
- *Cajanus lanceolatus* (W. Fitzg.) Maesen
Synonyms: *Atylosia lanceolata* W. Fitzg.
English: lanceolate cajanus
- *Cajanus lanuginosus* Maesen
Synonyms: *Cajanus lanuginosus* (S. T. Reynolds & Pedley) Maesen
English: woolly cajanus
- *Cajanus latisepalus* (S. T. Reynolds & Pedley) Maesen
Atylosia latisepala S. T. Reynolds & Pedley
English: broad-sepalled cajanus
- *Cajanus lineatus* (Wight & Arn.) Maesen

Synonyms: *Atylosia lawii* Wight; *Atylosia lineata* Wight & Arn.; *Cantharospermum lineatum* (Wight & Arn.) Raizada
English: lined cajanus
Hindi: jangal tur
- *Cajanus mareebensis* (S. T. Reynolds & Pedley) Maesen
Synonyms: *Atylosia mareebensis* S. T. Reynolds & Pedley
English: Mareeben's cajanus
- *Cajanus marmoratus* (R. Br. ex Benth.) F. Muell.
Synonyms: *Atylosia marmorata* R. Br. ex Benth.
English: marbled cajanus
- *Cajanus mollis* (Benth.) Maesen
Synonyms: *Atylosia mollis* Benth.; *Cantharospermum molle* (Benth.) Taub.; *Cantharspermum molle* (Benth.) Taub.; *Collaea mollis* Wall.
English: tender cajanus; wild pigeon pea
Hindi: ban tur
- *Cajanus niveus* (Benth.) Maesen
Synonyms: *Atylosia nivea* Benth.; *Cantharospermum niveum* (Benth.) Raizada
English: snowy cajanus
- *Cajanus platycarpus* (Benth.) Maesen
Synonyms: *Atylosia geminiflora* Dalzell; *Atylosia platycarpa* Benth.; *Cantharospermum distans* Baker; *Cantharospermum geminiflorum* (Dalzell) Raizada; *Cantharospermum geminifolium* (Dalzell) Raizada; *Cantharospermum platycarpum* (Benth.) Raizada
English: broad-fruited cajanus
Hindi: sukli sengha
- *Cajanus pubescens* (Ewart & J. L. Morrison) Maesen
Synonyms: *Atylosia pubescens* (Ewart & Morrison) S. T. Reynolds & Pedley; *Tephrosia pubescens* Ewart & J. L. Morrison
English: hairy cajanus; ripening cajanus
- *Cajanus reticulatus* (Aiton) F. Muell.
Synonyms: *Atylosia grandifolia* Benth.; *Cajanus reticulatus* (Dryand.) F. Muell.
English: net-like cajanus
- *Cajanus reticulatus* (Aiton) F. Muell. var. *grandifolius* (F. Muell.) Maesen
Synonyms: *Cajanus grandifolius* F. Muell.
English: large-leafed cajanus
- *Cajanus reticulatus* (Aiton) F. Muell. var. *maritimus* (S. T. Reynolds & Pedley) Maesen
Synonyms: *Atylosia reticulata* subsp. *maritima* S. T. Reynolds & Pedley
English: maritime cajanus
- *Cajanus reticulatus* (Aiton) F. Muell. var. *reticulatus*
Synonyms: *Dolichos reticulatus* Aiton
English: common net-like cajanus
- *Cajanus rugosus* (Wight & Arn.) Maesen
Synonyms: *Atylosia rugosa* Wight & Arn.; *Cantharospermum rugosum* (Wight & Arn.) Alston
English: wrinkled cajanus

Sinhalese: wal-kollu
- *Cajanus scarabaeoides* (L.) Thouars
Synonyms: *Atylosia pauciflora* (Wight & Arn.) Druce; *Atylosia scarabaeoides* (L.) Benth.; *Cantharospermum pauciflorum* Wight & Arn.; *Cantharospermum scarabaeoides* (L.) Baill.; *Cantharospermum scarabaeoideum* (L.) Baill.; *Dolichos medicagineus* Roxb.; *Dolichos minutus* Wight & Arn.; *Dolichos scarabaeoides* L.; *Rhynchosia biflora* DC.; *Rhynchosia scarabaeoides* (L.) DC.; *Stizolobium scarabaeoides* (L.) Spreng.
Chinese (Cantonese): shui kom ts'o
Chinese (Mandarin): man cao chong duo
Chinese (Yunnan): jia yan pi guo
English: scarab-like cajanus
French (Mascarene Islands): fausse pistache marronne; pistache marronne
Hindi: jangal tor; jangli tur; ram kurti
Malagasy: vahi-tsokona
Nepali: ban bhartha; ban kurthi
Sinhalese: wal-kollu
Thai: K̄hî̂h̄nxn t̄heā
- *Cajanus sericeus* (Benth. ex Baker) Maesen
Synonyms: *Atylosia sericea* Benth. ex Baker; *Cantharospermum sericeum* (Baker) Raizada
English: silken cajanus
Hindi: rantur
- *Cajanus trinervius* (DC.) Maesen
Synonyms: *Atylosia candollei* Wight & Arn.; *Atylosia major* Wight & Arn.; *Atylosia trinervia* (DC.) Gamble; *Atylosia trinervia* (DC.) Gamble var. *major* (Wight & Arn.) Gamble; *Cantharospermum trinervium* (DC.) Taub.; *Collaea trinervia* DC.; *Odonia trinervia* (DC.) Spreng.
English: jungle pea; three-veined cajanus
Sinhalese: atta-tora; et-tora
Toda: tifiry
- *Cajanus villosus* (Benth. ex Baker) Maesen
Synonyms: *Atylosia villosa* Benth. ex Baker
English: hairy cajanus
- *Cajanus viscidus* Maesen
English: glutinous cajanus
- *Cajanus volubilis* (Blanco) Blanco
Synonyms: *Cytisus volubilis* Blanco
English: whirling cajanus

4.2 ORIGIN OF SCIENTIFIC AND POPULAR TAXA NAMES

An immense diversity of the words denoting *Cajanus cajan*, especially in the modern Dravidian languages, may be considered quite supportive to the claims that pigeon pea was domesticated in the Indian subcontintent and used as one of the main components in everyday diets (Southworth 2006, Kassa et al. 2016). According to the available etymological databases, there are two attested Proto-Dravidian roots that are either directly or indirectly related to pigeon pea (Burrow and Emeneau 1998, Krishnamurti 2003, Starostin 2006). The first of them is **kāj-*, which served as the scientific name of the genus *Cajanus* by the French botanist and naturalist Adanson (1763). Its primeval meaning was *fruit* and remained the same in the direct descendants of the Proto-Dravidian, namely Proto-Northern, Proto-South, and Proto-South-Central, with **qāj-nǯ-*, **kāj-*, and **kāj-*, respectively, as well as in the majority of the modern Dravidian languages, such as Kolami, Kota, Kurukh, Malto, Naiki, Ollari, Toda, Tamil, or Telugu (Starostin 2006, 2009). At the same time, this meaning was shifted to semantically associated terms in certain languages, such as *to ripe* in Malayalam, *pod* in Kannada, *seed* in Tulu, *a pulse crop* or *dal* in Kui, Kuvi, Manda or Pengo, and precisely *pigeon pea* in Konda (Starostin 2006, Mikić 2016).

Overall, the earliest history of the peoples of South Asia and Oceania remains far from satisfactorily studied, analyzed, explained, and presented (Winters 2014). One of the issues in this remarkably complex topic is a possible Dravidian influence eastward away from its current speaking area, especially the western islands of Oceania, including Indonesia, where certain basic elements of human society, such as kinship, were assessed (Hage 2001). Following such results, there could be at least a slight possibility that the Proto-Dravidian root **kāj-* is responsible for certain names connected with various bean crops, including pigeon pea, such as in Balinese, Sundanese, or Tomia Wakatobi (Table 4.1).

Among numerous terms relating to pigeon pea in the Central-South Dravidian languages, Tamil and Telugu have several words possibly descending from the said root **kāj-*, namely *kanti* with its cognates, such as *kalvayam*, *karkai,* or *kaycci*, and *kandi* with its derivations, such as *kandul* or *kandulu*, respectively (Table 4.1). These words, either from Tamil or Telugu or from both could be borrowed with slight modifications of the initial consonant *k-* by both neighboring languages of other ethnolinguistic families. Such may comprise Indo-European, with a rather peculiar Indo-Aryan language Sinhalese, surrounded by Dravidian and with as strong Vedda substratum (Blundel 2006), with its *wal-kollu*, equally possible to be a borrowing of the Tamil word denoting pigeon pea and being of the Vedda origin and descriptive nature, where the element *kollu* means *leaf.* Other examples may be found in Sino-Tibetan, with both the Mandarin Chinese *huang dou shu* and Myanmar *pellhcainnngone*, and Austroasiatic, with the Vietnamese *cay dau chieu*.

Other and more widely known examples of embracing the pigeon pea crop together with its Dravidian-originating names, such as the aforementioned Tamil *kanti* or the Telugu *kandi*, are the consequence of the contacts of various European naval powers with the inhabitants of the Indian subcontinent during past several centuries. Often, they resulted in transferring both to new homelands in Africa and the Americas. Their traces are visible in the primary names designating pigeon pea in the Indo-European languages, such as Asturian, Dutch, English, with its local speeches in the some of the Caribbean islands, such as the Caymans and Jamaica, Portuguese, in both mother country and the Portuguese-speaking world, especially Brazil, where pigeon pea has been successfully domesticated and intensively grown for various purposes (Ceccon et al. 2013), Spanish, with its dialects in North, Central, and South Americas, and the creoles, such as Papiamento. Similar cases may be found in Chewa, Kinyarwanda, or Kirundi of the African Niger-Congo ethnolinguistic family (van der Maesen 1995).

The scientific name for the whole genus became a word in the languages of the peoples of mostly temperate climates, where pigeon pea has never been either cultivated or used, such as Czech, French, Indonesian, Italian, Russian, Serbian, or Waray (Table 4.1). The same is valid for the constructed languages, such as Esperanto.

Another attested Proto-Dravidian root with a link to pigeon pea is **toɣar-* (Starostin 2006), which the primeval meaning is *dal*, a grain legume soup or, in the case of many regions across the Indian subcontinent, a meal prepared specifically by cooking this native pulse crop. This represents an additional proof of the constant and essential place of grain legumes, to which we may rather reliably add pigeon pea, in the human consumption throughout the history of mankind (Kassa et al. 2012). The proto-word **toɣar-* is considered to have a complex evolution in two of its three main branches, namely Central and Southern, where its direct derivatives, **togar-* and **tUvar-*, retained the same meaning, but also gained another one, namely *pigeon pea* (Figure 4.1). The latter had prevailed in most of the contemporary Dravidian languages and acquired an immensely abundant diversity, as may be seen in the cases of Adilabadi, Gondi, Kannada, Kodava, Kolami, Konda, Madiya, Malayalam, Naiki, Tamil, Telugu, Tulu, or Yavatmal (Table 4.1). In Toda, its word *tifiry* denotes jungle pea, providing evidence that there have been and still are other *Cajanus* species save pigeon pea regarded as edible and suitable for using in the form of dal, although constantly neglected and underutilized (Smartt 1985).

The Dravidian words denoting pigeon pea, originating from the Proto-Dravidian **toɣar-* were borrowed by numerous close and distant languages, in both a geographical and linguistic sense. Its reflections may easily be visible in the Indo-European languages, such as neighboring Indo-Aryan, with Gujarati, Hindi, Kinwatkar, Maldivian, Marathi, Punjabi, Sanskrit or Sinhalese, and colonial Germanic English (Table 4.1); Afroasiatic languages, such as Chadic Hausa, Sino-Tibetan languages, with Tibetan, the Tai-Kadai languages, including Lao or Thai; and the Austronesian languages, such as Rotanese or Timorese.

As a non-native crop in various environments and in diverse unrelated languages, pigeon pea was named in a similar mode, having caused the nearly identical associations, primarily with various kinds of locally cultivated grain legumes. Such are pea (*Pisum sativum* L.), due to the roundish shape of pigeon pea grains, and soybean (*Glycine max* [L.] Merr.), hyacinth bean (*Lablab purpureus* [L.] Sweet), common bean (*Phaseolus* spp.), faba bean (*Vicia faba* L.) and *Vigna* beans. Among the collected names denoting pigeon pea (Table 4.1), it may be seen that this crop is connected with pea in Afrikaans, Catalan, English, Estonian, Fijian, Finnish, French, German, Haitian Creole, Hawaiian, Hungarian, Ibanag, Italian, Japanese, Portuguese, Russian, Spanish, Swedish, and Tongan. On the other hand, pigeon pea is reminiscent of various legume beans in Afrikaans, Arabic, Bikol, Bontoc, Cebuano, Filipino, Galician, Georgian, German, Icelandic, Igorot, Ilocano, Indonesian, Javanese, Khmer, Malay, Mangyan, Portuguese, Spanish, or Tagalog. An interesting kind of hybrid word, merging pea with bean, is present in Danish and both Bokmål and Nynorsk Norwegian. Linking pigeon pea specifically with soyabean, lentil (*Lens*

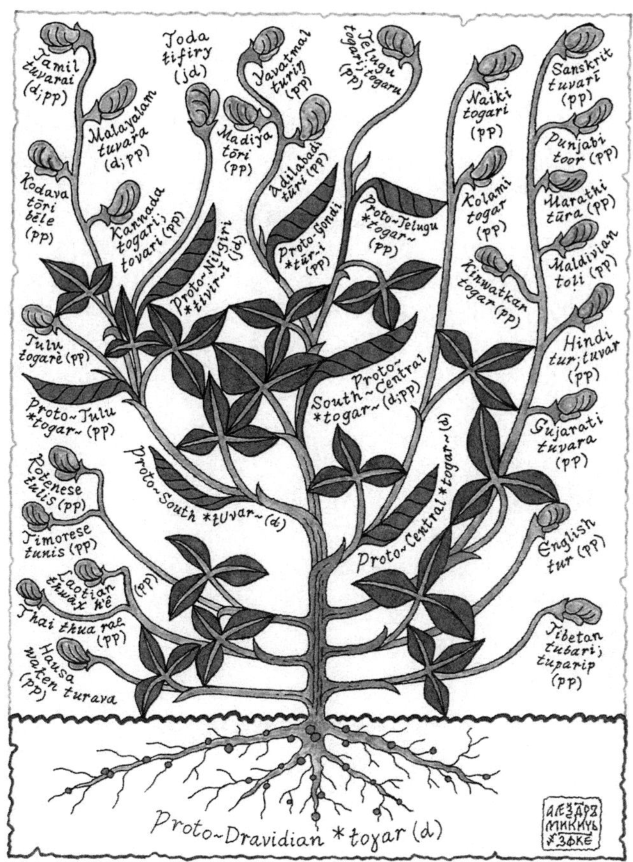

FIGURE 4.1 **(See color insert.)** One of the possible evolutions of the Proto-Dravidian root **toɣar-*, denoting dal (Starostin 2006), into its direct derivatives, represented as pods, and contemporary descendants, portrayed as flowers, in the Central and Southern Dravidian languages; the meaning of each proto- and modern word is given within the brackets, namely (d) for *dal*, (jd) for *jungle pea*, and (pp) for *pigeon pea.*

culinaris Medik.), and chickpea (*Cicer arietinum* L.) is demonstrated in Chinese, Chamorro, and Azerbaijani with Persian, respectively. The Swahili word denoting pigeon pea, *mbaazi*, is a curious African cognate of the Indian dal, referring to a protein-rich pea soup made exactly from pigeon pea (Yeakel et al. 2009).

The attested Proto-Indo-European root **arenko-*, **arn k(')-*, meaning both a kind of cereal (Pokorny 1959) and a leguminous plant in general (Nikolayev 2012) and being responsible for the scientific genus name *Arachis* spp. (Chapter 3), produced numerous derivatives in modern languages of its Indo-Aryan branch, such as in Assamese, Bengali, or Hindi, with its borrowing into Fijian, Maithili, Nepali, Odia (Table 4.1). In Dutch and German, there are the references to the bushy growth habit of pigeon pea (Mudaraddi et al. 2013), while some of the names in Galician, Portuguese, and Spanish describe it as a kind of tree, due to its advancing stem lignifications along with maturity (Elzaki et al. 2012). One of the names in Brazilian Portuguese points out its perennial nature, fully expressed in its native warm climates, where pigeon pea may live as much as seven years (Waldman et al. 2017).

The sole animal with which the pigeon pea crop is associated is, as its English name says, pigeon (Columbidae Leach). Like many other birds of similar size, such as poultry, pigeons prefer the pulses of a smaller seed size (Cacan et al. 2016), making them one of the most harmful pests in the grain production of lentil, pea, vetches (*Vicia* spp.), and various cereals (Patel 2016). On the other hand, pigeon pea and its pulse relatives are desirable in feeding pigeons (Xie et al. 2016), and it is no wonder that the name of this bird consists of names denoting *Cajanus cajan* in many languages of diverse ethnolinguistic families, such as in Afrikaans, Asturian, Azerbaijani, Estonian, Finnish, German, Hungarian, Icelandic, Japanese, Portuguese, Spanish, or Swedish (Table 4.1).

Pigeon pea is another prominent pulse crop globetrotter: its names in numerous mutually non-related languages postulate its country of origin in many a corner of the world. It is Angola in Danish, French, Italian, Portuguese, or Spanish; Congo in Afrikaans, English, or Haitian Creole; France in Chamorro; India in Arabic, Czech, or Polish; Puerto Rico in English or Hawaiian; and the Sudan in Arabic, as well as, by some reason, simply some mountains in Spanish and tropics in general in Catalan and Italian (Table 4.1).

5 *Cicer* L.

5.1 LIST OF TAXA SCIENTIFIC AND POPULAR NAMES

The following paragraphs provide a list of the species of the genus *Cicer* L. and their subtaxa, as regarded by several currently most relevant botanical databases, as well as some of their systematic synonyms and vernacular names, given alphabetically and according to the official or most used language designations (ISTA 1982, Rehm 1994, van der Maesen et al. 2007, Gledhill 2008, Porcher 2008, Hannan et al. 2013, The Plant List 2013, Ladizinsky and Abbo 2015, Ecocrop 2017, EPPO 2017, Ethnologue 2017, IBIS 2017, ILDIS 2017, Logos 2017, NPGS 2017, Wikipedia 2017, Wiktionary 2017).

- *Cicer acanthophyllum* Boriss.
Synonyms: *Cicer garanicum* Boriss.
English: spiny-leafed chickpea
Russian: nut garanskii; nut iglolistnyi; nut kolyuchelistnyi
- *Cicer anatolicum* Alef.
Synonyms: *Cicer glutinosum* Alef.; *Cicer songaricum sensu* Jaub. & Spach
English: Anatolian chickpea
Russian: nut anatoliiskii
- *Cicer arietinum* L. (Table 5.1)
Synonyms: *Cicer album* hort.; *Cicer arientinium* L.; *Cicer arientinum* L.; *Cicer edessanum* Bornm.; *Cicer grossum* Salisb.; *Cicer nigrum* hort.; *Cicer physodes* Rchb.; *Cicer rotundum* Alef.; *Cicer sativum* Schkuhr; *Cicer sintenisii* Bornm.; *Ononis crotalarioides* M. E. Jones
- *Cicer atlanticum* Coss. ex Maire
Synonyms: *Cicer maroccanum* Popov
English: Atlantic chickpea
- *Cicer balcaricum* Galushko
English: Balkarian chickpea
Russian: nut balkarskii
- *Cicer baldshuanicum* (Popov) Lincz.
Synonyms: *Cicer flexuosum* subsp. *baldshuanicum* Popov
English: Baldshuanian chickpea
Russian: nut baldzhuanskii
- *Cicer bijugum* Rech. f.
English: doubled chickpea; paired chickpea
- *Cicer canariense* A. Santos & G. P. Lewis

TABLE 5.1
Popular Names Denoting *Cicer arietinum* L. in Some World Languages and Other Linguistic Taxa

Language/Taxon	Name
Afar	sabbar-ḗ
Afrikaans	kikkererwt
Albanian	qiqër; qiqra
Albanian (Arvanitika)	k'ik'ere
Alemannic	Kichererbse
Amharic	shimibira
Arabic	amazir; hammous; hhimmass; ḥimmiṣ; ḥimmaṣ; hommos; hommos malana; homs shayie; hummus
Aramaic	ḥ rṭwmn'
Araucanian	kalfan
Argrobba	šumbura
Armenian	siseṛ
Armenian (Classical)	siseṛn
Aromanian	tseatsiri
Assamese	butmah
Asturian	garbanzu
Awngi	šəmbər-i
Azerbaijani	qoyunnoxudu
Bashkir	noqot borsaꝗь
Basque	garbantzua; txitxirioa
Belarusian	baranoŭ garoch; nut kul̓turnyj
Bengali	butakala; chanaabatulaa; chanabartula; chōlā; chotobata; chotobut
Berber	himz; ikiker
Bihari	chana
Bosniak	leblebija; slanutak
Bulgarian	leblebiya; nahutat; slanutak; stragali
Burji	šimbur-a
Burmese	kalabèh
Calabrian	ciciaru
Catalan	ceirons; cigró; cigronera; cigrons; ciurons
Caterisani	ciciru
Cebuano	garbanso
Chinese (Cantonese)	sānjiăo dòu
Chinese (Mandarin)	yīng zuĭ dòu
Corsican	cece
Croatian	slanac grah; slanac graj; slanutak
Czech	cizrna beraní
Daasanach	šumbur-a
Dalmatian	cič
Danish	garbanzobønner; kikært

(*Continued*)

TABLE 5.1 (*Continued*)
Popular Names Denoting *Cicer arietinum* L. in Some World Languages and Other Linguistic Taxa

Language/Taxon	Name
Dutch	keker kekererwt; kikkererwt; garbanzo
English	Bengal gram; ceci; cece; chana; chick pea; chickpea; common chickpea; Egyptian pea; garbanzo; garbanzo bean; gram; Kabuli chana
English (16th century)	cich; ciche pease
English (U.S.)	calavance; garbanzo; garbanzo bean
English (U.S., 17th century)	garvance
Esperanto	kikero
Estonian	kikerhernes
Filipino	tsikpi
Finnish	kahviherne; kikherne; kikkahviherne
Flemish	sissererwt
French	cicer tete-de-belier; cicérole; gairance; pois chiche; pois cornu
French (Old)	ceire; cice
Friulian	piçûl
Galician	garavanzo; grao de bico
Genoese	çeixao; seixao; seixau
Georgian	mukhudo
German	Echte Kicher; Felderbse; Kichererbse; Römische Kicher; Venuskicher
Greek	erévinthos o koinós; erévinthos o kriómorfos; revithia
Greek (Ancient)	kryós
Greek (Old)	ἐrévinthos
Gujarati	caṇā; chanaa; chaniaa
Gurage	šumbura
Hadiyya	šimbur-a
Harari	šumbura
Hebrew	khimtsá; khúmus
Hindi	but; chana
Hungarian	bagolyborsó; csicseriborsó
Icelandic	kjúklingabaun
Ido	garbanzo
Indonesian	kacang arab
Irish	piseánach
Istriot	pisiòl; sesmanarin; seʃemanarìn
Italian	cece
Japanese	hiyokomame
Kabyle	afercic; lh'emmez'
Kambaata	šumbur-a
Kannada	kaḍale
Kazakh	noqat; nut
K'iche'	karawan
Korean	byeong-alikong

(*Continued*)

TABLE 5.1 (*Continued*)
Popular Names Denoting *Cicer arietinum* L. in Some World Languages and Other Linguistic Taxa

Language/Taxon	Name
Kurdish (Central)	nŭk
Kurdish (Northern)	nahk; nok; nuk
Kyrgyz	buurçaktay; koybuurçak
Lak	nuxuţ qjuruv
Lao	kan ka kia
Latin	cicer
Latvian	aunazirņi; garbanzo pupas; turku zirņi
Ligurian	cêxe
Lithuanian	nutas; sėjamasis avinžirnis
Luxembourgish	käicher
Macedonian (Ancient, Hellenic)	kíkerroi
Macedonian (Modern, Slavic)	leblebija; naut; slanutok
Malagasy	karazana; voanjobory asiana
Malay	kacang kuda
Malayalam	kaṭala
Maldivian	sanā mugu
Maltese	ċiċri
Māori	tikipī
Marathi	lahāna ākārācā vāṭāṇā
Mari (Hill)	nut
Mari (Meadow)	nut
Mauritian Creole	dhale
Moldavian	nekhut
Mongolian	vandui
Mozarabic	chíchar
Neapolitan	cicero
Nepali	chana
Niçard	cee
Norwegian (Bokmål)	kikert
Norwegian (Nynorsk)	kikert
Odia	caṇā
Oromo	sabbar-ë; shumbur-ä
Otomi (Querétaro)	garbanzo
Papiamento	barbados
Pashto	nukhud
Persian	noẖud
Polish	ciecierzyca pospolita; cieciorka
Portuguese	ervanço; ervilha-de-bengala; ervilha-de-galinha; grão-de-bico; gravanço

(*Continued*)

TABLE 5.1 (*Continued*)
Popular Names Denoting *Cicer arietinum* L. in Some World Languages and Other Linguistic Taxa

Language/Taxon	Name
Portuguese (Brazil)	chícharo; ervanço; grão-de-bico; grão-de-cavalo; gravanço
Punjabi (Eastern)	chōlē
Punjabi (Western)	chuna
Q'eqchi'	karwans
Romagnol	runden
Romanian	năut
Russian	baranii gorokh; nagat; nakhat; nakhut; nut; puzyrnik; turetskii gorokh; shish
Saho	sabbar-ḗ
Salentino	cìcerh
Sanskrit	caṇakaḥ; harimantha; salealpriya; vajimantha
Sardinian (Campidanese)	cìxiri
Scots	chickpea
Serbian	leblebija; naut; slani grah; slani pasulj; slanutak
Serbian (Gallipoli)	slanȕtak
Serbian (Pčinja)	leblebíja; slanútak
Sicilian	cìciru
Sidamo	šimbur-a
Sinhalese	kaḍala; kondi
Slovak	cícer
Slovenian	čičerika; čičerka
Somali	shunburo
Spanish	garbanzo
Spanish (Old)	arvanço
Swahili	mnjegere mkubwa
Swedish	kikärt
Tagalog	garbanzo; garbansos
Tajik	nahūd
Tamil	koṇṭaik kaṭalai
Tatar	bären nogıtı; naxat; nut; puzırnik; törek borçagı; şiş borçagı
Telugu	śanagalu
Thai	t̄hạw lūkkị̀; t̄hạw h̄aw c̄hāng
Tigrinya	šəmbəra
Turkish	nohut
Turkmen	nokhud
Udmurt	nut
Ukrainian	baraniachyi gorokh; nut; turets'kyi gorokh
Urdu	tcna
Uzbek	no'xatday; no'xot; turk no'xati
Valencian	cigro

(*Continued*)

TABLE 5.1 (*Continued*)
Popular Names Denoting *Cicer arietinum* L. in Some World Languages and Other Linguistic Taxa

Language/Taxon	Name
Venetian	cexarìna
Vietnamese	đậu gà
Walloon	garvane; poes d' souke
Waray	cicer
Welsh	gwygbysen gwygbys
Yiddish	arbes; nahit
Yucatec	garbanzo
Zaza	nehe

English: Canarian chickpea
Spanish: garbancera
- *Cicer chorassanicum* (Bunge) Popov
Synonyms: *Cicer trifoliatum* Bornm.; *Orobus chorassanica* Bunge
English: Khorasan chickpea
Russian: nut khorasanskii
- *Cicer cuneatum* Hochst. ex A. Rich.
Amharic: ait shembra
English: wedge-like chickpea
Tigrinya: anchoa ater; ater quasot
- *Cicer echinospermum* P. H. Davis
English: spiny-seeded chickpea
- *Cicer fedtschenkoi* Lincz.
Synonyms: *Cicer fedtchenkoi* Lincz.; *Cicer songaricum* DC. var. *pamiricum* Paulsen; *Cicer songaricum* DC. var. *schugnanicum* Popov
English: Fedchenko's chickpea
Russian: nut Fedchenko
Tajik: tashkurut
- *Cicer flexuosum* Lipsky
English: winding chickpea
Russian: nut izvilistyi
Uzbek: togburchok
- *Cicer floribundum* Fenzl
English: flower-abundant chickpea
- *Cicer graecum* Orph.
Synonyms: *Cicer graecum* Boiss.
English: Greek chickpea
Greek: revithi
- *Cicer grande* (Popov) Korotkova
Synonyms: *Cicer flexuosum* subsp. *grande* Popov

English: large chickpea
Russian: nut bolshoi
- *Cicer heterophyllum* Contandr. et al.
English: different-leafed chickpea
- *Cicer incanum* Korotkova
Synonyms: *Cicer pungens* Boiss. var. *horridum* Popov
English: silver-colored chickpea
Russian: nut sedoi
- *Cicer incisum* (Willd.) K. Maly
Synonyms: *Anthyllis incisa* Willd.; *Cicer adonis* Nyman; *Cicer caucasicum* Bornm.; *Cicer ervoides* (Sieber) Fenzl; *Cicer incisum* (Willd.) K. Maly var. *libanoticum* (Boiss.) Bornm.; *Cicer minutum* Boiss. & Hohen; *Cicer pimpinellifolium* Jaub. & Spach; *Cicer pimpinellifolium* Jaub. & Spach subsp. *minutum* (Boiss. & Hohen.) Ponert
English: cleft-leafed chickpea
Russian: nut kavkazskii; nut kroshechnyi; nut nadrezannyi
- *Cicer isauricum* P. H. Davis
English: Isaurian chickpea
- *Cicer judaicum* Boiss.
Synonyms: *Cicer pinnatifidum* var. *judaicum* Popov
English: Judean chickpea
- *Cicer kermanense* Bornm.
English: Kermani chickpea
- *Cicer korshinskyi* Lincz.
English: Korshinsky's chickpea
Russian: nut Korzhinskogo
- *Cicer laetum* Rassulova & Sharipova
English: rattling chickpea
Russian: nut svetlyi
- *Cicer luteum* Rassulova & Sharipova
English: yellow chickpea
- *Cicer macracanthum* Popov
Synonyms: *Cicer songaricum* (DC.) Bunge var. *spinosum* Aitch.
English: large-thorned chickpea
Russian: nut dlinnokolyuchii
- *Cicer microphyllum* Benth.
Synonyms: *Cicer jacquemontii* Jaub. & Spach; *Cicer songaricum* auct. non (DC.) Bunge
English: small-leafed chickpea
- *Cicer mogoltavicum* (Popov) Korol.
Synonyms: *Cicer flexuosum* Lipsky subsp. *mogoltavicum* (Popov) Popov; *Cicer flexuosum* Lipsky var. *mogoltavicum* Popov
English: Mogoltauian chickpea
Russian: nut mogoltavskii
Tajik: nakhudi kukhi-mugul
- *Cicer montbretii* Jaub. & Spach

English: Montbret's chickpea
- *Cicer multijugum* Maesen
Synonyms: *Cicer jacquemontii* sensu auct.; *Cicer microphyllum* sensu auct.
English: compound chickpea
Russian: nut mnogoparnyi, nut zhakemonta
- *Cicer nuristanicum* Kitam.
English: Nuristani chickpea
- *Cicer oxyodon* Boiss. & Hohen.
English: bluish chickpea
- *Cicer paucijugum* Nevski
Synonyms: *Cicer songaricum* DC. var. *paucijugum* Popov
English: few-leaflet chickpea
Russian: nut maloparnyi
- *Cicer pinnatifidum* Jaub. & Spach
English: lobe-leafed chickpea
- *Cicer pungens* Boiss.
Synonyms: *Cicer spinosum* Popov
English: puncturing chickpea
Russian: nut kolyuchii
- *Cicer rassuloviae* Lincz.
Synonyms: *Cicer multijugum* Rassulova & Sharipova
English: Rasulova's chickpea
Russian: nut Rasulovoi
- *Cicer rechingeri* Podlech
English: Rechinger's chickpea
- *Cicer reticulatum* Ladiz.
English: net-like chickpea
- *Cicer songaricum* Stephan ex DC.
Synonyms: *Cicer popovii* Nevski; *Cicer songoricum* DC.; *Cicer soongaricum* DC.
English: Dzungarian chickpea
Mongolian: Zuungar banduukhai
Russian: nut dzhungarskii
Tajik: bide
- *Cicer spiroceras* Jaub. & Spach
English: apiral-horned chickpea
- *Cicer stapfianum* Rech. f.
English: Stapf's chickpea
- *Cicer subaphyllum* Boiss.
English: upright-leafed chickpea
- *Cicer tragacanthoides* Jaub. & Spach
Synonyms: *Cicer kopetdaghense* Lincz.; *Cicer straussii* Bornm.; *Cicer tragacanthoides* Jaub. & Spach var. *turcomanicum* Popov
English: tragacanth-like chickpea
Russian: nut kopetdagskii
- *Cicer uludereensis* Donmez

English: Uluderean chickpea
- *Cicer yamashitae* Kitam.
English: Yamashita's chickpea

5.2 ORIGIN OF SCIENTIFIC AND POPULAR TAXA NAMES

There are many archaeobotanical records relating to *Cicer* species, especially in the area of the Fertile Crescent. Among many, a noteworthy one is from Tell-el-Kerkh, in northwest Syria, where a whole range of forms, from wild relative *Cicer reticulatum*, to *proper* common chickpea, was found and dated to the late tenth millennium BP (Tanno and Willcox 2006).

The proto-words relating to cultivated chickpea (*Cicer arietinum*) have been attested in the Proto-Afroasiatic and Proto-Indo-European language families, which may be quite expected, since the geographic proximity of the crop's core environment (Redden and Berger 2007) and the families' supposed homelands (Huehnergard 2004, Bouckaert et al. 2012).

The Proto-Afroasiatic root **ŝV(m)bar-* originally denoted both kernel of corn and chickpea (Militarev and Stolbova 2007). This second meaning was preserved in some of the direct derivatives of the Proto-Afroasiatic language, such as the Proto-Agaw **ŝa(m)bar-*, the Proto-Highland-East-Cushitic **šimbur-*, the Proto-Lowland-East-Cushitic **šumbur-* and the Proto-Saho-Afar **sabbar-*, as well as the Proto-Semitic **ŝxabar-*, primarily denoting grain in general. In this way, it represents one of the most remarkable evidences of how agriculture played a significant role among its speakers, more than 10,000 years ago (Militarev 2002). It is also a part of the vocabulary of the languages belonging to Agaw or Central Cushitic, such as Awngi, to Highland East Cushitic, such as Burji, Hadiyya, Kambaata, and Sidamo, and to Lowland East Cushitic, such as Afar, Daasanach, Oromo, Saho, and Somali (Table 5.1). In its Semitic branch, its meaning related to chickpea is not present either in Arabic or Hebrew, as its most renown representatives, but in several African languages, such as Amharic, Gurage, Harari, or Tigrinya (Figure 5.1). There is a possibility that the words in the above-mentioned Semitic languages denoting chickpea were influenced by the Cushitic ones, within a wider cultural impact of the latter on the former (Pagani et al. 2012). It is noteworthy that the Proto-Afroasiatic root **ŝV(m)bar-* also produced a few more words in modern languages relating to other grain legumes, such as *säbbära*, denoting grass pea (*Lathyrus sativus* L.) in Tigre, and *sumburo*, designating pea (*Pisum sativum* L.) in Gollango, a dialect of Gawwada (Militarev and Stolbova 2007).

There are Afroasiatic languages demonstrating different forms of the word denoting chickpea, such as the Berber, with Kabyle, and the Semitic, with Arabic or Hebrew, well-known as anglicized *hummus* and with a yet insufficiently clarified etymology (Table 5.1).

The primeval meaning of the Proto-Indo-European root **kek-*, **k'ik'-* was equally oat (*Avena sativa* L.) and pea (Nikolayev 2012). This second original meaning was preserved only in the extinct Baltic language, Old Prussian. In all other attested direct descendants of Proto-European, this root gradually began to shift its meaning to chickpea. One of them is Old Armenian, with *siseṛn*, that gave the Modern Armenian

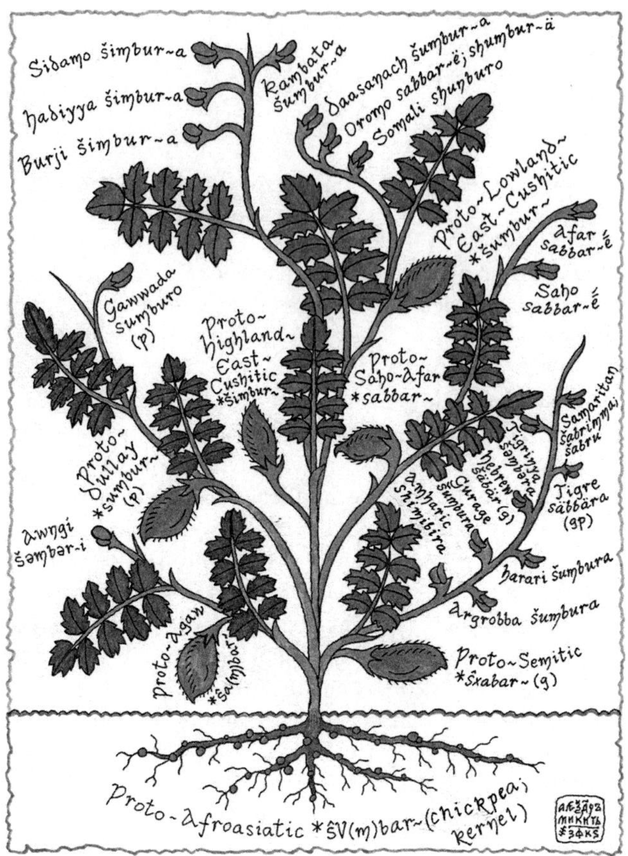

FIGURE 5.1 **(See color insert.)** One of the possible evolutions of the Proto-Afroasiatic root **ŝV(m)bar-*, denoting both chickpea and kernel of corn (Militarev and Stolbova 2007), into its direct derivatives, interpreted as pods, and contemporary descendants, illustrated as flowers, in the Cushitic and Semitic languages; the meanings of each proto- and modern word are *chickpea*, if given without brackets, as well as *grain* (g), *grass pea* (gp), and *pea* (p).

siser, while another is the Old Macedonian *kíkerroi*, also denoting chickpea and, possibly, being derived from the Proto-Hellenic **kikriós*, but with no attested forms in Greek and other descendants. The most remarkable outcome of a long evolution of the Proto-Indo-European **kek-*, **k'ik'-* is surely the Latin *cicer*, which brought forth countless descendants in nearly all groups of both extinct and living Romance languages. Such are the Iberian Romance, with Mozarabic; the Occitano-Romance, with Catalan, Niçard, or Valencian; the Gallo-Romance, with French or Walloon; the Gallo-Italic, with Genoese or Ligurian; the Venetian with the Italo-Dalmatian, with Calabrian, Caterisani, Corsican, Dalmatian, Istriot, Italian, Neapolitan, Salentino, or Sicilian; the Sardinian and the Eastern Romance, with Aromanian (Table 5.1). This Latin word was borrowed by numerous geographically neighboring languages, such as the Germanic, including English, as well as non-Indo-European, such as Basque and the Semitic Maltese (Mikić 2014d).

Another Proto-Indo-European root with important derivatives is **erəgʷ[h]-*, *eregw(h)o-*, *erogw(h)o-*, denoting both the kernel of a leguminous plant and pea (Pokorny 1959, Nikolayev 2012). This word is responsible for a number of the words denoting chickpea in modern languages of European origin, such as the Germanic, with U.S. English or Yiddish; the Hellenic, with Greek; the Iberian Romance, with Asturian, Galician, Portuguese, or Spanish; and several others, both Indo-European and non-Indo-European. The Portuguese and Spanish forms were imported into numerous languages of mutually distant language families across the world, beginning with the West European conquests from the early sixteenth century, as may be witnessed by the words denoting chickpea in Cebuano, K'iche', Malagasy, Otomi, Papiamento, Q'eqchi', Tagalog, or Yucatec (Table 5.1).

There is a possibility that the Persian word denoting chickpea, *nohud*, originated from the Proto-Indo-European root **k(')now-*, designating nut (Pokorny 1959, Nikolayev 2012) and, specifically, hazel (*Corylus* spp.). Thus, it may have an obviously descriptive nature, since a great similarity in both size and shape between hazelnuts and chickpea grains. This Persian word, probably together with the crop itself, was embraced by a number of neighboring languages of the same Iranian branch of the Indo-European family, such as Central and Northern Kurdish, Pashto, Tajik, or Zaza; as well as by certain Germanic, such as Yiddish; Romance, such as Moldavian and Romanian; and Slavic languages, such as Belarusian, Bulgarian, Macedonian, Russian, Serbian, or Ukrainian, with subsequent borrowings into the Baltic, such as Lithuanian (Table 5.1). By different paths, the Persian *nohud* became present in different non-Indo-European languages, such as the Altaic, with Azerbaijani, Bashkir, Kazakh, Tatar, Turkish, Turkmen, or Uzbek; the Caucasian, with Lak; the Kartvelian, with Georgian; and the Uralic, with both Hill and Meadow Mari or Udmurt (Nişanyan 2017).

In the languages developed and spoken in the regions without native chickpea cultivation, especially those with moderate or cold climates, the words related to it are most often associated with pea. Among the most prominent examples are all the modern Germanic languages, where chickpea is regularly formed by a compound word. The first segment is derived from the aforementioned Latin *cicer*, which was used by Linnaeus to denote the eponymous genus (Linnaeus 1753, 1758), while the second part of the word is the native word denoting pea (Table 5.1). If we remember that the Latin word is derived from the Proto-Indo-European root **kek-*, **k'ik'-*, denoting pea, then chickpea in the contemporary Germanic languages, such as the English *chickpea*, the German *Kichererbse* or the Norwegian *kikert*, represent a kind of pleonasm, since a literate translation of these compound words would be, interestingly enough, *pea-pea*.

Among the words connecting chickpea to pea are those in the other Indo-European languages, such as the Baltic, with Latvian and Lithuanian; the Celtic, with Irish and Welsh; the Romance, with French, Friulian, Istriot, Portuguese, and Walloon; and the Slavic, with Belarusian, Russian, Serbian, and Ukrainian (Table 5.1). The same association with pea is present in non-Indo-European languages, such as the Altaic, with Kyrgyz, Mongolian, or Tatar; the Sino-Tibetan, with Chinese; and the Uralic, with Estonian, Finnish, or Hungarian. In Serbian, there is a resemblance to common bean (*Phaseolus vulgaris* L.), while in Welsh to common vetch (*Vicia sativa* L.).

The Linnean epithet *arietinum*, *ram-like*, is a part of the names denoting chickpea in diverse languages, such as Belarusian, Czech, French, Lithuanian, Russian, or Ukrainian, while its peculiar grain feature resembling beak may be found in Galician and Portuguese names (Table 5.1). The imagined place of chickpea's origin varies from Egypt, in Indonesian, and Arabia, in English, over Turkey, in Latvian, Russian, Tatar, Ukrainian or Uzbek, to Kabul, in English, and Bengal, in English and Portuguese. Perhaps the most curious of all the associations regarding chickpea is the one to Venus, as seen in one of many names in German: perhaps for its supposed properties or for the delicacy of the sweet dishes made of its grains (Çekal et al. 2012).

6 *Ervum* L.

Synonyms: *Ervilia* Opiz; *Orobus* L.

6.1 LIST OF TAXA SCIENTIFIC AND POPULAR NAMES

This chapter is somewhat peculiar in comparison to the others since it deals with only one species, namely *Vicia ervilia* (L.) Willd., one of the first domesticated plants in the world and one of the two species used for the first successful and attested extraction of ancient DNA from any legume ever (Mikić et al. 2015b). During the past few centuries, there were numerous revisions of pioneering botanical classifications, which, as a consequence and specifically to the said species, resulted in dissolving the Linnean genera *Ervum* and *Orobus* (Linnaeus 1753, 1758) and distributing their taxa to several other categories. Purely out of historical reasons, as will be seen in Section 6.2, this section will present an extensive list of the taxa belonging to the former genera of *Ervilia* Opiz, *Ervum* L., and *Orobus* L., with their contemporary consensus names. There is also a list of vernacular names for the sole memory of theirs, that is, *Vicia ervilia* (ISTA 1982, Rehm 1994, Gledhill 2008, Porcher 2008, The Plant List 2013, Ecocrop 2017, EPPO 2017, Ethnologue 2017, IBIS 2017, ILDIS 2017, Logos 2017, NPGS 2017, Wikipedia 2017, Wiktionary 2017).

- *Ervilia caesarea* Alef. (unresolved)
- *Ervilia cassubica* Schur (unresolved)
- *Ervilia hirsuta* Opiz > *Vicia hirsuta* (L.) Gray
- *Ervilia monanthos* (L.) Opiz (unresolved)
- *Ervilia orobos* Schur (unresolved)
- *Ervilia sativa* Link > *Vicia ervilia* (L.) Willd.
- *Ervilia sylvatica* Schur (unresolved)
- *Ervilia tetrasperma* (L.) Opiz (unresolved)
- *Ervilia vulgaris* Godr. > *Vicia hirsuta* (L.) Gray
- *Ervum amoenum* (Fisch.) Trautv. > *Vicia amoena* Fisch.
- *Ervum bithynicum* > *Vicia bithynica* (L.) L.
- *Ervum calcaratum* Trautv. > *Vicia monantha* Retz. subsp. *monantha*
- *Ervum cappadocicum* (Boiss.) Stank. > *Vicia cappadocica* Boiss. & Balansa
- *Ervum cassubicum* (L.) Peterm. > *Vicia cassubica* L.
- *Ervum cyanea* Boiss. & Hohen. > *Lens culinaris* Medik. subsp. *orientalis* (Boiss.) Ponert
- *Ervum cyaneum* Boiss. & Hohen. > *Lens culinaris* Medik. subsp. *orientalis* (Boiss.) Ponert
- *Ervum erectum* Walter > *Galactia erecta* Vail
- *Ervum ervilia* L. > *Vicia ervilia* (L.) Willd.
- *Ervum ervoides* (Brign.) Hayek > *Lens ervoides* (Brign.) Grande

- *Ervum filiforme* Roxb. > *Vicia hirsuta* (L.) Gray
- *Ervum gracile* (Loisel.) DC. > *Vicia parviflora* Cav.
- *Ervum hirsutum* L. > *Vicia hirsuta* (L.) Gray
- *Ervum hohenakeri* Fisch. & C. A. Mey. > *Lens ervoides* (Brign.) Grande
- *Ervum kotschyanum* Boiss. > *Vicia montbretii* Fisch. & C. A. Mey.
- *Ervum lathyroides* > *Vicia lathyroides* L.
- *Ervum lens* L. > *Lens culinaris* Medik.
- *Ervum lenticula* Schreb. ex Sturm > *Lens ervoides* (Brign.) Grande
- *Ervum loiseleurii* M. Bieb. > *Vicia loiseleurii* (M. Bieb.) Litv.
- *Ervum monanthos* L. > *Vicia articulata* Hornem.
- *Ervum multiflorum* Pursh > *Astragalus tenellus* Pursh
- *Ervum nigricans* M. Bieb. > *Lens nigricans* (M. Bieb.) Godr.
- *Ervum orientale* Boiss. > *Lens culinaris* Medik. subsp. *orientalis* (Boiss.) Ponert
- *Ervum paucijugum* Trautv. > *Vicia cappadocica* Boiss. & Balansa
- *Ervum pictum* (Fisch. & C. A. Mey.) Alef. > *Vicia biennis* L.
- *Ervum pilosum* Alef. > *Vicia ludoviciana* Nutt.
- *Ervum pisiforme* (L.) Peterm. > *Vicia pisiformis* L.
- *Ervum pubescens* DC. > *Vicia pubescens* (DC.) Link
- *Ervum sardoum* Spreng. > *Vicia loiseleurii* (M. Bieb.) Litv.
- *Ervum soloniense* L. > *Vicia lathyroides* L.
- *Ervum sylvaticum* Fisch. > *Lens nigricans* (M. Bieb.) Godr.
- *Ervum tenuissimum* Pers. > *Vicia tetrasperma* (L.) Schreb.
- *Ervum terronii* Ten. > *Vicia loiseleurii* (M. Bieb.) Litv.
- *Ervum tetraspermum* L. > *Vicia tetrasperma* (L.) Schreb.
- *Ervum tridentatum* Alef. > *Vicia ludoviciana* Nutt.
- *Ervum tsydenii* (Malyschev) Stank. > *Vicia tsydenii* Malyschev
- *Ervum unijugum* > *Vicia unijuga* A. Br.
- *Ervum volubile* Walter > *Galactia regularis* (L.) Britton & al.
- *Ervum woronowii* (Bornm.) Stank. > *Lathyrus woronowii* Bornm.
- *Orobus alatus* Maxim. > *Lathyrus komarovii* Ohwi
- *Orobus albus* L. f. > *Lathyrus pannonicus* (Jacq.) Garcke
- *Orobus alpestris* Waldst. & Kit. > *Lathyrus alpestris* (Waldst. & Kit.) Celak.
- *Orobus alpestris* Ledeb. > *Lathyrus frolovii* Rupr.
- *Orobus angustifolius* L. > *Lathyrus pallescens* (M. Bieb.) K. Koch
- *Orobus anomalus* K. Koch > *Vicia abbreviata* Spreng.
- *Orobus aphaca* (L.) Doll > *Lathyrus aphaca* L.
- *Orobus atropatanus* Grossh. > *Lathyrus atropatanus* (Grossh.) Sirj.
- *Orobus aurantius* Steven > *Vicia crocea* (Desf.) Fritsch
- *Orobus aureus* Fisch. & C. A. Mey. > *Lathyrus aureus* (Steven) D. Brandza
- *Orobus austriacus* Crantz > *Lathyrus pannonicus* (Jacq.) Garcke subsp. *pannonicus* (Jacq.) Garcke
- *Orobus baicalensis* (Turcz.) Stank. & Roskov > *Vicia venosa* (Link) Maxim.
- *Orobus californicus* (Douglas) Alef. > *Lathyrus japonicus* Willd. subsp. *maritimus* (L.) P. W. Ball
- *Orobus canescens* L. f. > *Lathyrus filiformis* (Lam.) Gay

- *Orobus caucasicus* Spreng. > *Vicia abbreviata* Spreng.
- *Orobus ciliatidentatus* (Czefr.) Avazneli > *Lathyrus ciliatidentatus* Czefr.
- *Orobus croceus* Desf. > *Vicia crocea* (Desf.) Fritsch
- *Orobus cyaneus* Steven > *Lathyrus digitatus* (M. Bieb.) Fiori
- *Orobus davidii* (Hance) Stank. & Roskov > *Lathyrus davidii* Hance
- *Orobus diffusus* Nutt. > *Vicia americana* Willd.
- *Orobus digitatus* M. Bieb. > *Lathyrus digitatus* (M. Bieb.) Fiori
- *Orobus dispar* Nutt. > *Astragalus tenellus* Pursh
- *Orobus dissitifolius* Alef. > *Lathyrus graminifolius* (S. Watson) T. G. White
- *Orobus emodi* Fritsch > *Lathyrus emodi* (Fritsch) Ali
- *Orobus ewaldii* Meinsh. > *Lathyrus laevigatus* (Waldst. & Kit.) Gren. subsp. *laevigatus* (Waldst. & Kit.) Gren.
- *Orobus faba* Brot. > *Vicia faba* L.
- *Orobus formosus* Steven > *Vavilovia formosa* (Steven) Fed.
- *Orobus frolovii* Ledeb. > *Lathyrus frolovii* Rupr.
- *Orobus fruticosus* (Cav.) Pers. > *Coursetia fruticosa* (Cav.) J. F. Macbr.
- *Orobus gmelinii* DC. > *Lathyrus gmelinii* Fritsch
- *Orobus hirsutus* L. > *Lathyrus laxiflorus* (Desf.) Kuntze
- *Orobus hispanicus sensu* Lacaita > *Lathyrus pannonicus* (Jacq.) Garcke subsp. *hispanicus* (Lacaita) Bässler
- *Orobus humilis* Ser. > *Lathyrus humilis* (Ser.) Spreng.
- *Orobus incurvus* (Roth) A. Braun > *Lathyrus incurvus* (Roth) Willd.
- *Orobus intermedius* C. A. Mey. > *Lathyrus ledebourii* Trautv.
- *Orobus kolenatii* K. Koch > *Lathyrus aureus* (Steven) D. Brandza
- *Orobus komarovii* (Ohwi) Stank. & Roskov > *Lathyrus komarovii* Ohwi
- *Orobus krylovii* > *Lathyrus krylovii* Serg.
- *Orobus lacaitae* (Cefr.) Stank. & Roskov > *Lathyrus pannonicus* (Jacq.) Garcke subsp. *hispanicus* (Lacaita) Bässler
- *Orobus lacteus* (M. Bieb.) Wissjul. > *Lathyrus pannonicus* (Jacq.) Garcke subsp. *collinus* (J. Ortmann) Soó
- *Orobus laevigatus* Waldst. & Kit. > *Lathyrus laevigatus* (Waldst. & Kit.) Gren. subsp. *laevigatus* (Waldst. & Kit.) Gren.
- *Orobus lathyroides* L. > *Vicia unijuga* A. Br.
- *Orobus laxiflorus* Desf. > *Lathyrus laxiflorus* (Desf.) Kuntze
- *Orobus ledebourii* (Trautv.) Roldugin > *Lathyrus ledebourii* Trautv.
- *Orobus linifolius* Reichard > *Lathyrus linifolius* (Reichard) Bässler
- *Orobus littoralis* (Torr. & A. Gray) A. Gray > *Lathyrus littoralis* (Torr. & A. Gray) Walp.
- *Orobus longifolius* (Pursh) Nutt. > *Astragalus ceramicus* E. Sheld. var. *filifolius* (A. Gray) F. J. Herm.
- *Orobus luteus* L. > *Lathyrus gmelinii* Fritsch
- *Orobus maritimus* (L.) Rchb. > *Lathyrus japonicus* Willd. subsp. *maritimus* (L.) P. W. Ball
- *Orobus muhlenbergii* Alef. > *Lathyrus venosus* Willd. subsp. *venosus*
- *Orobus multijugus* (Ledeb.) Stank. & Roskov > *Lathyrus pannonicus* (Jacq.) Garcke subsp. *multijugus* (Ledeb.) Bässler

- *Orobus myrtifolius* Alef. > *Lathyrus palustris* L.
- *Orobus myrtifolius* (Willd.) Hall > *Lathyrus palustris* L.
- *Orobus niger* L. > *Lathyrus niger* (L.) Bernh.
- *Orobus nipponicus* (Matsum.) Stank. & Roskov > *Vicia nipponica* Matsum.
- *Orobus nissolia* > *Lathyrus nissolia* L.
- *Orobus occidentalis* (Fisch. & C. A. Mey.) Fritsch > *Lathyrus laevigatus* (Waldst. & Kit.) Gren. subsp. *occidentalis* (Fisch. & C. A. Mey.) Breistr.
- *Orobus ochroleucus* (Hook.) A. Br. > *Lathyrus ochroleucus* Hook.
- *Orobus ochroleucus* Waldst. & Kit. > *Vicia sparsiflora* Ten.
- *Orobus ohwianus* (Hosok.) Stank. & Roskov > *Vicia ohwiana* Hosok.
- *Orobus orientalis* Boiss. > *Lathyrus aureus* (Steven) D. Brandza
- *Orobus pallescens* M. Bieb. > *Lathyrus pallescens* (M. Bieb.) K. Koch
- *Orobus pannonicus* Jacq. > *Lathyrus pannonicus* (Jacq.) Garcke subsp. *pannonicus* (Jacq.) Garcke
- *Orobus persicus* Boiss. > *Vicia iranica* Boiss.
- *Orobus polymorphus* (Nutt.) Alef. > *Lathyrus eucosmus* Butters & St. John
- *Orobus pratensis* > *Lathyrus pratensis* L.
- *Orobus pseudorobus* > *Vicia pseudorobus* Fisch. & C. A. Mey.
- *Orobus ramuliflorus* Maxim. > *Vicia ramuliflora* (Maxim.) Ohwi
- *Orobus roseus* (Steven) Ledeb. > *Lathyrus roseus* Steven
- *Orobus saxatilis* Vent. > *Lathyrus saxatilis* (Vent.) Vis.
- *Orobus semenovii* Regel & Herder > *Vicia semenovii* (Regel & Herder) B. Fedtsch.
- *Orobus sericeus* Sessé & Moc. > *Tephrosia sinapou* (Buc'hoz) A. Chev.
- *Orobus sessilifolius* Sibth. & Sm. > *Lathyrus digitatus* (M. Bieb.) Fiori
- *Orobus sphaericus* (Retz.) Avazneli > *Lathyrus sphaericus* Retz.
- *Orobus subalpinus* Herbich > *Lathyrus subalpinus* Beck
- *Orobus subrotundus* > *Vicia subrotunda* (Maxim.) Czefr.
- *Orobus subvillosus* Ledeb. > *Vicia subvillosa* (Ledeb.) Boiss.
- *Orobus tomentosus* Desf. > *Coursetia fruticosa* (Cav.) J. F. Macbr.
- *Orobus transsylvanicus* Spreng. > *Lathyrus transsilvanicus*
- *Orobus triflorus* Stapf > *Vicia subvillosa* (Ledeb.) Boiss.
- *Orobus trifoliatus* Sessé & Moc. > *Vigna luteola* (Jacq.) Benth.
- *Orobus tuberosus* L. > *Lathyrus linifolius* (Reichard) Bässler
- *Orobus variegatus* Ten. > *Lathyrus venetus* (Mill.) Wohlf.
- *Orobus venetus* Mill. > *Lathyrus venetus* (Mill.) Wohlf.
- *Orobus venosus* Braun > *Lathyrus venosus* Willd.
- *Orobus venosus* Link > *Vicia venosa* (Link) Maxim.
- *Orobus venosus* Link var. *willdenowianus* Turcz. > *Vicia venosa* (Link) Maxim.
- *Orobus vernus* L. > *Lathyrus vernus* (L.) Bernh.
- *Orobus versicolor* J. F. Gmel. > *Lathyrus pannonicus* (Jacq.) Garcke subsp. *varius* (Hill) P. W. Ball
- *Vicia ervilia* (L.) Willd. (Table 6.1)

Synonyms: *Ervilia sativa* Link; *Ervum ervilia* L.; *Lens pygmaea* Grossh.

TABLE 6.1
Popular Names Denoting *Vicia ervilia* (L.) Willd. in Some World Languages and Other Linguistic Taxa

Language/Taxon	Name
Albanian (Arvanitika)	ro'v
Arabic	biqiat marratan; kersannah
Armenian	k'urrushna
Azerbaijani	mərciməyəbənzər çölnoxudu
Basque	eru
Belarusian	garoshak patserkapadobny
Catalan	erb
Danish	perlevikke
Dutch	bittere wikke; erve; linzenwikke
English	bitter vetch; ervil; lentil vetch
Finnish	linssivirvilä
French	alliez; ers; ers lentille; erse; ervilier; ervilière; lentille bâtarde; lentille du Canada; lentille ers; lentille erviliaire; pois de pigeon; vesce amère; vesce bâtarde; vesce blanche; vesce ervilia
Georgian	ugrekheli
German	Bitter-Wicke; Erfe; Ervilie; Erwenlinse; Linsen-Wicke; Stachelwicke; Steinlinse; Wicklinse
Greek	róvi
Greek (Ancient)	ǒrobos
Hebrew	h'chrs'hn'h
Italian	capogirlo; ervil; ervo; fragellini; girlo; lente-girlo; lero; ruviglia; vecciola; veggiolo zirlo
Kurdish (Northern)	kizin
Latin	ervilia; ervum
Limburgish	linsewèk
Macedonian	urov
Norwegian (Bokmål)	linsevikke
Norwegian (Nynorsk)	linsewikke
Persian	gavdaneh
Polish	wyka soczewicowata
Portuguese	ervilha-de-pombo; gero; marroiço; orobo
Russian	frantsuzskaia chechevitsa; goroshek chetkoobraznyi
Serbian	urov
Slovak	vika šošovicová
Sorbian (Lower)	sočna wójka
Sorbian (Upper)	sočna woka
Spanish	alacarceña; alcaruna; alverja; ervilla; lenteja bastarda; yero
Swedish	linsvicker
Tajik	gomuk
Turkish	burçak
Welsh	ytbysen y coed

6.2 ORIGIN OF SCIENTIFIC AND POPULAR TAXA NAMES

Bitter vetch (*Vicia ervilia* [L.] Willd.) is considered one of the most ancient domesticated plants in the world, along with several other pulse, cereal, fiber, and oil crops. It is present already in the Pre-Pottery Neolithic period in the Near East, with the remains dated as early as 11,700 years BP (White 2013). Together with pea (*Pisum sativum* L.), bitter vetch is one of the first legume species in the world with the known and attested extraction of ancient DNA (Mikić 2015c). However, despite all these kinds of accolades, bitter vetch is still an underutilized and neglected crop, even in its native Mediterranean environments (Berger et al. 2003).

The Linnean genera *Ervum* L. and *Orobus* L. have much in common. Both comprise legume species, some of which are important pulses, and both were eventually dissolved, with their members attributed to other genera, as may be seen in the first section of this chapter. In addition, they have a common etymology: although Linnaeus used two names, at a first glance rather different, to name two legume genera, it is most likely that both had had the same ultimate linguistic origin (Figure 6.1).

What stands behind both Linnean genera names is the Proto-Indo-European root **erəgʷ[h]-*, *eregw(h)o-*, *erogw(h)o-*, denoting both the kernel of a leguminous plant and pea (Pokorny 1959, Nikolayev 2012). The evolution of this word is not as complex as it may seem, because of the frequent shift of meaning from pea to bitter vetch and *vice versa* and even including chickpea (*Cicer arietinum* L.). The most obvious development is attested in the Germanic languages, where the initial meaning of *pea* was kept in the form of **arwait-*, **arwīt-*, and nearly all of its modern descendants (see Chapter 14).

The Proto-Hellenic **ĕ́rovos* produced the Old Greek *órovos*, denoting bitter vetch, and *érévinthos*, denoting chickpea. The former maintained its basic form and essential meaning in the Modern Greek *róvi*, in the neighboring languages, such as the Slavic Macedonian and Serbian, and more or less geographically distant Indo-European languages, such as the Romance Italian and Portuguese (Table 6.1). On the other hand, the latter Old Greek word evolved into the Modern Greek word for chickpea *revithiá* (Mikić 2009).

In a similar way, the hypothetical Proto-Italic word **erou̯om* is considered a direct ancestor of the well-known Latin *ervum*, denoting bitter vetch, from which, in turn, derive contemporary descendants denoting pea in certain Romance languages, such as Portuguese (see Chapter 14). The same Latin word also produced a certain number of less widely used and almost obsolete names designing bitter vetch in various Indo-European languages, such as in the Germanic, with Dutch, English or German, and the Romance, with Catalan, French, Italian, Portuguese, or Spanish (Table 6.1). It was also borrowed into some adjacent non-Indo-European languages, such as Basque.

It is noteworthy that Turkish is the sole Altaic language where the attested Proto-Altaic root **bŭkrV*, ultimately denoting pea, nut, and cone and through its direct derivative, the Proto-Turkic **burčak*, designing pea, shifted its meaning from *pea* to *bitter vetch* (Starostin et al. 2003).

The vast majority of the collected words signifying bitter vetch demonstrate that it is by far considered a kind of vetch (Table 6.1). At the same time and in many

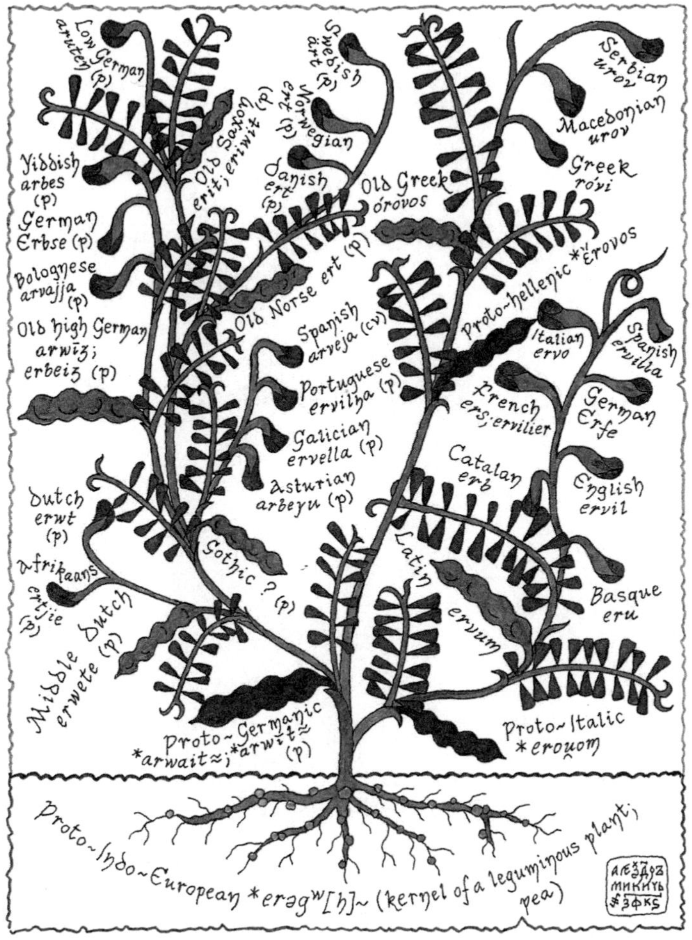

FIGURE 6.1 **(See color insert.)** One of the possible evolutions of the Proto-Indo-European root **erəgʷ[h]-*, denoting both the kernel of a leguminous plant and pea (Pokorny 1959, Nikolayev 2012), into its direct derivatives, drawn as pods, and contemporary descendants, rendered as flowers, in the Indo-European languages; the meanings of the proto- and modern words are *bitter vetch*, if given without brackets, as well as *common vetch* (cv) and *pea* (p).

languages of diverse ethnolinguistic families, bitter vetch bears a strong resemblance to lentil (*Lens culinaris* Medik.), especially because of their common growth habit and morphology of leaves and flowers. Such examples may be seen in the Indo-European, encompassing the Germanic, with Dutch, English, German Limburgish, Norwegian, or Swedish; the Romance, with French, Italian, or Spanish; and the Slavic, with Polish, Russian, Slovak, Lower or Upper Sorbian, as well as the Altaic, with Azerbaijani, and the Uralic languages, with Finnish. In some languages, bitter vetch is primarily linked to chickpea, such as in Azerbaijani, or to pea, such as in Belarusian or Russian.

Among numerous other associations relating to bitter vetch, we may also mention those with forest in Welsh, with pigeon in French and Portuguese, or pearls in Danish, most likely due to the shape of its mature pods (Table 6.1). It is very interesting that one of the numerous metropolitan French names makes a connection with Canada, as its proposed region of origin and for the reason probably already fallen into oblivion, while one of the Russian names links bitter vetch to no other country but France: just one more among uncountable circle-like pathways that have been used by pulse crops for their indiscernible journeys across the land and sea.

7 *Faba* Mill.

Synonyms: *Orobus* L.; *Vicia* L.

7.1 LIST OF TAXA SCIENTIFIC AND POPULAR NAMES

Similar to Chapter 6, which is devoted to one species, *Vicia ervilia* (L.) Willd., this chapter will also deal with one pulse species, namely *Faba vulgaris* Moench or *Vicia faba* L. The first segment of the present chapter will give an insight to its subtaxa and demonstrate a remarkable richness of vernacular names of this species (ISTA 1982, Rehm 1994, Gledhill 2008, Porcher 2008, The Plant List 2013, Ecocrop 2017, EPPO 2017, Ethnologue 2017, IBIS 2017, ILDIS 2017, Logos 2017, NPGS 2017, Wikipedia 2017, Wiktionary 2017).

- *Vicia faba* L. (Table 7.1)

Synonyms: *Faba bona* Medik.; *Faba equina* Medik.; *Faba faba* (L.) House; *Faba major* Desf.; *Faba minor* Roxb.; *Faba sativa* Bernh.; *Faba sativa* Moench; *Faba vulgaris* Moench; *Faba vulgaris* var. *paucijuga* Alef.; *Orobus faba* Brot.; *Vicia esculenta* Salisb.; *Vicia faba* subsp. *paucijuga* Muratova; *Vicia vulgaris* Gray

- *Vicia faba* L. var. *equina* St.-Amans (Table 7.2)

Synonyms: *Vicia faba* subsp. *equina* (Pers.) Schubl. & G. Martens; *Vicia faba* var. *minor* Peterm.

- *Vicia faba* L. var. *faba* (Table 7.3)

Synonyms: *Faba vulgaris* var. *major* Harz; *Vicia faba* [unranked] *major* (Harz) Beck

- *Vicia faba* L. var. *minuta* (hort. ex Alef.) Mansf. (Table 7.4)

Synonyms: *Faba vulgaris* var. *minor* Harz; *Faba vulgaris* var. *minuta* hort. ex Alef.; *Vicia faba* [unranked] *minor* (Harz) Beck

7.2 ORIGIN OF SCIENTIFIC AND POPULAR TAXA NAMES

From many a point of view, faba bean (*Vicia faba*, syn. *Faba vulgaris*) has a specific and unique place among the pulse crops. Its growth habit and general morphology have never ceased to inspire generations of taxonomists to assess its proper status in relation to other botanically close species and, so far, only to end where it all began, that is, with its Linnean designation as the most widely accepted within the global legume research community (Linnaeus 1753, 1758). Furthermore, even though it is officially considered a vetch and that has few closer relatives, with a possibility of interspecific hybridization, faba bean remains one of the very few crops in the whole world without identified primitive forms in wild floras, although very recent discoveries seem to identify its

TABLE 7.1
Popular Names Denoting *Vicia faba* L. in Some World Languages and Other Linguistic Taxa

Language/Taxon	Name
Achterhooks	grote boon
Adyghe	ceshä
Afrikaans	breë boontjie
Ainu	mame
Akhvakh	hali
Akkadian	luppu
Aknadian	hale
Albanian	bathë
Albanian (Arvanitika)	bathe
Amharic	bak'ēla
Andi	holi
Arabic	full
Aragonese	faba
Aramaic	dagr-
Archi	bex:`é čaq
Armenian	bakla; lobi
Asturian	fabona
Avar	holo
Awngi	adangwari
Aymara	jawasa
Azerbaijani	at paxlası
Bashkir	baqsa borsağı
Basque	baba
Basque (Baztanese)	illar
Basque (Biscayan)	irar
Basque (Lapurdian)	illar
Basque (Lower Navarrese)	illar
Basque (Upper Navarrese)	illar
Basque (Roncalese)	ilhar-xúri
Basque (Souletin)	ílhar; ilhar-biríbil
Belarusian	bob
Berg en Terblijt	bwoan
Beurla-Reagaird	pònair
Bezhta	holo
Bilen	adängʷal; adogur; baldanguā́
Bilzen	baun
Boyko	bib
Brakel (Gelderland)	boèn
Bree	buu-uun
Brescian	faa
Breton	fav

(*Continued*)

TABLE 7.1 (*Continued*)
Popular Names Denoting *Vicia faba* L. in Some World Languages and Other Linguistic Taxa

Language/Taxon	Name
Budels	boén
Bulgarian	bakla; cher bob
Buli	ḅaal
Bunjevac	bob
Catalan	fava; favera
Catanian	favjana
Caterisani	vajana
Chamalal	hal
Chechen	qo
Chewa	nyemba yotakata
Chinese	cán dòu
Chuvash	nímëş parşí
Cineni	ngüre
Colognian	Bunne
Cornish	fav
Corsican	fava
Crimean Tatar	eşek baqlası; pasle
Croatian	bob
Cumbric	favenn
Czech	bob koňský; bob obecný; vikev bob
Dalmatian	fua
Damot	adangwari
Danish	favabønne; hestebønne; vælsk bønne
Dutch	labboon; tuinboon
Dutch Low Saxon	platte peters; tuunbonen
Egyptian (Coptic)	aro
Egyptian (Old)	puyr
Elfdalian	byöna
English	bell bean; broad bean; Celtic bean; English bean; faba bean; fava bean; field bean; horse bean; pigeon bean; tic bean; tick bean; Windsor bean
English (Old)	béan
Erzya	kuvtjol
Esperanto	fabo
Estonian	põlduba
Eys	boeën
Faroese	bøna
Finnish	härkäpapu
Finnish (Helsinki)	bōōna
Finnish (South-West)	pōōnā
Flemish	boon
French	fève; fèverole; vesce feve

(*Continued*)

TABLE 7.1 (*Continued*)
Popular Names Denoting *Vicia faba* L. in Some World Languages and Other Linguistic Taxa

Language/Taxon	Name
French (Ontario)	pois blanc
French (Quebec)	fagioli; fève des marais; gourgane
French (Rwanda)	fève des marais
Frisian (North)	ääkerbuan; eekerbuun; hingstbuan; sjok buan; swinbuan
Frisian (West)	bean; beane; beanne
Friulian	fave
Galambu	ándì
Galician	faba loba; fabaca; faballón; fabón
Gawwada	äläło
Geffen	bôn
Geji	ḅáali
Genk	boen
Genoese	bazann-a; fâva
Georgian	lobio; ts'erts'vi
Gera	handìmì
German	Ackerbohne; Acker-Wicke; Dicke Bohne; Faberbohne; Favabohne; Große Bohne; Pferdebohne; Puffbohne; Saubohne; Schweinsbohne; Viehbohne
Glavda	ʔagùra
Greek	kýamos; koukí
Greek (Old)	fakós
Gronings	peerdeboon; woalse boon
Guduf-Gava	ŋgurè
Hasselt	boeën
Hebrew	fūl; pool
Heusden (Belgium)	boen
Hindi	bakalā; baklasim; kala matar
Horst	boën
Hungarian	bab; lóbab
Hunsel	boeën
Hutsul	bîp
Icelandic	bóndabaunir; hestabaunir; velskar baunir
Ido	fabo
Indonesian	kacang babi; kara oncet
Ingrian	papu
Ingush	qeş
Irish	pónaire; pónaire leathan
Istriot	fava
Italian	fava; favino; favito
Japanese	sora mame
Joratian	favioûla
Judaeo-Spanish	ava; avika

(*Continued*)

TABLE 7.1 (*Continued*)
Popular Names Denoting *Vicia faba* L. in Some World Languages and Other Linguistic Taxa

Language/Taxon	Name
Kabyle	ibawen
Kalmyk	bob
Kampen	bone
Kanne	boen
Karachay-Balkar	hans qudoru
Karaim	baqla; burcacyk; bob
Karata	hale
Karelian	papu
Kashubian	bób
Kazakh	at burşaq; iri burşaq
Kempenlands (Eersel)	bón
Khamiri	adogur
Khanty	bob
Khwarshi	ħel
Koersel	Boen
Komi	bobijas
Korean	jamdu
Kumyk	burçaq
Kurdish (Northern)	baqil; beqle; beqilk; keşol
Lagwan	máágùréé́
Lak	luħi qjuru
Latin	faba
Latvian	pupa
Lebbeke	boeën
Lemko	bib
Lezgian	paxla; xaru
Ligurian	bazzan-a; fâva
Limburgish	akkerboean; boen; haofboean; paersboean
Lithuanian	pupa
Livonian	pubād
Lunteren	boan
Luxembourgish	boun
Macedonian	bakla; bob
Malay	kacang buncis besar; kacang faba; kacang fava; kacang loceng; kacang parang; kacang tik
Malayalam	avara
Malgwa	ŋ̀gurè
Maltese	fula
Mangas	ḅāl
Mansi	bob
Manx	poanrey

(*Continued*)

TABLE 7.1 (*Continued*)
Popular Names Denoting *Vicia faba* L. in Some World Languages and Other Linguistic Taxa

Language/Taxon	Name
Mapundungun	awar
Mari (Hill)	tarakan pərsa
Mari (Meadow)	neməçpursa
Mauritian Creole	bokla; fève de Mardis
Meänkieli	pööna
Mehri	dəgərät
Mofu-Gudur	mada aŋgar
Mogum	giri; gír(k)
Moksha	babanjsnav
Mongolian	morin shosh
Moravian (Czech Republic)	buny
Mwaghavul	mbwaalaa
Nahuatl	caxtillāntlahtōlli; caxtillān ayecohtli
Nederasselt	bôn
Nepali	bakulla simee
Ngas	gürm
Nijswiller	bòn
Norwegian (Bokmål)	åkerbønne; bondebønne; fababønne; favabønne; hestebønne
Norwegian (Nynorsk)	hestebønne
Occitan	fava
Old Prussian	babâ; babo
Onze-Lieve-Vrouw-Waver	Boeën
Oromo	adangwaaree; adunguri; daangulle; otongora
Ossetian	qædur
Persian	baqla
Picard	fèfe
Piedmontese	fava
Polish	bób
Portuguese	fava; fava-comum; fava-italiana; fava silvestre; feijão-fava
Quechua	chaqullu; habas hawas
Romagnol	feva
Romani	boba
Romanian	bob
Russian	bob konskii; bob obyknovennyi; bob russkii; bob sadovyi; boby
Rusyn (Carpathian)	bib
Rusyn (Pannonian)	bob
Saho	adagur
Sami (Inari)	páápu
Sami (Northern)	báhpu
Sami (Skolt)	kåårak
Sami (Southern)	bööna

(*Continued*)

TABLE 7.1 (*Continued*)
Popular Names Denoting *Vicia faba* L. in Some World Languages and Other Linguistic Taxa

Language/Taxon	Name
Samogitian	popa
Sanskrit	vartulakam; vrhulkm
Sardinian (Campidanese)	faa; fae
Sardinian (Logudorese)	fa
Scottish Gaelic	pònair
Serbian	bob
Shehri	dəgərät
Sicilian	fava; faviana
Silesian (Upper)	bober
Slovak	bôb
Slovenian	bob
Slovincian	bouna
Somali	digir
Somrai	giri; ǯìrī
Soqotri	dengo; dígir
Sorbian (Lower)	bob
Sorbian (Upper)	bobroch; konjacy bob
Spanish	haba; habichuela
Stein	boean
Sudovian	baba
Swahili	mharagwe-pana
Swedish	bondböna
Tabasaran	xaru
Tajik	boqilo; boqlī
Tatar	bakça borçagı
Thai	t̄hạ̀w pāk x̂ā
Tindi	beč'at'ub hali
Turkish	bakla
Udmurt	s'öd kóžy
Ukrainian	bib; bib zvichainyi; bib kins'kii; boby
Uzbek	bokla; boqla, burchoq
Veluws	platte peters
Venetian	fava
Venlo	boeën
Venray	boën
Vietnamese	đậu răng ngựa
Võro	uba
Waanrode	boën
Walloon	grosse feve; pitite feve
Walshoutem	boën
Wandala	gíre

(*Continued*)

TABLE 7.1 (*Continued*)
Popular Names Denoting *Vicia faba* L. in Some World Languages and Other Linguistic Taxa

Language/Taxon	Name
Welsh	ffa; ffeuen; ponar
Welsh Romani	rattlers
West Frisian Dutch	bòòòn; bòòòòn; bòòòòòn
Yiddish	bob
Yoruba	ẹwa
Zenaga	ti-d_īgi-d
Zul	ḅaali
Zwartebroek	boon; kruuper

TABLE 7.2
Popular Names Denoting *Vicia faba* L. var. *equina* St.-Amans in Some World Languages and Other Linguistic Taxa

Language/Taxon	Name
Dutch	molleboon; paardenboon; veldboon; wierboon
English	field bean; horse bean
French	fève à cheval
German	Pferdebohne
Italian	fava cavallina; favetta; favetta cavallina
Latvian	lauka pupa; vidējrupjsēklu pupa; zirgu pupa
Polish	bobik
Portuguese	fava-cavalinha; fava-da-holanda; fava-de-cavalo; favarola; faveira
Russian	konskie boby; kormovye boby; melkosemennye boby
Serbian	konjski bob; krmni bob; stočni bob
Spanish	haba caballar
Turkish	ufak bakla
Vietnamese	đậu ngựa

progenitor, existing 14,000 years ago (Caracuta et al. 2016). Faba bean appeared among the first cultivated plant species in the very dawn of agriculture in the Near East, as early as more than 10,200 years BP (Caracuta et al. 2017). The linguistic evidence on the role that faba bean had in everyday diets of our ancestors is remarkably rich, especially in the Afroasiatic, Caucasian, and Indo-European ethnolinguistic families.

There are several roots denoting faba bean in the Afroasiatic languages (Militarev and Stolbova 2007):

- **(ʔa-)da(n)g(w)ir-*, denoting exclusively faba bean, and with a remarkable evolution into the Proto-Central-Cushitic **ʔa-da(n)g^wVr-*, was responsible for the words in Awngi, Bilen, Damot, and Khamiri, the

TABLE 7.3
Popular Names Denoting *Vicia faba* L. var. *faba* in Some World Languages and Other Linguistic Taxa

Language/Taxon	Name
Burmese	sandusi
Chinese	bai hua can dou; hong hua can dou
Danish	agerbønne; hestebønne; valsk bønne; valskbønne; vølskbønne
Dutch	duiveboon; paardenboon; tuinbonen; tuinboon; veldboon; waalse boon
English	broad bean; faba bean; fava bean; red-flowered broad bean; Windsor bean
English (U.S.)	fava-bean
French	fève à fleurs rouges; fève des marais; fève ornementales; grosse fève
French (Quebec)	gorgane; gourgane
German	Dicke Bohne; Puffbohne
Hindi	anhuri; bakla; kala matar
Indonesian	kacang babi
Italian	fagioli; fava; fava grossa comune
Japanese	sora mame
Latvian	cūku pupa; dārza pupa; rupjsēklu pupa
Malay	kacang babi
Norwegian	baunevikke
Polish	wyka bób
Portuguese	fava; faveira
Punjabi	bakla; chastang; kabli bakla
Russian	boby krupnosemennye; boby ovoshchnye
Saho	baldangā; bardanguā
Serbian	baštenski bob; krupnosemeni bob; ljudski bob; povrtarski bob
Spanish	faba haba; haba común; haba de huerta; haba mayor; habichuela; haboncillo
Thai	chiang mai; thua yang
Turkish	bakla

Proto-Lowland-East-Cushitic **ʔa-da(n)g*^{w}*Vr-* and Proto-Saho-Afar **ada-gur*, bringing forth the words in Oromo, Saho, and Somali, and the Proto-Semitic **dVgVr-*, producing the words in Arabic, Aramaic, Mehri, Shehri, and Soqotri;

- **ʕadas-*, with a primeval designation of faba bean, had few derivatives with shifted meaning to other pulse crops and plant-related terms;
- **ʔary/w-* ~ **(ʔV)w/yar-* denoted equally seed, corn, and faba bean and, through the Egyptian *ỉwry*, produced the word in Coptic;
- Designing corn and faba bean, **da/ingw-* gave the words denoting the latter in the Berber, with Zenaga, the Cushitic, with Bilen, and the Semitic, with Soqotri;
- **ḥVnbal-*, initially denoted corn, but subsequently evolved into the Proto-Semitic **ḥunbul-* and Proto-West-Chadic **mV-bwal-*, both of which began to design faba bean, as witnessed by the corresponding words in Arabic and in Buli, Geji, Mangas, Mwaghavul, and Zul;

TABLE 7.4
Popular Names Denoting *Vicia faba* L. var. *minuta* (hort. ex Alef.) Mansf. in Some World Languages and Other Linguistic Taxa

Language/Taxon	Name
Dutch	duivenboon
English	Celtic bean; field bean; tickbean
French	féverole
German	Ackerbohne
Italian	favetta; favino
Latvian	lopbarības pupa; sīksēklu pupas
Portuguese	favarola; favinha
Russian	boby melkosemennye
Serbian	krmni bob; stočni bob; sitnosemeni bob
Spanish	haba menor; haba pequeña

- *lap-, designed both faba bean and corn, and produced the Proto-Semitic *lupp-, denoting only the former, as may be seen in Akkadian;
- Denoting both millets (e.g., *Eleusine* spp., *Panicum* spp., *Pennisetum* spp., *Setaria* spp., or *Sorghum* spp.) and faba bean, **mang-* gave the Proto-Semitic **mang-/*magg-*, indicating equally lentil and faba bean;
- **pal-* denoted both corn and faba bean and was responsible for the Egyptian words *pry (n)* and *puyr* and the Semitic names in Arabic, Hebrew, and Maltese (Table 7.1).

The ancient Caucasian ethnolinguistic family, with the languages spoken on the northern slopes of the highest mountain range in Europe, from the Black Sea to the Caspian Sea, has two attested roots related to faba bean (Nikolayev and Starostin 1994). One of them is **hōwƚ[ā]*, denoting both lentil (*Lens culinaris* Medik.) and faba bean (Starostin 2005d, Mikić and Vishnyakova 2012). This root evolved into the Proto-Avar-Andi-Dido **ħoli* (Starostin 2003b), referring to both faba bean and pea (*Pisum sativum* L.), that produced the words related to faba bean in Akhvakh, Aknadian, Andi, Avar, Chamalal, Karata, and Tindi, and the Proto-Tsezic **hel(u) A*, also associated with faba bean and pea, with the word in Khwarshi (Table 7.1). On a higher level, that is, if the supposed Dené-Caucasian ethnolinguistic macrofamily is considered, this Proto-Caucasian root is corresponding to the Proto-Basque **iƚha-r̄*, designing equally pea, faba bean, and common vetch (*Vicia sativa* L.), and being reflected through several Basque dialects, such as Baztanese, Lapurdian, Lower and Upper Navarrese, Roncalese, and Souletin, in which it is related to faba bean (Mikić 2011b, Bengtson 2015). Both Proto-Basque *iƚha-r̄* and Proto-Caucasian **hōwƚ[ā]* descend from the Proto-Dené-Caucasian **hVwƚV*, with a primeval signifying faba bean (Figure 7.1). Another Proto-Caucasian root is **qŏrʔā (~-rħ-)*, originally denoting pea (Starostin 2005d) and directly deriving into the Proto-Nakh **qo(w)e ~ *qe(w)u*, which gave the words meaning faba bean in modern Chechen and Ingush, as well as in the Abkhazo-Adyghean, with Adyghe, the Lak and the Lezgic, with Lezgian and Tabasaran (Starostin 2003a).

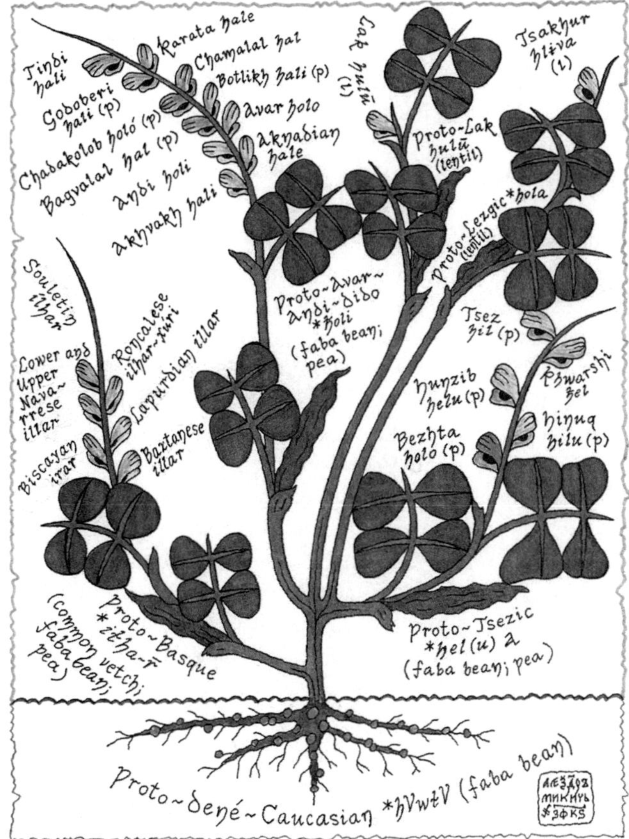

FIGURE 7.1 **(See color insert.)** One of the possible evolutions of the root **hVwƛV* of the proposed Proto-Dené-Caucasian language, denoting faba bean (Starostin 2015), into its direct derivatives, shown as pods, and contemporary descendants, depicted as flowers, in the Basque and Caucasian languages; the meanings of the proto-words are given within brackets, while the meanings of the modern words are *faba bean*, if given without brackets, as well as *lentil* (l) and *pea* (p).

Among many Proto-Indo-European roots related to pulse crops with a large number of attested direct derivatives, **bhabh-*, *bhabhā* is regarded as having the literal meaning of a descriptive one, that is, *swollen* or *swelling*, and was used to denote faba bean (Pokorny 1959, Nikolayev 2012). This was entirely preserved, although more than five millennia have passed between the time this Proto-Indo-European word-root was used in spoken form and its countless descendants in the contemporary Indo-European languages (Mikić 2011d). Its direct derivatives include (Tables 7.1 through 7.4):

- The Proto-Albanian **bhakā*; with its modern form;
- The unattested and hypothetic Proto-Baltic root, most likely similar to the extinct *baba*, *babo* in Old Prussian and *baba* in Sudovian and producing

the corresponding words in Latvian, Lithuanian, and Samogitian (Mikić 2014b);

- The Proto-Germanic **bau-nō(n-)*, giving the words in Afrikaans, Colognian, Danish, Dutch with its remarkably dialectal diversity, Dutch Low Saxon, English, Faroese, Flemish, Frisian, German, Icelandic, Luxembourgish, Norwegian and Swedish and being borrowed by the neighboring Goidelic Celtic, such as Beurla-Reagaird, Irish, Manx, and Scottish Gaelic (MacBain 1911), the Slavic Moravian and Slovincian, and the Uralic Helsinki Finnish, South-West Finnish, Meänkieli, and Southern Sami;
- The unattested Proto-Italic root, bringing forth the Faliscan *haba* and the Latin *faba*, with the latter being a direct ancestor of the words denoting faba bean in Aragonese, Asturian, Brescian, Catalan, Catanese, Caterisani, Corsican, Dalmatian, French, Friulian, Galician, Genoese, Istriot, Italian, Joratian, Judaeo-Spanish, Ligurian, Occitan, Picard, Piedmontese, Portuguese, Romagnol, Sardinian, Sicilian, Venetian, and Walloon, with a transfer from Latin to the Brythonic Celtic, with Breton, Cornish, Cumbric, and Welsh (Mikić 2014c), and from Spanish into the Amerind Aymara and Quechua;
- The Proto-Slavic **bobŭ*, producing the words relating to faba bean in Belarusian, Boyko, Bulgarian, Bunjevac, Croatian, Czech, Hutsul, Kashubian, Lemko, Macedonian, Polish, Russian, Carpathian and Pannonian Rusyn, Serbian, Upper Silesian, Slovak, Slovenian, Lower and Upper Sorbian, and Ukrainian (Mikić 2014a) and being borrowed by the Altaic Kalmyk and Karaim, the Germanic Yiddish, the Indo-Aryan Romani, the Romance Romanian and the Uralic Estonian, Finnish, Hungarian, Ingrian, Karelian, Khanty, Komi, Livonian, Mansi, Moksha, Inari Sami, Northern Sami, and Võro (Stoddard et al. 2009);
- The single descendant of Proto-Indo-European **bhabh-*, *bhabhā* with the meaning shifted was Old Greek, where, in the form of *fakós*, began and continued to denote lentil (Mikić 2010); on the other hand, one of the modern Greek words associated with faba bean has its roots in the Ancient Greek *kókkos*, referring to grain, kernel, and seed and with a still undetermined origin and, perhaps, borrowed from some of the Pre-Hellenic peoples (Liddell et al. 1940).

The Arabic root *baḳl*, generally denoting edible plants and vegetables, evolved into *baḳla(t)*, with a similar meaning (Nişanyan 2017), and was imported into other Semitic languages, such as Amharic, as well as in the Altaic, including Azerbaijani, Crimean Tatar, Karaim, Tatar, Turkish, and Uzbek (Mikić and Perić 2011), the Caucasian, with Lezgian, the Indo-European, such as Armenian, Bulgarian, Macedonian, Hindi, Kurdish, Nepali, Persian, and Tajik, and some creole languages, like Mauritian Creole (Table 7.1).

There are languages, belonging to diverse ethnolinguistic families, where faba bean is most often associated with pea. In other words, faba bean is regarded as a kind of pea and this may be a case in such circumstances where pea was an older and already well-established crop, whereas faba bean was introduced later. Such examples may be found in the Altaic languages Ainu, Bashkir, Chuvash, Crimean

Tatar, Japanese, Karachay-Balkar, Karaim, Kazakh, Korean, Kumyk and Tatar, the Caucasian Archi, and the Indo-European Hindi and Kurdish, and the Uralic Hill and Meadow Mari, Moksha, and Udmurt (Table 7.1).

Regarding animals, faba bean is, quite frequently, connected with horse, such as in Czech, English, German, Gronings, Mongolian, Norwegian, Russian, Serbian, Ukrainian, and Vietnamese, or, rarer, with pig, with Frisian and German (Tables 7.1 and 7.2).

Two names denoting faba bean may be considered remarkably curious: the first is the one in Welsh Romani, with an obvious onomatopoeic background, while the second one is the Upper Sorbian, as a hybrid word of *bob* (faba bean) + *hroch* (pea) = *bobroch* (Table 7.1).

8 *Glycine* Willd.

Synonyms: *Chrystolia* Montrouz.; *Dolichos* L.; *Glycine* L.; *Kennedia* Vent.; *Leptocyamus* Benth.; *Leptolobium* Benth.; *Phaseolus* L.; *Rhynchosia* Lour. *Soja* Moench; *Teramnus* P. Browne; *Zichya* Bolívar

8.1 LIST OF TAXA SCIENTIFIC AND POPULAR NAMES

Although moderately rich in species in comparison to an average genus of legume plants, the genus *Glycine* Willd. provides us with vernacular names for only two species, one of which is economically the most important grain legume crop in the world, while another is its important wild relative. The following lines will bring a list of the species and their subtaxa within this genus, as well as an extensive collection of compiled popular names (Broué et al. 1977, Hymowitz and Newell 1981, ISTA 1982, Rehm 1994, Gledhill 2008, Hymowitz et al. 1998, Porcher 2008, The Plant List 2013, Ecocrop 2017, EPPO 2017, Ethnologue 2017, IBIS 2017, ILDIS 2017, Logos 2017, NPGS 2017, Wikipedia 2017, Wiktionary 2017).

- *Glycine albicans* Tindale & Craven
English: whitening soybean
- *Glycine aphyonota* B. E. Pfeil
English: aphyonote soybean
- *Glycine arenaria* Tindale
English: sandstone soybean
- *Glycine argyrea* Tindale
English: silvery soybean
- *Glycine canescens* F. J. Herm.
Synonyms: *Glycine sericea* (F. Muell.) Benth.; *Glycine sericea* (F. Muell.) Benth. var. *orthotrica* J. M. Black; *Leptocyamus sericeus* F. Muell.
English: silky glycine
- *Glycine clandestina* J. C. Wendl.
Synonyms: *Leptocyamus clandestinus* (Wendl.) Benth.; *Leptocyamus microphyllus* Benth.; *Leptolobium clandestinum* (Wendl.) Benth.; *Leptolobium microphyllum* Benth.; *Teramnus clandestinus* (Wendl.) Spreng.
Chinese: pēnghú dàdòu
English: love creeper; twining glycine
- *Glycine curvata* Tindale
English: curved soybean
- *Glycine cyrtoloba* Tindale
English: basketlike-lobed soybean
- *Glycine dolichocarpa* Tateishi & H. Ohashi
Chinese: biǎn dòujiá dàdòu

English: long-fruited soybean
- *Glycine falcata* Benth.
English: sickle-like soybean
- *Glycine gracei* B. E. Pfeil & Craven
English: Grace's soybean
- *Glycine hirticaulis* Tindale & Craven
English: hairy-stemmed soybean
- *Glycine hirticaulis* Tindale & Craven subsp. *hirticaulis*
English: common hairy-stemmed soybean
- *Glycine hirticaulis* Tindale & Craven subsp. *leptosa* B. E. Pfeil
English: slender hairy-stemmed soybean
- *Glycine lactovirens* Tindale & Craven
English: milk-flourishing soybean
- *Glycine latifolia* (Benth.) C. A. Newell & Hymowitz
Synonyms: *Glycine latifolius* (Labill.) Benth. var. *latifolia* Benth.; *Glycine tabacina* (Labill.) Benth. var. *latifolius* Benth.; *Leptocyamus latifolius* Benth.
English: wide-leafed soybean
- *Glycine latrobeana* (Meisn.) Benth.
Synonyms: *Leptocyamus tasmanicus* Benth.; *Zichya latrobeana* Meisn.
English: Australian anchor plant; clover glycine
- *Glycine max* (L.) Merr. (Table 8.1)
Dolichos soja L.; *Glycine angustifolia* Miq.; *Glycine gracilis* Skvortsov; *Glycine hispida* (Moench) Maxim.; *Glycine hispida* var. *brunnea* Skvortsov; *Glycine hispida* var. *lutea* Skvortsov; *Glycine soja* (L.) Merr.; *Phaseolus max* L.; *Soja angustifolia* Miq.; *Soja hispida* Moench; *Soja japonica* Savi; *Soja max* (L.) Piper; *Soja soja* H. Karst.; *Soja viridis* Savi
- *Glycine microphylla* (Benth.) Tindale
Synonyms: *Leptolobium microphyllum* Benth.
English: small-leafed soybean
- *Glycine montis-douglas* B. E. Pfeil & Craven
English: Mount Douglas soybean
- *Glycine peratosa* B. E. Pfeil & Tindale
English: abundant soybean
- *Glycine pescadrensis* Hayata
English: peach-like soybean
- *Glycine pindanica* Tindale & Craven
English: Pindan sooybean
- *Glycine pullenii* B. E. Pfeil et al.
English: Pullen's soybean
- *Glycine rubiginosa* Tindale & B. E. Pfeil
English: red-ovary soybean
- *Glycine soja* Siebold & Zucc. (Table 8.2)
Synonyms: *Glycine formosa* Hosok.; *Glycine formosana* Hosok.; *Glycine javanica* Thunb.; *Glycine max* subsp. *soja* (Siebold & Zucc.) H. Ohashi; *Glycine ussuriensis* Regel & Maack; *Glycine ussuriensis* Regel & Maack var. *brevifolia* Kom.; *Rhynchosia argyi* H. Lev.

TABLE 8.1
Popular Names Denoting *Glycine max* (L.) Merr. in Some World Languages and Other Linguistic Taxa

Language/Taxon	Name
Afrikaans	sojaboon
Albanian	sojë
Alemannic	soja; sojabohne
Amharic	ye'ākurī āteri
Arabic	fawall alssawia; fûl sûyah
Aragonese	soya
Armenian	soya
Asturian	soya
Azerbaijani	ekme soya; soya
Bashkir	soja
Basque	soja
Bavarian	sojaboanl
Belarusian	soevye boby; soia; soia kul'turnaia; soia shchatzinistaia
Bengali	saẏābina
Bihari	sō'ābīna
Bosniak	soja
Breton	greunnen soja; soja
Bulgarian	soya
Bunjevac	soja
Catalan	soia; soya
Cebuano	utaw
Chinese (Cantonese)	huáng dòu; wong dau
Chinese (Gan)	sà dòu
Chinese (Hakka)	thai-theu; vòng-theu
Chinese (Mandarin)	dà dòu; huáng dòu; kuān yèmán dòu
Chinese (Taiwan)	mao dou
Chuvash	soya
Croatian	soja
Czech	sója luštinatá
Danish	sojabønne
Dumi	gya:ksi
Dutch	sojaboon
English	edamme; soya; soya-bean; soya bean; soybean
English (UK)	soya bean
English (U.S.)	soybean
Esperanto	sojfabo; sojglicino; sojherbo
Estonian	karvane sojauba; kultuur-sojauba; põld-sojauba; soja; sojauba
Filipino	toyo
Finnish	soija; soijapapu
French	haricot oléagineux; pois chinois; soja; soya
Frisian (North)	soojabuan

(Continued)

TABLE 8.1 (*Continued*)
Popular Names Denoting *Glycine max* (L.) Merr. in Some World Languages and Other Linguistic Taxa

Language/Taxon	Name
Galician	soia
Georgian	soia
German	Soja; Sojabohne
Greek	sógia; fasóli apó sógia
Guarani	sóha
Gujarati	sōyābīna
Haitian Creole	soja
Hausa	waken soya
Hebrew	soya
Hindi	bhat; bhatwar; bhetmas; soyaabeen
Hmong	taum hwv
Hungarian	szója; szójabab
Icelandic	sojabaun
Ido	flava soyo; soyo
Ilocano	bukel ti soya; soya
Indonesian	kacang kedelai; kedelai
Irish	pónaire soighe
Italian	soia; soja
Jamaican Patois	saibiin; sayabiin
Japanese	daizu
Javanese	dhelé
Kannada	sōyā avare
Kazakh	ekpe soya; mädenï soya; qıtayburşaq; tükti soya
Khaling	gipsi
Khmer	sândaèk an gen sar; sândaèk sieng
Kinyarwanda	soya
Korean	daedu; kong
Kurdish (Central)	es'yā
Kurdish (Northern)	fasulyeya soyayê; soya
Kyrgyz	soya
Lao	thwàx khôn; thwàx tê
Latvian	sarmatainā soja; soja
Lithuanian	gauruotoji soja; soja
Luxembourgish	soja
Macedonian	soja
Maithili	bhaṭamāsa
Malay	kacang soya
Malayalam	sēāyābīns
Maltese	sojja
Māori	meatia

(*Continued*)

TABLE 8.1 (*Continued*)
Popular Names Denoting *Glycine max* (L.) Merr. in Some World Languages and Other Linguistic Taxa

Language/Taxon	Name
Marathi	sōyābīna
Mari (Hill)	soia
Mari (Meadow)	soia
Min (Eastern)	uòng-dâu
Min (Southern)	n̂g-tāu
Mingrelian	mukhudo
Moldovan	soe mare
Mongolian	shar buurtzag; tarimal sharbuurtzag
Myanmar	lasi; pengapi; peryatpym
Nepali	bhaṭamāsa
Norwegian (Bokmål)	soyabønne
Norwegian (Nynorsk)	soyabønne
Occitan	sòja
Ossetian	kul'turon sojæ; sojæ
Otomi (Northwestern)	soja
Papiamento	e soja
Pashto	daal sabinui
Persian	suia
Polish	soja owłosiona; soja warzywna; soja zwyczajna
Portuguese	feijão-chinês; feijão-soja; soja
Portuguese (Brazil)	feijão-chinês; feijão-soja; soja
Punjabi (Eastern)	sō'i'ā bīna
Punjabi (Western)	soya bain
Quechua	suya
Romanian	soia
Russian	soevye boby; soia kul'turnaia; soia obyknovennaia
Sanskrit	sōyāmāṣaḥ
Scots	soya bean
Serbian	kineski pasulj; soja
Sindhi	sujhabjhn
Sinhalese	boo-mae
Slovak	sója fazuľová
Slovenian	soja
Somali	landkruusaro
Sorbian (Lower)	soja
Sorbian (Upper)	soja; sojabob
Spanish	frijol de soya; haba soya; soja; soya
Sundanese	kedelé
Swahili	maharage; soya
Swedish	sojaböna

(*Continued*)

TABLE 8.1 (*Continued*)
Popular Names Denoting *Glycine max* (L.) Merr. in Some World Languages and Other Linguistic Taxa

Language/Taxon	Name
Tagalog	balatong; utaw
Tajik	lūbijoi
Tamil	cōyā avarai
Tatar	soia
Telugu	sōyā cikkuḍu
Thai	t̄hạwh̄elụ̄xng; thuaa leuuang; thua lueang; thua phra lueang; thua rae
Thulung	kɛksi
Tibetan	rgya-sran; sran ser
Turkish	çin lubyası; soya; soya lubyası
Udmurt	soya
Ukrainian	soia kul'turna; soia shchetynysta
Urdu	soya phalli
Uyghur	dad'r
Uzbek	soya; so'ya
Vietnamese	đậu nành; đậu tương; đỗ tương; quantan
Võro	sojauba
Waray	soya
Welsh	soia
Xhosa	weembotyi
Yiddish	soye
Yucatec	soja
Zhuang	duh; duhhenj

TABLE 8.2
Popular Names Denoting *Glycine soja* Siebold & Zucc. in Some World Languages and Other Linguistic Taxa

Language/Taxon	Name
Chinese	dà dòu; shan huang dou; ye da dou
Czech	sója dyvoká
English	reseeding soybean; soja; wild soybean
French	soja sauvage
German	Wilde Sojabohne
Hungarian	vad szója
Indonesian	kedalai hitam
Italian	soia selvatica
Japanese	tsuru-mame
Korean	dolkong
Russian	glitsine soia; soia dikaya; soia ussuriiskaia
Serbian	divlja soja
Spanish	soja silvestre
Vietnamese	đậu tương leo; đậu tương núi

- *Glycine stenophita* B. E. Pfeil & Tindale
English: narrow-habitat soybean
- *Glycine syndetika* B. E. Pfeil & Craven
English: syndetic soybean
- *Glycine tabacina* (Labill.) Benth.
Synonyms: *Glycine koidzumii* Ohwi; *Kennedia tabacina* Labill.
Chinese: yān dòu
English: glycine-pea; pea glycine; variable glycine; wild soybean
Japanese: bōko-tsuru-mame
- *Glycine tomentella* Hayata
Synonyms: *Glycine tomentosa* (Benth.) Benth.; *Leptolobium tomentosum* Benth.
Chinese: duǎn róng yě dàdòu; kuò yè dàdòu
English: hairy glycine; Peak Downs-clover; rusty glycine; woolly glycine

8.2 ORIGIN OF SCIENTIFIC AND POPULAR TAXA NAMES

The name of the genus *Glycine* was derived from the Greek adjective *glykýs*, meaning *sweet tasting*. Linnaeus used it to denote a new genus, referring to the sweetness of edible tubers of one of its species, known mostly as American potato-bean (*Glycine apios* L.) (Linnaeus 1753, 1758). In subsequent classifications, this taxon gained new status and, as *Apios americana* Medik., was moved to the genus of its own, while the name *Glycine* remained to denote many other annual and perennial plants, including the most important grain legume crop in the world today (Shurtleff and Aoyagi 2004).

It is commonly considered that the domestication of soybean (*Glycine max* [L.] Merr.) began around 7000 BC in China, about 5000 BC in Japan, and approximately in 1000 BC in Korea (Lee et al. 2011). According to the Chinese tradition, emperor Sheng-Nung proclaimed soybean one of the five sacred crops, along with barley (*Hordeum vulgare* L.), rice (*Oryza sativa* L.), millet (*Panicum miliaceum* L.), and wheat (*Triticum aestivum* L.), in his opus *Ben Tsao Gang Mu*, dating from 2838 BC (Kalaiselvan et al. 2010). The most widely used of the Chinese names, *dà dòu*, literally means *large bean*, while another Chinese name, *huáng dòu*, means *yellow bean*. The Japanese and the Korean names are, in fact, slight modifications of the Chinese *dà dòu* (Table 8.1).

The first mention of soybean and its product in any European language is in the famous dictionary known as *Nippo Jisho* or *Vocabvlario da Lingoa de Iapam*, compiled by Jesuit missionaries in Nagasaki, Japan, in 1603, that was, at the same time, the first dictionary of Japanese and any European language (Cooper 1976). As may be obviously seen from the list of the compiled words denoting soybean in the world's languages, a vast majority of the names have the common segment, *soia*, *soja*, or *soy(a)*, regardless of the ethnolignuistic family to which they belong and the exact *Glycine* species (Tables 8.1 and 8.2). It is likewise in the Afroasiatic, with the Chadic and the Semitic subfamilies; in the Altaic, with its Turkic branch; in the Austronesian; in the Dravidian; and in the Indo-European, with its Armenian, Baltic, Celtic, Germanic, Hellenic, Indo-Aryan, Iranian, Italic, and Slavic groups. The word was also borrowed by the Amerind, with Guarani and Yucatec; by Basque; by the Niger-Congo; by the Uralic; by the creoles, such as Haitian creole and Papiamento; and by constructed languages, such as Esperanto, and Ido (Table 8.1).

The aforementioned worldwide spread morpheme *soy-* represents a corruption of the Mandarin Chinese word denoting soya sauce, *chĭyóu*, and its equivalents in Cantonese Chinese, *sihyàuh*, and Japanese, *shōyu*. The origin of these terms and the contemporary Mandarin Chinese word *sù*, denoting grain or seed, is the Proto-Chinese root **shok*, referring to the same terms and with a rather complex evolution (Mikić et al. 2013a). The ancestor of this Proto-Chinese root and the Proto-Kiranti **sV̀k-cɪə̀*, again designing grain and seed (Starostin 2005c), is the Proto-Sino-Tibetan root **sok*, with the same basic meaning (Starostin 2005e). Many Kiranti languages, such as Khaling, Kulung, Limbu, Thulung, and Yamphu, preserved it in their semantically corresponding words, in some of which these are also relating to lentil (*Lens culinaris* Medik.).

There is a hypothetical and unconventional linguistic theory about the Dené-Caucasian ethnolinguistic family, allowing a possibility of associating language taxa regarded as either independent groups or isolated by contemporary mainstream linguistics. One of its fundamental postulates is that the Basque, the Caucasian, the Burushaski, the Yenissenian, the Sino-Tibetan, and the Na-Dené families represent a continuum, ranging from the Pyrenees, over Central Asia and to North America, and that it was conceived somewhere in Central Asia, between 15,000 and 10,000 years ago and with certain genetic evidence (González et al. 2006). Also the speakers of this Proto-Dené-Caucasian were, in fact, the Eurasian hunter-gatherers, who, eventually with an advance of the Neolithic and agriculture, sought shelter in less-accessible areas, such as high mountains, valleys, or inhabited plains, and thus became a kind of islands in the sea of farmers (Ruhlen 2001). Following this minutely elaborated alternative proposal, with precisely exercised methods of comparative linguistics and abundant lexical evidence, it is possible to associate the said Proto-Sino-Tibetan **sok* with the related Proto-Basque **a=hoc*, denoting husk and wheat chaff (Bengtson 2015); the Proto-Caucasian **c̣HwekĔ (~ -k-)*, relating to chaff and straw (Starostin 2005d); the Proto-Burushaski **ṣiqá* (Starostin 2005a), referring to grass; the Proto-Yenisseian **TVKV* (Starostin 2005g), designing husk; and, very speculatively, some Na-Dené languages, such as the Mescalero-Chiricahua *yoo* and the Plains Apache with *zhoo*, both denoting a small round object; and, finally and extremely cautiously and quite conditionally, attempt to suggest that all of them originated from the reconstructed Proto-Dené-Caucasian root **sṭHwẹkĔ (~ -k-)*, basically denoting chaff (Starostin 2015), but also containing a germ of the subsequent word signifying the large and yellow and the most significant grain crop in East Asia: soybean (Figure 8.1).

There is another Sino-Tibetan root relating to soybean, namely the Proto-Kiranti **gèpsi ~ *gèsip (/ʔk-)*, retaining its primeval meaning in Dumi, with *gya:ksi*, Khaling, with *gipsi*, and Thulung, with *kɛksi* (Table 8.1), and with a shift to something bean-like in Limbu, with *khesippā* and Yamphu, with *käkkräŋma* (Starostin 2005c).

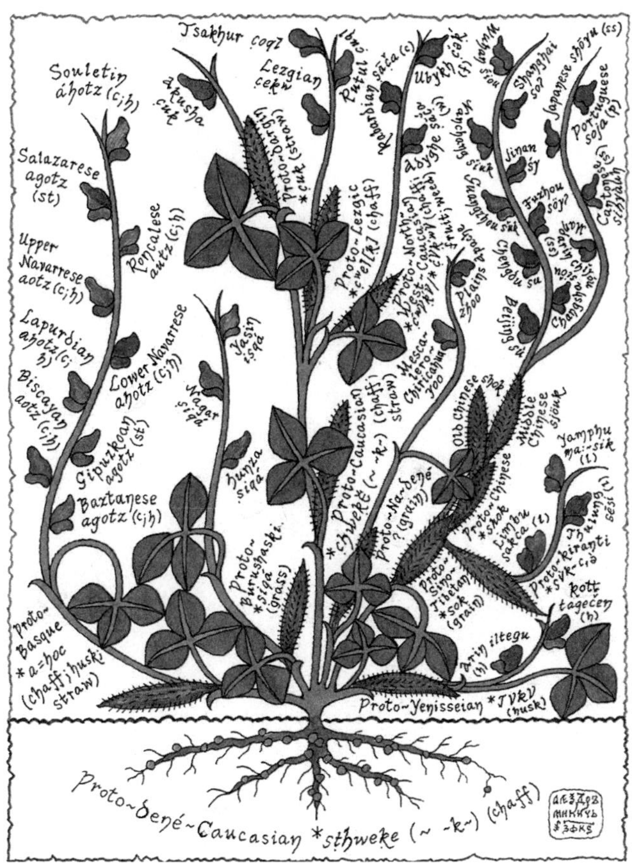

FIGURE 8.1 **(See color insert.)** One of the possible evolutions of the root **sṭHwekĔ (~ -k-)* of the proposed Proto-Dené-Caucasian language, denoting chaff (Starostin 2015), into its direct derivatives, represented as pods, and contemporary descendants, portrayed as flowers, in the Basque, Burushaski, Caucasian, Na-Dené, Sino-Tibetan, and Yenisseian languages; the meanings of the proto-words are given within brackets and the meaning of each modern word, if given without brackets, is the same in its proto-word, while with the bracketed abbreviations, such as (c) for *chaff*, (f) for *fruit*; (g) for *grain*, (h) for *husk*, (l) for *lentil*, (p) for *soybean plant*, (ss) for *soybean sauce,* and (w) for *weed*, are given to mark the distinction where needed.

Due to the shape of its grain, soybean has been, as its very name in English points out, associated with faba (*Vicia faba* L.) and other beans. Such examples may be easily found in the languages of the diverse world of ethnolinguistic families, such as the Afroasiatic, with Arabic and Hausa; the Austroasiatic, with Khmer and Vietnamese; the Austronesian, with Indonesian and Malay; the Altaic, with Turkish; the Indo-European, with nearly all of its representatives; the Tai-Kadai, with Lao and Thai; and the Uralic, with Estonian, Finnish, Hungarian, and Võro. Much rarer resemblances are those to pea (*Pisum sativum* L.), such as in the Altaic Mongolian and the Kartvelian Mingrelian languages (Table 8.1).

Among numerous other attributes of the names relating to soybean in various languages, we may find that it is cultivated, such as in Belarusian, Estonian, Kazakh, Ossetian, Russian, and Ukrainian; that it may be used immature, such as in Ido and Polish; that it is hairy, such as in Estonian, Polish, and Ukrainian; that its grain is large, such as in Moldovan; and that it is rich in oil, such as in French (Table 8.1).

It may be interesting to note that the sole imagined country of origin of soybean in linguistically more or less unrelated languages, such as French, Kazakh, Portuguese, Serbian, or Turkish, but with a common impression about its exotic nature, is China, and, without any doubt, the answer proves as fully correct.

9 *Lablab* Adans.

Synonyms: *Dolichos* L.; *Vigna* Savi

9.1 LIST OF TAXA SCIENTIFIC AND POPULAR NAMES

The present chapter deals with another monotypic genus and has a similar internal structure to those of Chapters 6 and 7. Along with a very brief list of the currently recognized taxa, there is an extensive table with vernacular names denoting the species *Lablab purpureus* (L.) Sweet, with a remarkable diversity in regional forms of English and Spanish languages across the world (Westphal 1975, ISTA 1982, Rehm 1994, Pengelly and Maass 2001, Maass et al. 2005, Wang et al. 2007, Gledhill 2008, Porcher 2008, The Plant List 2013, Sonari et al. 2015, Ecocrop 2017, EPPO 2017, Ethnologue 2017, IBIS 2017, ILDIS 2017, Logos 2017, NPGS 2017, Wikipedia 2017, Wiktionary 2017).

- *Lablab purpureus* (L.) Sweet (Table 9.1)

Synonyms: *Dolichos albus* Lour.; *Dolichos benghalensis* Jacq.; *Dolichos lablab* L.; *Dolichos purpureus* L.; *Glycine lucida sensu* Blanco, *non* J. R. Forst.; *Lablab cultratus* DC.; *Lablab lablab* (L.) Lyons; *Lablab niger* Medik.; *Lablab prostrata* R. Br.; *Lablab vulgaris* (L.) Savi; *Lablab vulgaris* var. *albiflorus* DC; *Vigna aristata* Piper

- *Lablab purpureus* (L.) Sweet subsp. *bengalensis* (Jacq.) Verdc.

Synonyms: *Dolichos bengalensis* Jacq.; *Dolichos lablab* L. var. *bengalensis* Nakai; *Dolichos uniflorus* Phamhoang

English: Bengalese hyacinth bean; horse gram

French: Dolique du Bengal; dolique sauvage

Japanese: Shiro fuji mame

Vietnamese: Đậu móng chim

- *Lablab purpureus* (L.) Sweet subsp. *purpureus*

Synonyms: *Dolichos lablab* L.; *Dolichos purpureus* L.; *Lablab leucocarpos* Savi; *Lablab niger* Medik.; *Lablab vulgaris* Savi

English: Annual hyacinth bean; cultivated hyacinth bean; eat-all hyacinth bean; garden lablab; hyacinth bean

- *Lablab purpureus* (L.) Sweet subsp. *purpureus* var. *albiflorus*

Synonyms: *Lablab leucocarpus* Savi; *Lablab vulgaris* (L.) Savi var. *albiflorus* DC.,

Chinese: Bai hua bian dou

English: White-flowered hyacinth bean; white-podded hyacinth bean

French: Dolique à fleur blanche; dolique beurre

Japanese: Shiro fuji mame

- *Lablab purpureus* (L.) Sweet subsp. *purpureus* var. *macrocarpon*

Synonyms: *Dolichos lablab* L. subsp. *macrocarpon*

Chinese: Shi jia bian dou

TABLE 9.1
Popular Names Denoting *Lablab purpureus* (L.) Sweet in Some World Languages and Other Linguistic Taxa

Language/Taxon	Name
Akkadian	libbu; luppu
Amharic	amora-guaya
Arabic	lablab; lupp
Aramaic	lev
Assamese	sim; urahi; urchi; uri; urshi
Asturian	chaucha xaponesa; fríjol d'Exiptu; poroto; xudía; zarandaja
Balochi	mohbel
Beja	kuashrengeig
Bengali	rajashimbi; shim
Bilen	gerenga
Catalan	mongeta egípcia; pèsol d'Austràlia
Chewa	kukuso; manbamba; mkhunguzu
Chinese (Cantonese)	pin tau; tseuk tau
Chinese (Mandarin)	biǎndòu; peng pi dou; que dou; rou dou
Czech	dlouhatec lablab; lablab purpurový
Danish	hjelmbønne; hyacintbønne
Dutch	komak
Dyula	agni guango ahura
Egyptian (Coptic)	gerenga
English	Australian pea; batao; batau; bataw; bonavist bean; bonavist pea; butter bean; dolichos bean; dolique lablab; Egyptian kidney bean; field bean; hyacinth bean; Indian bean; lablab; lablab bean; lablab-bean; papaya bean; poor-man's bean; seim bean; Tonga bean
English (Antilles)	black bean; bonavist
English (Australia)	papaya bean; poor man bean; Rongai dolichos; Tonga bean
English (Bahamas)	bonavist
English (Caribbean)	butter bean
English (Cayman Islands)	bonavist
English (Dominica)	butter bean
English (Fiji)	bonavist
English (Guyana)	butter bean
English (India)	country bean; field bean
English (Jamaica)	banner bean
English (Mauritius)	hyacinth bean
English (Niue)	bonavist
English (Seychelles)	lablab
English (Society Islands)	hyacinth bean
English (Trinidad)	bounavista pea; sem; seim bean
Esperanto	egipta fabo; hinda fabo; lablabo
Fijian	natoba; toba
Filipino	bàtau; itab; pardá

(*Continued*)

TABLE 9.1 (*Continued*)
Popular Names Denoting *Lablab purpureus* (L.) Sweet in Some World Languages and Other Linguistic Taxa

Language/Taxon	Name
Finnish	hyasinttipapu
French	dolique d'Égypte; lablab; pois antaque
French (French Antilles)	pois boucoussou
French (Martinique)	pois contour; pois coolis; pois d'un sou; pois en tout temps; pois indien
French (Mauritius)	haricot; pois antaque
French (Ontario)	carmelita; frijol caballero; pois de senteur; pois nourrice
French (Réunion)	antaque; pois gervais; pois gerville
French (Seychelles)	haricot rouge
German	Ägyptische Bohne; Faselbohne; Hyazinth-Bohne; Helmbohne; Indische Bohne
Greek	dóliho; dóliho tis Aigýptou
Greek (Cyprus)	louria; louvia
Grenadian Creole	bunabis
Guarani	cumana tupi
Gujarati	vāla
Guyanese Creole	bannabees
Hebrew	lvlv mtfs
Hindi	ballar; sem
Hungarian	bivalyvborsó
Indonesian	kacang bado; kacang biduk; kacang komak; komak
Italian	dolico egiziano; fagiolo d'Egitto; fagiolo del Cairo; lablab
Japanese	fuji mame; ingen
Javanese	kårå; kårå andhong; kårå usĕng; kårå wĕdhus; kekårå
Kannada	amare; avare; āvare; avari; chapparadavare; chikkadikai
Kazakh	sümbil burşaq
Konso	o-cala
Korean	pyeondu
Kota (India)	avr
Luri	boragada
Madurese	komak
Malagasy	macape
Malay	kara-kara; kekara
Malayalam	amara; amarakka; avara; avarakka
Marathi	pavta; wa
Mashi	muhulula
Meitei	haraā'ī uraī
Myanmar	pe-gyi
Nepali	hā'isintha bina; ijipsiyana bina; inḍiyana bina; ṭāṭē simī
Nheengatu	cumandiata
Norwegian (Bokmål)	hjelmbønne
Norwegian (Nynorsk)	hjelmbønne

(*Continued*)

TABLE 9.1 (*Continued*)
Popular Names Denoting *Lablab purpureus* (L.) Sweet in Some World Languages and Other Linguistic Taxa

Language/Taxon	Name
Nyanja	chinkamba
Persian	labalab
Polish	chropawiec pospolity; fasolnik egipski; wspięga pospolita
Portuguese	dólico-do-Egito; feijão cutelinho; lablab
Portuguese (Brazil)	ablabe; cumandá-açu; cumandália; cumandatiá; fava-cumandália; feijão-colubrino; feijão-cutelinho; feijão-da-Índia; feijão-lablabe; guar; labe-labe; lablabe; luz-do-dia; mangavi
Prakrit	simvā
Rotenese	loto; ndoto; roto
Russian	dolikhos lablab; dolikhos obyknovennyi; egipetskie boby; giatsintovye boby; lobiia
Sanskrit	nispavah; rājaśimbī
Serbian	australijski grašak; čarobni pasulj; lablab; ljubičasti pasulj; ukrasni pasulj
Shona	chizembera
Sinhalese	ho-dhambala; kiri-dambala; kos-ata-dambala; ratu-peti-dambala; sudu-peti-dambala
Spanish	chaucha japonesa; dólico gigante; fríjol de Egipto; judía; poroto; zarandaja
Spanish (Argentina)	poroto de Egipto
Spanish (Chile)	poroto bombero
Spanish (Colombia)	frijol jacinto
Spanish (Costa Rica)	chimbolo verde
Spanish (Cuba)	frijol de la tierra
Spanish (El Salvador)	frijol caballero; frijol de adorno
Spanish (Mexico)	gallinita
Spanish (Peru)	frijol bocón; frijol chileno
Spanish (Puerto Rico)	chichaso; frijol caballo
Spanish (Venezuela)	caraota chivata; chwata; gallinazo blanco; quiquaqua; tapirucusu
Sundanese	kacang jĕriji; kacang pĕda; roay katopès
Swahili	mfiwi mafuta; mnjahe
Swedish	hjälmböna; hjälmbönssläktet
Tagalog	abitsuwelas; bataw; bitsuwelas; habitsuwelas; sibatse
Tamil	avarai; minni; motchai; tatta-payaru
Thai	t̄hạwpæb; thua nang
Toda	efīr
Tulu	abadhè; abarè; avadhè; ăvadhè; avarè
Turkish	adi misir börülccesi; lablab
Umbundu	maculululu
Urdu	sam
Vietnamese	đậu ván
Zande	mukangi
Zulu	ossangue

English: Annual hyacinth bean; cultivated hyacinth bean; eat-all hyacinth bean; garden lablab; hyacinth bean
French: Dolique d'Egypte; dolique mangetout; dolique pourpre
- *Lablab purpureus* (L.) Sweet subsp. *purpureus* var. *purpureus*
Synonyms: *Dolichos purpureus* L.; *Lablab niger* Medikus var. *typica*
Chinese: Hei zi bian dou; hong jia bian dou; zi hong se bian dou; zi jia bian dou
English: Annual hyacinth bean; cultivated hyacinth bean; garden lablab; hyacinth bean; shelling hyacinth bean
French: Dolique à écosser; dolique d'Egypte; dolique pourpre du Soudan
German: Purpurbohne
Korean: Kkachikong
Vietnamese: Đậu vản bạchbiển
- *Lablab purpureus* (L.) Sweet subsp. *uncinatus* Verdc.
English: Hooked hyacinth bean
- *Lablab purpureus* (L.) Sweet subsp. *uncinatus* Verdc. var. *rhomboïdeus* (Schinz).
English: Lobe-leafed hyacinth bean
- *Lablab purpureus* (L.) Sweet subsp. *uncinatus* Verdc. var. *uncinatus* Verdc.
English: Common hooked hyacinth bean

9.2 ORIGIN OF SCIENTIFIC AND POPULAR TAXA NAMES

Apart from its many roles relating to food and feed, the only species of the genus *Lablab* Adans, hyacinth bean (*Lablab purpureus* [L.] Sweet) is also regarded as a favorite ornamental plant in contrasting environments, and it is no wonder that its name is derived from the noun denoting a gemstone known as *jacinth*, a variety of zircon. The whole plant, encompassing stems and lateral branches, compound leaves, flowers, pods, and grains, is dominantly colored by anhtocyanin, which gives this species a remarkably handsome outlook and provides a solid basis for breeding exclusively and literally for beauty (Mihailović et al. 2016).

It has been debated for a long time where exactly hyacinth bean had originated and had been domesticated, that is, either in Africa or India (Maass 2016). Quite recently, a complex analysis using molecular genomic tools postulated that hyacinth bean could originate in Africa and, by means of early domesticated escapes, reached India, where it underwent an advanced process of improvement (Maass et al. 2010). Charred seeds of hyacinth bean may be found across the Indian subcontinent, usually with a partial preservation of hilum, an important archaeobotanical feature. Among such sites, where the remains of hyacinth bean were found, we may mention the Neolithic site of Sanganakallu, in southern India, dating to either the mid- or late second millennium BC (Fuller and Harvey 2006).

Apart from *hyacinth bean*, the English language has a genuine treasury of the names relating to this pulse (Table 9.1). These words designate quite diverse terms, such as the imagined country from which hyacinth bean was introduced, with the islands of Tonga, a visual resemblance to some specific crop, such as pea (*Pisum sativum* L.), common bean (*Phaseolus* spp.), and faba bean (*Vicia faba* L.), and specific local names. The name *poor-man's bean* may be treated as a kind of evidence of a low-input production of hyacinth bean and its essential contribution to the everyday diets among the peoples in many regions (Mikić and Perić 2016).

The Latinized word *dolichos* was first introduced into plant taxonomy by Linnaeus (Linnaeus 1753, 1758) to denote a genus with the same name, *Dolichos* L., still existing today and belonging to the tribe Phaseoleae (Bronn) DC. and comprising about 60 species of herbaceous plants and shrubs. The first scientific name of hyacinth bean was *Dolichos lablab* L. and is still in use by numerous researchers as the most frequent synonym of *Lablab purpureus* Sweet (Schaaffhausen 1963). The origin of the Linnean genus name is the Ancient Greek adjective *dolihós*, meaning *long* and pointing out the species' vining stems and elongated pods. The word *dolihós* shares the same ultimate source with its equivalents in other ancient Indo-European languages, such as the Avestan *darəγa*, the Hittite *daluki*, the Latin *longus*, the Proto-Germanic **langaz*, the Proto-Slavic **dlŭgŭ*, the Sanskrit *dīrgha*, and the Proto-Indo-European root *dḷh_1g^hós, also denoting something long (Liddell et al. 1940, Ringe 2006). The memory of the Linnean genus name may be found mostly in the Indo-European languages, such as English, French, Greek, Italian, Portuguese, Russian, and Spanish (Table 9.1).

There is a Semitic word denoting hyacinth bean, present in Akkadian, Arabic, Aramaic, and Hebrew (Table 9.1), all descended from the Proto-Semitic **lupp-*, denoting faba bean (*Vicia faba* L.), and, ultimately from the Proto-Afroasiatic root **lap-*, designing both faba bean and corn (Militarev and Stolbova 2007, Nişanyan 2017). At the same time, there is a morphologically similar Semitic word, *lūbiyā*, with the cognates in Aramaic and Hebrew, and with an almost identical meaning, *faba bean*, and referring to some not sufficiently identified kind of legume bean from Egypt (Nişanyan 2017). This word was borrowed into some geographically close Altaic, Caucasian, and Iranian and Indo-Aryan languages, where it began to denote common bean (*Phaseolus vulgaris* L.) that, in the meanwhile, had been introduced from the Americas (see Chapter 13). At the same time, *lūbiyā* is still denoting hyacinth bean in a number of modern Eurasiatic languages, such as Cypriot Greek or Russian (Table 9.1). The spread of this name and its derivatives across the Eastern Mediterranean and in the Caucasus is still without any firm testimony, while the relationship between this word and *lablab* is also rather uncertain, despite their mutual and multiple similarities and a possibility of having an identical direct predecessor (Figure 9.1).

What may make the issue of the origin and diversity of the aforementioned words denoting hyacinth bean additionally interesting is a subsequent derivation of the Turkish *lablab* into *leblebi*, where it was also attributed to the meaning of common bean (*Phaseolus vulgaris* L.), especially in Anatolia (Nişanyan 2017). One of the proposed paths of the odyssey of this intriguing word may be that *lablab* was transferred from Arabic into Persian, after the invasion of Iran in the mid-seventh century; then it was embraced, along with numerous Persian words, by the Seljuk Turks after their conquest of the Persian lands in the eleventh century and ending in the vocabulary of the Ottoman Turkish. The derivative *leblebi*, along with some other words, began to denote chickpea (*Cicer arietinum* L.) and, especially, its roasted grain. This should not be surprising, since chickpea has already been

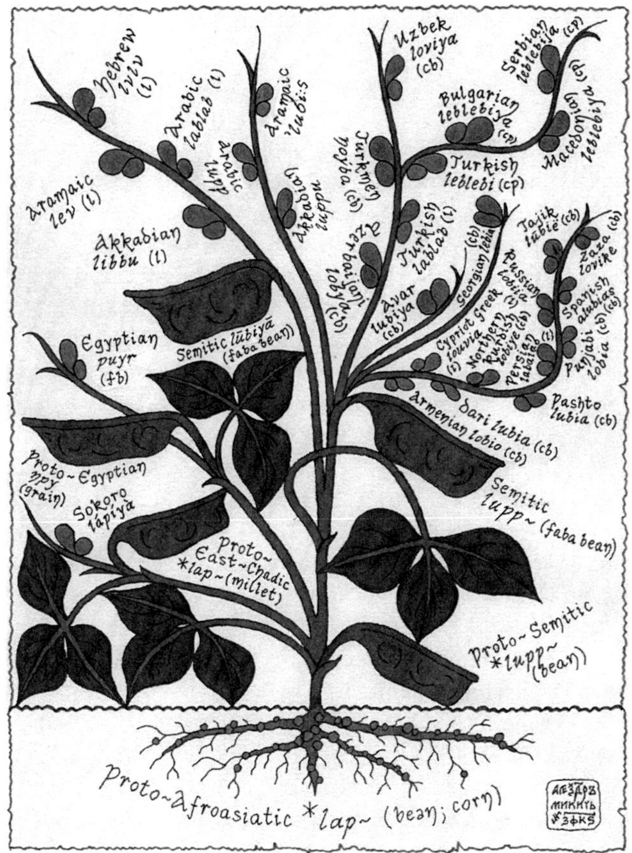

FIGURE 9.1 **(See color insert.)** One of the possible evolutions of the Proto-Afroasiatic root **lap-* originally denoting a bean-like pulse and corn (Militarev and Stolbova 2007), into its direct derivatives, interpreted as pods, and contemporary descendants, illustrated as flowers, in the Chadic, Egyptian, and Semitic languages and with borrowings in other ethnolinguistic families; the meanings of the proto-words are given within brackets, while the meanings of the modern words are the same as in their own proto-words, if given without brackets, as well as *chickpea* (cp), *common bean* (cb), *faba bean* (fb), and *lablab* (l).

present in Asia Minor for almost ten millennia, as one of the very first domesticated crops in the world. The Ottoman enslavement of the vast European territories introduced their culture to the local peoples, including their delicate and rich cuisine with numerous chickpea-based dishes. The Turkish *leblebi* was imported by some Slavic languages in the Balkan Peninsula, such as the Bulgarian and Macedonian *leblebiya* and Serbian *leblebija*, also with a primary meaning of *roasted chickpea grain* and sometimes referring to the chickpea crop itself (see Chapter 5).

One more etymological curiosity, which may make the issue of a real origin of the word *lablab* even more complex, is the existence of the Proto-Indo-European root **leb-*, initially denoting blade or something elongated, concave, and edge-sharp, produced the Ancient Greek *lobó-s*, which was used to denote legume beans in general and, more specifically, faba bean (Pokorny 1959, Nikolayev 2012). It is noteworthy that this Proto-Indo-European root and the Ancient Greek word are also present in the Russian noun *lopast'*, denoting, among few other related terms, a legume pod (Nikolayev 2012). In the sixth and fifth centuries BC, Milesian Greek established their first colonies in the Black Sea coastal region below the Southern Caucasus, then known as Colchis and today being a border country between Russia and Georgia (Giorgadze and Inaishvili 2016). Colchis, like the other Greek Pontic colonies, played an important role in supplying the Hellenic city-states with agricultural products and it is probable that faba bean was among them. It could be easily borrowed by local languages, such as Georgian and Armenian, as *lobio* and *lobi*, respectively, with the same meaning and then transferred southwards, where it came in contact with the northernmost Arabic communities and the Persians. The meaning could be shifted from *Vicia faba* to *Lablab purpureus* and later to Seljuk Turkish and other surrounding languages (Mikić and Perić 2016).

Whatever the origin of the name *lablab* may be, it has become spread worldwide, denoting this crop mainly in the Indo-European languages, with English, French, Italian, Portuguese, Russian, or Serbian, and the constructed ones that were based upon them, such as Esperanto (Table 9.1).

As mentioned in the introductory passages of this chapter segment, hyacinth bean became fully domesticated in India before the arrival of the Indo-Aryan tribes, witnessed by one of the Proto-Dravidian roots, **ávarai*, denoting exactly this crop (Starostin 2006). This produced a descendant in its Southern branch, the Proto-Southern root **avarai*, which had an evolution of its own in the Tamil-Kanadda and Tulu groups. In the first group, it gave the modern words denoting hyacinth bean in Kanadda, in Malayalam, and Tamil and, via the Proto-Nilgiri **avîrä*, in Kota and Toda. In the second group, the Proto-Southern **avarai* brought forth several corresponding names in Tulu (Table 9.1).

As may be seen in the first section of this chapter, there are few hyacinth bean subtaxa with the vernacular names referring that this crop is consumed in the form of immature pods, such as *eat-all* in English and *mangetout* in French. This way of use is also common to the *Phaseolus* and *Pisum* species, where the etymology of related terms is explained in more details (see Chapters 13 and 14).

Since listing all the names containing a reference to *Lablab purpureus* Sweet in various languages would be too excessive, we are taking liberty to decline it in favor of pointing out a few features. Hyacinth bean resembles pea in few languages, such as Chinese, some variants of English, Kazakh, Korean, or Serbian, and a pulse crop, mostly among the Indo-Aryan languages, with the examples in Assamese, Hindi, Nepali, Prakrit, or Urdu (Table 9.1). It is also associated with horse, such as in Ontario French or several local variants of Spanish.

Egypt is, by far, the supposed (and true) homeland of hyacinth bean in the greatest number of examined languages, with the relating names in Asturian, Catalan, English, Esperanto, French, German, Greek, Italian, Polish, Portuguese, Spanish, or Turkish. Then, there follow India in Esperanto, French, German, Nepali, Portuguese, and Russian; Australia in Catalan, English, and Serbian; and Japan in some Asturian and Spanish names (Table 9.1).

Hyacinth bean is well known as an ornamental plant (Anderson et al. 1996), and it is no wonder that there is a reference to its sweet fragrance in Ontario French. Moreover, hyacinth bean received one of the greatest compliments for a grain crop, as may be seen in Bengali and Sanskrit, where its name literally means *royal legume*.

10 *Lathyrus* L.

Synonyms: *Anurus* Presl; *Aphaca* L.; *Aphaca* Mill.; *Cicercula* Medik.; *Clymenum* L.; *Ervum* L.; *Graphiosa* Alef.; *Konxikas* Raf.; *Lastila* Alef.; *Menkenia* Bubani; *Navidura* Alef.; *Nissolia* Jacq.; *Ochrus* Mill.; *Orobus* L.

10.1 LIST OF TAXA SCIENTIFIC AND POPULAR NAMES

In this extensive chapter, the reader will find diverse information on one of the most abundant genera containing several economically significant species. It comprises a list of the majority of the most widely accepted taxa within the global taxonomic community, as well as several tables with rich data on vernacular names in numerous world's languages and other linguistic categories designing important pulse crops of the genus *Lathyrus* L. (ISTA 1982, Rehm 1994, Davies and Jones 1995, Asmussen and Liston 1998, Kenicer et al. 2005, Kenicer 2007, 2008, Gledhill 2008, Porcher 2008, Leht 2009, The Plant List 2013, Ecocrop 2017, EPPO 2017, Ethnologue 2017, IBIS 2017, ILDIS 2017, Logos 2017, NPGS 2017, Wikipedia 2017, Wiktionary 2017).

- *Lathyrus alatus* (Maxim.) Kom.
English: winged vetchling
- *Lathyrus allardii* Batt.
English: Allard's vetchling
- *Lathyrus alpestris* (Waldst. & Kit.) Kit.
Synonyms: *Lathyrus friedrichsthallii* (Griseb.) Prain; *Orobus alpestris* Waldst. & Kit.
English: alpine vetchling
Polish: groszek alpejski
- *Lathyrus amphicarpos* L.
Synonyms: *Lathyrus quadrimarginatus* Bory & Chaub.
English: amphicarpic vetchling
Italian: cicerchia con quattro ali
- *Lathyrus angulatus* L.
Synonyms: *Lathyrus leptophyllus* M. Bieb.
Catalan: guixó angulós; veça cantelluda
English: angled pea; angular pea; slender wild pea
French: gesse anguleuse
Italian: cicerchia angolosa
Spanish: pluma de angel
Swedish: vinkelvial
- *Lathyrus anhuiensis* Y. J. Zhu & R. X. Meng
Chinese: Ānhuī xiāng wāndòu shǔ
English: Anhui vetchling

- *Lathyrus annuus* L. (Table 10.1)
Synonyms: *Lathyrus trachyspermus* Webb ex J. J. Rodr.
- *Lathyrus aphaca* L. (Table 10.2)
Synonyms: *Aphaca marmorata* Alef.; *Lathyrus affinis* Guss.; *Lathyrus aphaca* var. *affinis* (Guss.) Arcang.; *Lathyrus aphaca* var. *biflorus* Post; *Lathyrus aphaca* var. *floribundus* (Velen.) K. Maly; *Lathyrus aphaca* var. *modestus* P. H. Davis; *Lathyrus aphaca* var. *pseudoaphaca* (Boiss.) P. H. Davis; *Lathyrus floribundus* Velen.; *Lathyrus polyanthus* Boiss. & Blanche; *Lathyrus pseudoaphaca* Boiss.
- *Lathyrus arizonicus* Britton
Synonyms: *Lathyrus lanszwertii* Kellogg var. *arizonicus* (Britton) S. L. Welsh
English: Arizona vetchling
- *Lathyrus atropatanus* (Grossh.) Sirj.
Synonyms: *Lathyrus nivalis* Hand.-Mazz. subsp. *atropatanus*; *Orobus atropatanus* Grossh.
English: Azerbaijani vetchling
Russian: China azerbaidzhanskaia; sochevichnik azerbaidzhanskii
- *Lathyrus aureus* (Steven ex Fisch. & C. A. Mey.) D. Brandza
Synonyms: *Orobus aureus* Steven ex Fisch. & C. A. Mey.; *Orobus kolenatii* K. Koch; *Orobus orientalis* Boiss.
Armenian: sovorakan tap'volorr
English: golden pea; golden vetchling
Polish: groszek złocisty
Russian: china zolotistaia; sochevichnik zolotistyi
Swedish: gyllenärt
Turkish: altuni mürdümük
Ukrainian: chyna zolotysta; horoshok zolotystyi
- *Lathyrus azureus* Dean
English: azure vetchling
- *Lathyrus basalticus* Rech. f.
English: basaltic vetchling
- *Lathyrus bauhini* Genty
Synonyms: *Lathyrus ensifolius* (Lapeyr.) Gay; *Lathyrus filiformis* var. *bauhini* (Genty) Beck; *Lathyrus filiformis* (Lam.) Gay var. *ensifolius* (Lapeyr.) Hayek
English: Bauhin's vetchling
German: Schwert-Platterbse; Schwertblättrige Platterbse
Sorbian (Upper): banćikaty hróšik
- *Lathyrus belinensis* Maxted & Goyder
English: Belin vetchling
French: gesse de Belin
Swedish: flamvial
- *Lathyrus berteroanus* Savi
Synonyms: *Lathyrus debilis* Clos; *Lathyrus gracilis* Phil.
English: Bertero's vetchling
- *Lathyrus biflorus* T. W. Nelson & J. P. Nelson
English: twoflower pea; twoflower vetchling

Turkish: ıki çiçekli mürdümük
- *Lathyrus bijugatus* T. G. White
English: Drypark pea; Latah tule-pea; pinewoods sweetpea
- *Lathyrus bijugus* Boiss. & Noe
English: twofold vetchling
- *Lathyrus binatus* Pancic
English: paired vetchling
- *Lathyrus blepharicarpus* Boiss.
Arabic: aljulban jaffani alththimar
English: eyelid-fruited vetchling; Syrian vetchling
Hebrew: tvfch rsn
- *Lathyrus boissieri* Sirj.
English: Boissier's vetchling
- *Lathyrus brachycalyx* Rydb.
Synonyms: *Lathyrus brachycalyx* Rydb. var. *brachycalyx*
English: Bonneville pea; Bonneville vetchling; Rydberg's sweet Pea; short-calyx vetchling
- *Lathyrus brachyodon* Murb.
English: short-tooth vetchling
- *Lathyrus bungei* Boiss.
English: Bunge's vetchling
- *Lathyrus cabreranus* Burkart
Synonyms: *Lathyrus dumetorum* Burkart; *Lathyrus pubescens* Clos
English: Cabrera vetchling
- *Lathyrus campestris* Phil.
Synonyms: *Lathyrus debilis* Vogel var. *campestris* (Phil.) Reiche; *Lathyrus gracillimus* Reiche
English: field vetchling
- *Lathyrus caudatus* Wei & H. P. Tsui
Chinese: yèshān lí dòu
English: tailed vetchling
- *Lathyrus cassius* Boiss.
Arabic: aljulban alkasiusi
English: Cassian vetchling
Hebrew: tvfch ksvs
- *Lathyrus chloranthus* Boiss.
English: greenish vetchling
Finnish: turkinnätkelmä
Russian: china zheltozelenaia
Swedish: grönvial
- *Lathyrus chrysanthus* Boiss.
English: Golden-flowered vetchling
- *Lathyrus cicera* L. (Table 10.3)
Synonyms: *Lathyrus aegaeus* Davidov.
- *Lathyrus cicera* L. var. *cicera*
English: common red vetchling

- *Lathyrus cicera* var. *lineatus* Post
English: linear red vetchling
- *Lathyrus cicera* var. *negevensis* Plitmann
English: Negev red vetchling
- *Lathyrus cicera* var. *patagonica* Speg.
English: Patagonian red vetchling
- *Lathyrus ciliatidentatus* Czefr.
Synonyms: *Orobus ciliatidentatus* (Czefr.) Avazneli
English: cilia-toothed vetchling
Russian: china resnitchatozubchataia
- *Lathyrus ciliolatus* Sam. ex Rech. f.
English: ciliolate vetchling
Hebrew: tvfch hshlvchvt
- *Lathyrus cirrhosus* Ser.
Catalan: pèsol bord
English: orange vetchling
- *Lathyrus clymenum* L. (Table 10.4)
Synonyms: *Lathyrus articulatus* L.; *Lathyrus articulatus* subsp. *clymenum* (L.) Maire; *Lathyrus clymensum* L.; *Lathyrus purpureus* Desf.
- *Lathyrus coerulescens* Boiss. & Reut.
English: darkish blue-green vetchling
- *Lathyrus colchicus* Lipsky
English: Colchis vetchling
Russian: china kolkhidskaia
- *Lathyrus crassipes* Gillies ex Hook. & Arn.
Synonyms: *Lathyrus stipularis* C. Presl
English: arvejilla; thicky vetchling
Spanish: arvejilla; arvejilla de campo
- *Lathyrus cyaneus* (Steven) K. Koch
Synonyms: *Orobus cyaneus* Steven
English: dark blue vetchling
- *Lathyrus davidii* Hance
Synonyms: *Orobus davidii* (Hance) Stank. & Roskov
Chinese: chin yin hua; dàshān lí dòu; jiang mang xiang wan dou; da shan li dou; jiang mang shan li dou; shan chiang tou; shan jiang dou
English: David's vetchling
Japanese: itachi sasage
Russian: china Davida; sochevichnik Davida
- *Lathyrus decaphyllus* Pursh
Synonyms: *Lathyrus ornatus* Nutt.; *Lathyrus polymorphus* Nutt.
English: prairie vetchling
- *Lathyrus decaphyllus* Pursh var. *decaphyllus*
English: common prairie vetchling
- *Lathyrus decaphyllus* Pursh var. *incanus* (J. G. Sm. & Rydb.) Broich
English: grey-haired prairie vetchling
- *Lathyrus delnorticus* C. L. Hitchc.

English: Del Norte pea
- *Lathyrus dielsianus* Harms
Chinese: zhōng huà shān lí dòu
English: Diels' vetchling
- *Lathyrus digitatus* (M. Bieb.) Fiori
Synonyms: *Lathyrus cyaneus* (Steven) K. Koch; *Lathyrus cyaneus* (Steven) K. Koch subsp. *digitatus*; *Lathyrus cyanus* (Steven) K. Koch; *Lathyrus sessilifolius* (Sibth. & Sm.) Ten.; *Lathyrus tempskyanus* (Freyn & Sint.) K. Maly; *Orobus cyaneus* Steven; *Orobus digitatus* M. Bieb.; *Orobus sessilifolius* Sibth. & Sm.
English: finger-like vetchling
Italian: cicerchia digitata
Russian: china golubaia; china palchataia; sochevichnik palchatyi
Ukrainian: chyna pal'chasta; horoshok pal'chastyi
- *Lathyrus dominianus* Litv.
English: Domin's vetchling
Russian: china Domina
- *Lathyrus elodes* Link ex Colmeiro
English: whorled vetchling
- *Lathyrus emodii* (Wall. ex Fritsch) Ali
Synonyms: *Lathyrus laevigatus sensu* Sanjappa; *Lathyrus laevigatus* (Waldst. & Kit.) Gren. subsp. *emodi* (Fritsch) H. Ohashi; *Lathyrus luteus* Baker; *Orobus emodi* Fritsch
English: Emodian vetchling
Swedish: stor guldärt
- *Lathyrus eucosmus* Butters & H. St. John
Synonyms: *Lathyrus brachycalyx* Rydb. subsp. *eucosmus* S. L. Welsh; *Lathyrus brachycalyx* Rydb. var. *eucosmus* S. L.Welsh; *Lathyrus decaphyllus* sensu auct.; *Lathyrus polymorphus* sensu auct.; *Orobus polymorphus* (Nutt.) Alef.
English: bush vetchling; semmly vetchling
Navajo: e'e'aahjí'ígíí
- *Lathyrus filiformis* (Lam.) J. Gay
Synonyms: *Lathyrus canescens* (L. f.) Godr. & Gren.; *Orobus canescens* L. f.; *Orobus filiformis* Lam.
Catalan: pesolina borda
English: threadlike-leafed vetchling
Italian: cicerchia filiforme
- *Lathyrus fissus* Ball
English: cloven vetchling
- *Lathyrus frolovii* Rupr.
Synonyms: *Orobus alpestris* Ledeb.; *Orobus frolovii* Ledeb.
English: Frolov's vetchling
Russian: china Frolova; sochevichnik Frolova
- *Lathyrus glandulosus* Broich
English: glandular vetchling
- *Lathyrus gloeospermus* Warb. & Eig
English: sticky-seeded vetchling

Hebrew: tvfch dvk
- *Lathyrus gmelinii* Fritsch
Synonyms: *Orobus gmelinii* DC.; *Orobus luteus* L.
Chinese: xīnjiāng shān lí dòu
English: Gmelin's vetchling
Finnish: keltalinnunherne
Kazakh: sarı qoyanburşaq
Russian: china Gmelina; sochevichnik zheltyi
- *Lathyrus golanensis* Cohen & Plitmann
English: Golan vetchling
Hebrew: tvfch hgvln
- *Lathyrus gorgoni* Parl.
Synonyms: *Lathyrus amoenus* Fenzl; *Lathyrus gorgonei* Parl.; *Lathyrus gorgonei* Parl. subsp. *tiriopolitanus* (Davidov) Ponert; *Lathyrus gorgonii* Parl.
Arabic: aljulban alkarih
English: gorgon-like vetchling
Hebrew: tvfch rch-'mvd
Italian: cicerchia gorgonio
- *Lathyrus graminifolius* (S. Watson) T. G. White
Synonyms: *Lathyrus palustris* var. *graminifolius* S. Watson
English: grassleaf pea
- *Lathyrus grandiflorus* Sm.
English: everlasting pea; large-flowered vetchling; two-flower everlasting pea
Finnish: isonätkelmä
Italian: cicerchia a fiori grandi
Japanese: oorenrisou
Polish: groszek wielkokwiatowy
Swedish: jättevial
- *Lathyrus grimesii* Barneby
English: Grimes' pea; Grimes' vetchling
- *Lathyrus hallersteinii* Baumg.
Synonyms: *Lathyrus pratensis* L. subsp. *hallersteinii* (Baumg.) Nyman
English: Hallerstein's vetchling
- *Lathyrus hasslerianus* Burkart
English: Hassler's vetchling
- *Lathyrus heterophyllus* L.
Czech: hrachor různolistý
English: Norfolk everlasting pea
Esperanto: latiro diversfolia
Finnish: pallenätkelmä; pikkuruusunätkelmä
German: Verschiedenblättrige Platterbse
Italian: cicerchia a foglie variate
Polish: groszek różnolistny
Sorbian (Upper): wšelakołopjenaty hróšik
Swedish: vingvial
- *Lathyrus hierosolymitanus* Boiss.

English: Jerusalem vetchling
Hebrew: tvfch rvshlm
- *Lathyrus hirsutus* L. (Table 10.5)
- *Lathyrus hirticarpus* J. Mattatia & Heyn
English: hairy-fruited vetchling
Hebrew: tvfch sh'r-fr
- *Lathyrus hitchcockianus* Barneby & Reveal
English: Bullfrog Mountain pea; Bullfrog Mountain wild pea; Hitchcock's peavine; Hitchcock's sweet pea
- *Lathyrus holochlorus* (Piper) C. L. Hitchc.
Synonyms: *Lathyrus ochropetalus* subsp. *holochlorus* Piper
English: thinleaf pea; thin-leaved peavine
- *Lathyrus humilis* (Ser.) Spreng.
Synonyms: *Lathyrus altaicus* Ledeb.; *Orobus humilis* Ser.
Chinese: ǎi shān lí dòu
English: low vetchling
Russian: china nizkaia; china prizemistaia; sochevichnik prizemistyi
- *Lathyrus hygrophilus* Taub.
Synonyms: *Lathyrus kilimandsharicus* Taub.
English: Kilimanjaro peavine; Kilimanjaro vetchling; moisture-loving vetchling
- *Lathyrus ibicuiensis* Abruzzi de Oliveira
English: Ibicuí vetchling
- *Lathyrus inconspicuus* L. (Table 10.6)
Synonyms: *Lathyrus erectus* Lag.; *Lathyrus hispidulus* Boiss.
- *Lathyrus incurvus* (Roth) Willd.
Synonyms: *Orobus incurvus* (Roth) A. Braun; *Vicia incurva* Roth
English: bent vetchling
Polish: groszek odgięty
Russian: china izognutaia; china sognutaia
Swedish: sabelvial
Ukrainian: chyna zignuta; horoshok zignenyi; horoshok zignutyi
- *Lathyrus japonicus* Willd. (Table 10.7)
Synonym: *Lathyrus aleuticus* (T. G. White) Pobed.; *Lathyrus japonicus* Willd. var. *aleuticus* (T. G. White) Fernald; *Lathyrus maritimus* L. var. *aleuticus* T. G. White.
- *Lathyrus japonicus* Willd. subsp. *japonicus*
English: Japanese vetchling; northern vetchling
Norwegian (Bokmål): nordlig strandflatbelg
- *Lathyrus japonicus* Willd. subsp. *maritimus* (L.) P. W. Ball
Synonyms: *Lathyrus japonicus* var. *glaber* (Ser.) Fernald; *Lathyrus maritimus* Bigelow; *Pisum maritimum* L.; *Pisum maritimum* var. *glabrum* Ser.
Danish: strand-fladbælg; strandært
English: maritime vetchling; sea pea; sea vetchling
Estonian: rand-seahernes
French: gesse maritime; pois de mer

Lithuanian: pajūrinis pelėžirnis
Norwegian (Bokmål): sørlig strandflatbelg
- *Lathyrus japonicus* Willd. subsp. *maritimus* (L.) P. W. Ball var. *acutifolius* (Bab.) Bassler
English: thorny-leafed vetchling
- *Lathyrus japonicus* Willd. subsp. *maritimus* (L.) P. W. Ball var. *pellitus* Fernald
English: skinny vetchling
- *Lathyrus jepsonii* Greene
English: Jepson's vetchling
- *Lathyrus jepsonii* Greene subsp. *californicus* (S. Watson) C. L. Hitchc.
Synonym: *Lathyrus venosus* var. *californicus* S. Watson; *Lathyrus watsonii* T. G. White
English: Californian vetchling
- *Lathyrus jepsonii* Greene subsp. *jepsonii*
English: delta tule pea; Jepson's pea
- *Lathyrus jordanii* (Ten.) Ces., Pass. et Gib.
English: Jordan vetchling
Italian: cicerchia di Giordano
- *Lathyrus komarovii* Ohwi
Synonyms: *Lathyrus alatus* (Maxim.) Kom.; *Orobus alatus* Maxim.; *Orobus komarovii* (Ohwi) Stank. & Roskov
Chinese: sān mài shān lí dòu
English: Komarov's vetchling
Russian: china komarova; china krylataia; sochevichnik komarova
- *Lathyrus krylovii* Serg.
Synonyms: *Lathyrus laevigatus* Fritsch subsp. *krylovii* (Serg.) Hendrych; *Orobus krylovii* (Serg.) Stank. & Roskov
Chinese: xiá yèshān lí dòu
English: Krylov's vetchling
Russian: china krylova; sochevichnik krylova
- *Lathyrus laevigatus* (Waldst. & Kit.) Gren.
Synonyms: *Orobus laevigatus* Waldst. & Kit.
Catalan: guixó groc
English: smooth-surfaced vetchling
German: gelbe Platterbse
Lithuanian: geltonžiedis pelėžirnis
Polish: groszek wschodniokarpacki
Swedish: guldärt
Ukrainian: chyna gladen'ka; horoshek gladen'kyi; zaiachyi horokh rzhavyi
- *Lathyrus lanszwertii* Kellogg
Synonyms: *Lathyrus coriaceus* T. G. White; *Lathyrus goldsteinae* Eastw.
English: Nevada peavine; Nevada sweet pea
Turkish: Nevada mürdümüğü
- *Lathyrus lanszwertii* subsp. *aridus* (Piper) Bradshaw
Synonyms: *Lathyrus lanszwertii* var. *aridus* (Piper) Jeps.

English: Nevada pea
- *Lathyrus lanszwertii* var. *sandbergii* (T. G. White) Broich
Synonyms: *Lathyrus sandbergii* (T. G. White) Howell
- *Lathyrus latidentatus* Jelen.
English: wide-toothed vetchling
Russian: china shirokozubchataia
- *Lathyrus latifolius* L. (Table 10.8)
Synonyms: *Lathyrus megalanthus* Steud.; *Lathyrus membranaceus* C. Presl; *Lathyrus sylvestris* L. subsp. *latifolius* Bonnier & Layens.
- *Lathyrus latifolius* L. forma *albiflorus* Moldenke
English: white-flowered everlasting pea
French: pois vivace à fleur blanche
- *Lathyrus latifolius* L. forma *lanceolatus* Freyn
English: lance-leafed everlasting pea
- *Lathyrus latifolius* L. forma *rubicundus* Moldenke
English: ruddy everlasting pea
- *Lathyrus laxiflorus* (Desf.) Kuntze
Synonyms: *Lathyrus inermis* Rochel ex Friv.; *Orobus hirsutus* L.; *Orobus laxiflorus* Desf.
Arabic: laljulban rakhu al'azhar
English: loose-flowered vetchling
Italian: cicerchia laxiflora
Russian: china redkotsvetkovaia; sochevichnik redkotsvetkovyi
Ukrainian: china ridkokvitkova; horoshok ridkotsvityi
- *Lathyrus ledebourii* Trautv.
Synonym: *Lathyrus pannonicus* (Jacq.) Garcke subsp. *ledebourii* (Trautv.) Bässler; *Orobus intermedius* C. A. Mey.; *Orobus ledebourii* (Trautv.) Roldugin
English: Ledebour's vetchling
Russian: china Ledebura; sochevichnik promezhutochnyi
- *Lathyrus lentiformis* Plitmann
English: lentil-like vetchling
Hebrew: tvfch dsht
- *Lathyrus leptophyllus* M. Bieb.
Synonyms: *Lathyrus angulatus* sensu auct.
Armenian: tap'volorr neghaterev
English: Angular vetchling
Russian: china uglovataia
- *Lathyrus leucanthus* Rydb.
Synonyms: *Lathyrus laetivirens* Rydb.; *Lathyrus lanszwertii* Kellogg var. *laetivirens* (Rydb.) S. L. Welsh; *Lathyrus lanszwertii* Kellogg var. *pallescens* Barneby; *Lathyrus leucanthus* Rydb. var. *laetivirens* (Rydb.) C. L. Hitchc.
English: white-flowered vetchling
- *Lathyrus linearifolius* Vogel
English: aspen pea; line-leafed vetchling
- *Lathyrus libani* Fritsch

English: Lebanon vetchling
- *Lathyrus linifolius* (Reichard) Bassler (Table 10.9)
Synonyms: *Lathyrus linifolius* (Reichard) Bässler var. *montanus* (Bernh.) Bässler; *Lathyrus macrorrhizus* Wimm.; *Lathyrus montanus* Bernh.; *Orobus linifolius* Reichard; *Orobus tuberosus* L.
- *Lathyrus littoralis* (Nutt.) Endl.
Synonyms: *Astrophia littoralis* Torr. & A. Gray; *Lathyrus littoralis* (Torr. & A. Gray) Walp.; *Orobus littoralis* (Torr. & A. Gray) A. Gray
English: silky beach pea
- *Lathyrus lomanus* I. M. Johnst.
English: Lomanus' vetchling
- *Lathyrus lusitanicus* Mart. ex Ser., nom. nud.
English: Portuguese vetchling
- *Lathyrus macropus* Gillies ex Hook. & Arn.
Synonyms: *Lathyrus linearifolius* Griseb.
English: large-based vetchling
- *Lathyrus macrostachys* Vogel
English: large-peduncle vetchling
- *Lathyrus magellanicus* Lam.
Synonyms: *Lathyrus gladiatus* Hook.; *Lathyrus hookeri* G. Don var. *tricho-calyx* (Phil.) Burkart; *Lathyrus megellanicus* Lam.; *Lathyrus patagonicus* Hauman; *Lathyrus pterocaulos* Phil.
English: Magellan's vetchling
- *Lathyrus marmoratus* Boiss. & Blanche
English: marble vetchling
Hebrew: tvfch nh
- *Lathyrus miniatus* M. Bieb. ex Steven
Synonyms: *Lathyrus rotundifolius* subsp. *miniatus* (M. Bieb. ex Steven) P. H. Davis
English: small vetchling
- *Lathyrus mulkak* Lipsky
English: mulkak vetchling
Russian: china mulkak
Tajik: myulkak
Ukrainian: chyna miul'kak
Uzbek: mulkak
- *Lathyrus multiceps* Clos
Synonyms: *Lathyrus ecirrhosus* Phil.; *Lathyrus eurypetalus* Phil.; *Lathyrus multiceps* Clos var. *normalis* Burkart; *Lathyrus multiceps* Clos var. *setiger* (Phil.) Acevedo; *Lathyrus setiger* Phil.
English: many-headed vetchling
- *Lathyrus nervosus* Lam.
Synonyms: *Lathyrus americanus* (Mill.) Kupicha; *Lathyrus armitageanus* Loudon
English: veined vetchling
Swedish: safirvial
- *Lathyrus neurolobus* Boiss. & Heldr.
English: vein-lobed vetchling

- *Lathyrus nevadensis* S. Watson
Synonyms: *Vicia nana* Kellogg
English: Cusick's pea; Nevada vetchling; purple peavine; Sierra pea
- *Lathyrus nevadensis* S. Watson subsp. *cusickii* (S. Watson) C. L. Hitchc.
Synonyms: *Lathyrus cusickii* S. Watson
English: Cusick's vetchling
- *Lathyrus nevadensis* S. Watson subsp. *nevadensis*
English: Nevada peavine; Nevada vetchling
- *Lathyrus niger* (L.) Bernh. (Table 10.10)
Synonyms: *Orobus niger* L.
- *Lathyrus nigrivalvis* Burkart
English: black-folding vetchling
Spanish: alverjilla; choreque
- *Lathyrus nissolia* L. (Table 10.11)
Synonyms: *Orobus nissolia.*
- *Lathyrus nissolia* L. subsp. *nissolia*
Czech: hrachor trávolistý pravý
English: common grass vetchling
- *Lathyrus nissolia* subsp. *pubescens* (Beck) Soják
Czech: hrachor trávolistý pýřitý
English: maturing grass vetchling
- *Lathyrus nitens* Vogel
English: glittering vetchling
- *Lathyrus nivalis* Hand.-Mazz.
English: snow-loving vetchling
- *Lathyrus numidicus* Batt.
English: Numidian vetchling
- *Lathyrus occidentalis* (Fisch. & C. A. Mey.) Fritsch
Synonyms: *Orobus luteus* var. *occidentalis* Fisch. & C. A. Mey.
English: occidental vetchling
Sorbian (Upper): žołty hróšik
- *Lathyrus ochraceus* Kitt.
Synonyms: *Lathyrus gmelinii* Fritsch; *Lathyrus laevigatus* subsp. *occidentalis* auct.; *Lathyrus luteus* auct.; *Orobus gmelinii* Fisch. ex DC., nom. inval.; *Orobus luteus* L.
English: ochre-colored vetchling
Italian: cicerchia gialla
Swedish: stor guldärt
- *Lathyrus ochroleucus* Hook.
Synonyms: *Lathyrus albidus* Eaton; *Lathyrus glaucifolius* Beck; *Lathyrus ochroleucous* Hook.; *Orobus ochroleucus* (Hook.) A. Br.
English: cream pea; pale vetchling; yellow vetchling
- *Lathyrus ochrus* (L.) DC. (Table 10.12)
Synonyms: *Pisum ochrus* L.
- *Lathyrus odoratus* L. (Table 10.13)
- *Lathyrus pallescens* (M. Bieb.) K. Koch

Synonyms: *Orobus angustifolius* L.; *Orobus canescens* sensu auct.; *Orobus filiformis* sensu auct.; *Orobus pallescens* M. Bieb.
English: fading-colored vetchling; pale vetchling
Hungarian: sápadt lednek
Russian: china bledneiushchaia; china blednovataia; sochevichnik bledneiushchii; sochevichnik uzkolistnyi
Ukrainian: horoshok blida; horoshok blidyi; horoshok pol'ovyi; horoshok syvavyi; horoshok siven'kyi
- *Lathyrus palustris* L. (Table 10.14)
Synonyms: *Lathyrus incurvus* Rchb.; *Lathyrus macranthus* (T. G. White) Rydb.; *Lathyrus myrtifolius* Willd.; *Lathyrus occidentalis* Torr. & A. Gray; *Lathyrus paluster* sensu auct.; *Lathyrus palustris* L. var. *linearifolius* Ser.; *Lathyrus palustris* L. var. *macranthus* (T. G. White) Fernald; *Lathyrus palustris* L. var. *myrtifolius* (Willd.) A. Gray; *Lathyrus palustris* L. var. *pilosus* (Cham.) Ledeb.; *Lathyrus palustris* L. var. *retusus* Fernald & St. John; *Lathyrus pilosus* Cham.; *Orobus myrtifolius* Alef.; *Orobus myrtifolius* (Willd.) Hall.
- *Lathyrus palustris* L. subsp. *nudicaulis* (Willk.) P. W. Ball
Synonyms: *Lathyrus nudicaulis* (Willk.) Amo; *Lathyrus palustris* var. *nudicaulis* Willk.
English: hairless-stemmed marsh vetchling
- *Lathyrus palustris* L. subsp. *palustris*
English: common marsh vetchling
Norwegian (Bokmål): snau myrflatbelg
- *Lathyrus palustris* L. subsp. *pilosus* (Cham.) Hulten
Synonyms: *Lathyrus palustris* var. *pilosus* (Cham.) Ledeb.; *Lathyrus pilosus* Cham.
Chinese: rou mao xiang wan dou; shān lí dòu
English: hairy marsh vetchling
Japanese: ezo no renri sou
Norwegian (Bokmål): håret myrflatbelg
- *Lathyrus pancicii* (Jurisic) Adamovic
Synonym: *Orobus pancicii* Jurisic
Bulgarian: Panchichevo sekirche
English: Pančić's vetchling
Serbian: Pančićev grahor
- *Lathyrus pannonicus* (Jacq.) Garcke (Table 10.15)
Synonyms: *Lathyrus albus* (L. f.) Kitt.; *Orobus albus* L. f.
- *Lathyrus pannonicus* (Jacq.) Garcke subsp. *collinus* (Ortmann) Soo
Synonyms: *Orobus pannonicus* var. *collinus* Ortmann
Czech: hrachor panonský chlumní
English: hilly Pannonian vetchling
- *Lathyrus pannonicus* (Jacq.) Garcke subsp. *pannonicus*
Synonyms: *Orobus pannonicus* Jacq.
Czech: hrachor panonský pravý
English: common Pannonian vetchling
- *Lathyrus pannonicus* (Jacq.) Garcke subsp. *varius* (Hill) P. W. Ball
Synonyms: *Orobus varius* Hill; *Orobus versicolor* J. F. Gmel.

English: versatile Pannonian vetchling
- *Lathyrus paraguariensis* Hassl.
English: Paraguayan vechling
- *Lathyrus paranensis* Burkart
Synonyms: *Lathyrus magellanicus* Arechav.
English: Paraná vetchling
- *Lathyrus parodii* Burkart
English: Parodi's vetchling
- *Lathyrus parvifolius* S. Watson
Synonyms: *Lathyrus pauciflorus* Fernald subsp. *schaffneri* (Rydb.) Piper; *Lathyrus schaffneri* Rydb.
English: small-leafed vetchling
- *Lathyrus pastorei* (Burkart) Rossow
Synonyms: *Lathyrus multiceps* Clos var. *pastorei* Burkart
English: Pastore's vetchling
- *Lathyrus pauciflorus* Fernald
English: Few-flowered vetchling
- *Lathyrus pisiformis* L. (Table 10.16)
- *Lathyrus plitmannii* Greuter & Burdet
English: Plitmann's vetchling
Hebrew: tvfch chd-'vrk
- *Lathyrus polymorphus* Nutt.
English: hoary vetchling
Navajo: ha'a'aahjí'ígíí
- *Lathyrus polyphyllus* Nutt.
Synonyms: *Lathyrus ecirrhosus* A. Heller
English: leafy pea; multi-leaflet vetchling
Turkish: yapraksı mürdümük
- *Lathyrus pratensis* L. (Table 10.17)
Synonyms: *Orobus pratensis* L.
- *Lathyrus pseudocicera* Pamp.
Synonyms: *Lathyrus gorgonei* Parl. subsp. *lineatus* (Post) Ponert
English: false-chickpea vetchling
Hebrew: tvfch dvm
- *Lathyrus pubescens* Hook. & Arn.
Synonyms: *Lathyrus acutifolius* Vogel; *Lathyrus andicolus* Gand.; *Lathyrus dumetorum* Phil.; *Lathyrus petiolaris* Vogel; *Lathyrus purpureo-coeruleus* Knowles & Westc.
English: ripening vetchling
- *Lathyrus pulcher* J. Gay
Synonyms: *Lathyrus elegans* Porta & Rigo; *Lathyrus tremolsianus* Pau
Catalan: guixa borda; pèsol de marge; pèsol valencià; pesolera de pastor
English: beautiful vetchling
- *Lathyrus pusillus* Elliott
Synonyms: *Lathyrus arvensis* Phil.; *Lathyrus cicera* Hauman; *Lathyrus cicera* L. var. *patagonica* Speg.; *Lathyrus crassipes* Hook. & Arn.; *Lathyrus*

crassipes Phil.; *Lathyrus debilis* Vogel var. *arvensis* (Phil.) Reiche; *Lathyrus dicirrhus* Clos; *Lathyrus engelmanni* Bisch.; *Lathyrus guaraniticus* Hassl.; *Lathyrus lancifolius* Rchb.; *Lathyrus montevidensis* Vogel; *Lathyrus stipularis* C. Presl var. *patagonica* (Speg.) Speg.
English: petty vetchling; singletary vetchling
- *Lathyrus quinquenervius* (Miq.) Litv.
Synonyms: *Lathyrus palustris* L. var. *sericea* Franch.; *Vicia quinquenervia* Miq.; *Vicia quinquenervius* Miq.
Chinese: shān lí dòu; wu mai shan li dou; wu mai ye xiang wan dou
English: five-veined vetchling
Japanese: renri sou
Russian: china piatizhilkovaia
- *Lathyrus rigidus* T. G. White
Synonyms: *Lathyrus albus* A. Gray
English: stiff pea; stiff vetchling
- *Lathyrus roseus* Steven
Synonyms: *Orobus roseus* (Steven) Ledeb.
English: pink vetchling
Finnish: kaukasiannätkelmä
Georgian: ardzhakeli
Hebrew: tvfch hchrmvn
Polish: groszek różowy
Russian: china rozovaia; sochevichnik rozovyi
- *Lathyrus rotundifolius* Willd.
Synonyms: *Lathyrus drummondii* hort.; *Lathyrus latifolius* L. var. *rotundifolius*; *Lathyrus litvinovii* Iljin; *Lathyrus peduncularis* Poir.
English: Persian everlasting pea; round-leaf vetchling
Polish: groszek okrągłolistny
Russian: china kruglolistnaia; china Litvinova
Ukrainian: chyna kruhlolista; horoshok kruhlolistyi
- *Lathyrus sativus* L. (Table 10.18)
Synonyms: *Lathyrus asiaticus* (Zalkind) Kudr.; *Lathyrus sativus* L. subsp. *asiaticus* Zalkind.
- *Lathyrus saxatilis* (Vent.) Vis.
Synonyms: *Lathyrus ciliatus* Guss.; *Orobus saxatilis* Vent.; *Vicia saxatilis* (Vent.) Tropea
Bulgarian: skalno sekirche
Catalan: guixera de roca; guixera rupestre
Hebrew: tvfch hsl'm
Italian: cicerchia rupestre
English: rock-loving vetchling
Russian: china skalistaia; goroshek shchebnistyi
Ukrainian: china skel'na; horoshok skel'nyi
- *Lathyrus setifolius* L.
Arabic: aljulban daeif al'awraq
Catalan: guixera fina

English: brown vetchling; narrow-leaved; red vetchling
Hebrew: tvfch chchll
Italian: cicerchia capillare
Russian: china shchetinistaya
Ukrainian: chyna shchetynolista; horoshok shchetynolystyi
- *Lathyrus spathulatus* Celak.
English: spatulate vetchling
Hebrew: tvfch hgll
- *Lathyrus sphaericus* Retz. (Table 10.19)
Synonyms: *Lathyrus coccineus* All.; *Lathyrus hygrophilus* sensu Robyns; *Lathyrus viciodes* DC.; *Orobus sphaericus* (Retz.) Avazneli.
- *Lathyrus splendens* Kellogg
English: Campo pea; pride of California
- *Lathyrus stenophyllus* Boiss. & Heldr.
English: narrow-leafed vetchling
- *Lathyrus subalpinus* Beck
Synonyms: *Orobus subalpinus* Herbich
English: subalpine vetchling
Russian: china subalpiiskaia; sochevichnik subalpiiski
Ukrainian: chyna subal'piis'ka
- *Lathyrus subandinus* Phil.
English: sub-Andean vetchling
- *Lathyrus sulphureus* W. H. Brewer ex A. Gray
English: snub pea; sulphurous vetchling
- *Lathyrus sylvestris* L. (Table 10.20)
Synonyms: *Lathyrus variegatus* Gilib., nom. Inval.
- *Lathyrus szowitsii* Boiss.
English: Szowits' vetchling
- *Lathyrus tingitanus* L. (Table 10.21)
Synonyms: *Lathyrus coruscans* Emb. & Maire; *Lathyrus mexicanus* Schlect.
- *Lathyrus tomentosus* Lam.
Synonyms: *Lathyrus sericeus* Lam.
English: rough-haired vetchling
- *Lathyrus torreyi* A. Gray
Synonyms: *Lathyrus villosus* Torr.
English: redwood pea; Torrey's peavine; Torrey's vetchling
Turkish: Torrey mürdümüğü
- *Lathyrus tracyi* Bradshaw
Synonyms: *Lathyrus lanszwertii* Kellogg var. *tracyi* (Bradshaw) Isely
English: Tracy's vetchling
- *Lathyrus transsylvanicus* (Spreng.) Rchb.
Synonyms: *Lathyrus laevigatus* (Waldst. & Kit.) Gren. subsp. *transsylvanicus* (Spreng.) Breistr.; *Lathyrus laevigatus* (Waldst. & Kit.) Gren. subsp. *transsylvanicus* (Spreng.) Soó; *Lathyrus luteus* (L.) Peterm. subsp. *transsylvanicus* (Spreng.) Dostal; *Orobus transsylvanicus* Spreng.
Bulgarian: transilvansko sekirche

English: Transylvanian vetchling
Hungarian: Erdélyi lednek
Polish: groszek transylwański
Russian: china transilvanskaia; sochevichnik transilvanskii
Swedish: karpaterärt
Ukrainian: chyna transyl'vans'ka; horoshek transil'vans'kyi
- *Lathyrus tremolsianus* Pau
Synonyms: *Lathyrus elegans* Porta & Rigo; *Lathyrus latifolius* L. var. *angustifolius sensu* Willk.
English: elegant everlasting pea
- *Lathyrus tropicalandinus* Burkart
English: tropical Andean vetchling
- *Lathyrus tuberosus* L. (Table 10.22)
Synonyms: *Lathyrus festivus* Sennen.
- *Lathyrus undulatus* Boiss.
Synonyms: *Lathyrus rotundifolius* sensu auct.
English: wavy vetchling
Russian: china volnistaia
- *Lathyrus vaniotii* H. Lev.
Synonyms: *Vicia venosa* Maxim. var. *willdenowiana* Miura
Chinese: dōngběi shān lí dòu
English: Korean mountain vetchling; Vaniot's vetchling
- *Lathyrus venetus* (Mill.) Wohlf.
Synonyms: *Lathyrus variegatus* (Ten.) Gren. & Godr.; *Orobus variegatus* Ten.; *Orobus venetus* Mill.
English: Venetian vetchling
German: Bunte Platterbse; Venezianische Platterbse
Hungarian: tarka lednek
Italian: cicerchia veneta
Polish: groszek błękitny
Russian: china golubaia; china sinevataia; sochevichnik goluboi; sochevichnik sinevatyi
Ukrainian: chyna riaba; chyna siniuvata; chyna venetsians'ka; horoshok riabyi; horoshok sorokatyi; horoshok strokatyi
- *Lathyrus venosus* Muhl. ex Willd.
Synonyms: *Lathyrus multiflorus* Torr. & A. Gray; *Lathyrus oreophyllus* Wooton & Standl.; *Lathyrus rollandii* Vict. & J. Rousseau; *Lathyrus venosus* Willd. var. *intonsus* Butters & St. John; *Lathyrus venosus* Willd. var. *meridionalis* Butters & St. John; *Orobus venosus* Braun
English: bushy vetchling; forest pea; smooth veiny pea; veiny pea
Russian: goroshek zhilkovatyi; sochevichnik zhilkovatyi
- *Lathyrus venosus* Muhl. ex Willd. var. *intonsus* Butters & H. St. John
Synonyms: *Lathyrus venosus* var. *meridionalis* Butters & H. St. John
English: long-haired vetchling
- *Lathyrus venosus* Muhl. ex Willd. var. *venosus*

English: common forest pea
- *Lathyrus vernus* (L.) Bernh. (Table 10.23)
Synonyms: *Lathyrus vernus* f. *roseus* Beck; *Orobus vernus* L.
- *Lathyrus vernus* (L.) Bernh. subsp. *flaccidus* (Ser.) Arcang.
Synonyms: *Orobus vernus* var. *flaccidus* Ser.
English: pendulous spring vetchling
- *Lathyrus vernus* (L.) Bernh. subsp. *vernus*
Synonyms: *Orobus vernus* L.
English: common spring vetchling
- *Lathyrus vestitus* Nutt.
English: covered vetchling; Pacific pea
Russian: tikhookeanskii goroshek
Turkish: Pasifik mürdümüğü
- *Lathyrus vestitus* Nutt. subsp. *alefeldii* (T. G. White) Broich
Synonyms: *Lathyrus alefeldii* T. G. White; *Lathyrus strictus* Nutt.
English: Alefeld's vetchling
- *Lathyrus vestitus* Nutt. subsp. *laetiflorus* (Greene) Broich
Synonyms: *Lathyrus laetiflorus* Greene; *Lathyrus laetiflorus* subsp. *barbarae* (T. G. White) C. L. Hitchc.
English: shaky-flowered vetchling
- *Lathyrus vestitus* Nutt. subsp. *vestitus*
Synonyms: *Lathyrus puberulus* T. G. White ex Greene; *Lathyrus vestitus* subsp. *puberulus* (T. G. White ex Greene) C. L. Hitchc.; *Lathyrus vestitus* var. *puberulus* (T. G. White ex Greene) Jeps.
English: common covered vetchling
- *Lathyrus vinealis* Boiss. & Noe
Arabic: aljulban alkarmi
Armenian: tap'volorr khaghoghanmany
English: vinyard vetchling
Russian: china vinogradnikovaia
- *Lathyrus vivantii* P. Monts.
English: Vivantius' vetchling
- *Lathyrus whitei* Kupicha
Synonyms: *Lathyrus longipes* T. G. White; *Lathyrus venosus* Hemsl.
English: White's vetchling
- *Lathyrus woronowii* Bornm.
Synonyms: *Ervum woronowii* (Bornm.) Stank.; *Lathyrus furtivus* Woronow
English: Voronov's vetchling
Russian: china Voronova
- *Lathyrus zalaghensis* Andr.
English: Zalagh vetchling
- *Lathyrus zionis* C. L. Hitchc.
Synonyms: *Lathyrus brachycalyx* Rydb. var. *zionis* (C. L. Hitchc.) S. L. Welsh
English: Zion vetchling

TABLE 10.1
Popular Names Denoting *Lathyrus annuus* L. in Some World Languages and Other Linguistic Taxa

Language/Taxon	Name
Arabic	aljulban alhuali
Catalan	guixera borda; guixera pàllida
Czech	hrachor roční
English	annual vetchling; annual yellow vetchling; fodder pea; red fodder pea
Estonian	suvik-seahernes; üheaastane seahernes
Finnish	välimerennätkelmä
French	gesse annuelle
Georgian	kanis matkvartzana
German	Einjährige Platterbse
Greek (Ancient)	ǎrakos; árako-s; árak-s
Hebrew	tvfch chd-shnt
Italian	cicerchia pallida
Russian	china odnoletniaia
Serbian	jednogodišnji grahor
Spanish	cicércula annual
Swedish	fodervial

TABLE 10.2
Popular Names Denoting *Lathyrus aphaca* L. in Some World Languages and Other Linguistic Taxa

Language/Taxon	Name
Arabic	bîqîyah; hhamâm el burg; jalban eafaqa
Catalan	fesolet; gerdell; tapissot bord
Czech	hrachor pačočkový
Danish	bladløs fladbælg
Dutch	naakte lathyrus
English	yellow pea; yellow vetch; yellow vetchling
Esperanto	rampa latiro; senfolia latiro
Finnish	korvakenätkelmä
French	gesse aphylle; gesse sans feuilles
German	Ranken-Platterbse
Hebrew	tvfch mtzv
Hindi	janglimatar; pili matter
Hungarian	levéltelen lednek
Istriot	biʃèto
Italian	afaca; cicerchia bastarda
Japanese	takuyourenrisou

(*Continued*)

TABLE 10.2 (*Continued*)
Popular Names Denoting *Lathyrus aphaca* L. in Some World Languages and Other Linguistic Taxa

Language/Taxon	Name
Norwegian (Bokmål)	mølleflatbelg
Norwegian (Nynorsk)	mølleskolm
Occitan	gèissa sens fuèlhas
Persian	xlr bibrg
Polish	groszek bezlistny
Portuguese	chícharo amarelo; ervilha olho de boneca
Punjabi (Western)	jungli mutter
Russian	china bezlistochkovaia
Serbian	bezlisni grahor
Sorbian (Upper)	wobwitkowy hróšik
Spanish	afaca; alverja silvestre
Swedish	spjutvial
Turkish	sarı burçak; sarı mürdümük
Ukrainian	chyna bezlystochkova; horoshok bezlistyi
Welsh	ydbysen felen; ytbysen felen

TABLE 10.3
Popular Names Denoting *Lathyrus cicera* L. in Some World Languages and Other Linguistic Taxa

Language/Taxon	Name
Arabic	aljulban alhumusy
Armenian	tap'volorr karmiry
Asturian	almorta de monte
Catalan	guixó; guixó ver; guixonera; herba de forca
Chinese	dui ye xiang wan dou; xiang wan dou
Czech	hrachor cizrnový
Dutch	kekerlathyrus
English	chickling vetch; dwarf chickling-vetch; flatpod peavine; lesser chickpea; red pea; red vetchling
Finnish	etelännätkelmä; purppuranätkelmä
French	gesse chiche; gesse garosse; gesse pois-chiche; gessette; jarosse; jarousse; petite gesse
German	Futterplatterbse; Platterbse; Kleine Platterbse; Rote Platterbse
Hungarian	csicserilednek
Italian	cicerchia cicerchiella; mochi; moco
Latin (Medieval)	geissas caninas
Norwegian (Bokmål)	purpurflatbelg

(*Continued*)

TABLE 10.3 (*Continued*)
Popular Names Denoting *Lathyrus cicera* L. in Some World Languages and Other Linguistic Taxa

Language/Taxon	Name
Norwegian (Nynorsk)	purpurskolm
Occitan	gèissa chicha
Persian	xlr nxudi
Polish	groszek cicierzycowaty
Portuguese	chícharo-miudo
Russian	china nutovaya
Serbian	grašasti grahor; orašasti grahor
Spanish	alcaballares; almorta; almorta de monte; almorta salvaje; almorta silvestre; alvejana; alverja caballar; alverja de guija; alverjón; aracus; arbejones; chícharo; chícharo-miudo; chícharos; chícharos de Lisboa; cicercha; cicercula; cicércula; cicérula; cirésula; cuchillejos; diente de muerto; galbana; galgana; galgana de Nebrija; galgarra; galgarria; garbanzo negro; garbanzos gitanos; gríjoles; guija; guija silvestre; guijas; guijas bordes; guijilla; guijillas; guisa silvestre; guisante salvaje; guixons; gálgana; habilla; latiro; lenteja forrajera; pitos silvestre; sabillones; titarro; titarros; veza loca
Swedish	rödvial
Turkish	kırmızı mürdümük
Ukrainian	chyna nutova; horoshok nutovyi

TABLE 10.4
Popular Names Denoting *Lathyrus clymenum* L. in Some World Languages and Other Linguistic Taxa

Language/Taxon	Name
Arabic	aljulban alkalimini
Czech	hrachor popínavý
English	Spanish vetchling
French	gesse clymène; gesse d'Espagne; gesse pourpre
German	Purpur-Platterbse
Greek	ispanikós víkos; láthyros to klýmenon
Hebrew	tvfch sfrd
Hungarian	borsóbükköny
Italian	cicerchia articolata; cicerchia porporina
Japanese	otomerenrisou
Neapolitan	doleca
Portuguese	chicharão-de-Torres; cisirão-de-Torres; ervilhaca púrpura
Russian	china chlenistaia
Spanish	abejaquilla; alvehón; arvejana; arvejones; arvejón; chícharos menores de España; cicercha purpurina; conejitos; cuchillejo; garbaneta; guijas; guisantera; guisantes silvestres; pelailla; presulillo; présule zorrero
Swedish	bandvial; ledvial

TABLE 10.5
Popular Names Denoting *Lathyrus hirsutus* L. in Some World Languages and Other Linguistic Taxa

Language/Taxon	Name
Arabic	aljulban almashear
Catalan	veçó eriçat
Czech	hrachor chlupatý
Dutch	ruige lathyrus
English	austrian winterpea; Caley pea; hairy vetchling; rough pea; singletary pea; winterpea
Estonian	karvane seahernes
Finnish	karvanätkelmä
French	gesse hérissée; gesse velue
German	Behaarte Platterbse; Behaartfrüchtige Platterbse; Rauhhaarige Platterbse
Hebrew	tvfch sh'r
Hungarian	borzas lednek
Italian	cicerchia pelosa
Norwegian (Bokmål)	lodneflatbelg
Norwegian (Nynorsk)	lodneskolm
Persian	xlr krki
Polish	groszek kosmatostrąkowy
Russian	china shershavaia; china zhestkovolosistaia; sochevichnik zhestkovolosistyi
Sorbian (Upper)	kosmaty hróšik
Spanish	guija velluda
Swedish	luddvial
Turkish	tüylü mürdümük
Ukrainian	chyna shorstka; horoshok mokhnatyi; horoshok sherstkovolosyi; horoshok shortskyi
Welsh	ytbysen flewgodog; ytbysen flewog

TABLE 10.6
Popular Names Denoting *Lathyrus inconspicuus* L. in Some World Languages and Other Linguistic Taxa

Language/Taxon	Name
Albanian	vingjër
Bulgarian	nevzrachno sekirche
Croatian	sitna graholika; sitna kukavičica
English	inconspicuous vetchling
Finnish	vähänätkelmä
German	Kleinblütige Platterbse
Hebrew	tvfch zkvf
Italian	cicerchia a fiore piccolo
Japanese	suzumenorenrisou
Norwegian (Bokmål)	sveltflatbelg
Norwegian (Nynorsk)	sveltskolm
Polish	groszek liściakowy
Russian	china nezametnaia
Serbian	sitni grahor; sitnocvetni grahor
Swedish	dvärgvial
Turkish	yılan mürdümüğü

TABLE 10.7
Popular Names Denoting *Lathyrus japonicus* Willd. in Some World Languages and Other Linguistic Taxa

Language/Taxon	Name
Azerbaijani	yapon gülülcəsi
Chinese	hǎibīn shān lí dòu
Danish	strand-fladbælg
Dutch	zeelathyrus
English	beach pea; circumpolar pea; sea pea; sea vetchling
Estonian	rand-seahernes
Finnish	merinätkelmä
French	gesse maritime
French (Canada)	pois de mer
Frisian (North)	dünemeert
German	Strand-Platterbse
Gullah	peawine
Icelandic	baunagrass
Japanese	hama endou
Nivkh	tur; turi

(Continued)

TABLE 10.7 (*Continued*)
Popular Names Denoting *Lathyrus japonicus* Willd. in Some World Languages and Other Linguistic Taxa

Language/Taxon	Name
Norwegian (Bokmål)	strandflatbelg
Norwegian (Nynorsk)	strandskolm
Polish	groszek nadmorski
Russian	china yaponskaia
Serbian	primorski grahor
Sorbian (Upper)	přimórski hróšik
Swedish	strandvial
Ukrainian	chyna yapons'ka
Welsh	pysen gochlas ar môr; pysen gochlas arfor; ytbysen arfor; ytbysen y môr

TABLE 10.8
Popular Names Denoting *Lathyrus latifolius* L. in Some World Languages and Other Linguistic Taxa

Language/Taxon	Name
Arabic	al gulbân el kabîr
Armenian	dzmerrayin tap'volorr
Asturian	albejana
Catalan	llapissots; pèsol bord; pesolera de camp; pesolera silvestre
Chinese	kuān yèshān lí dòu
Czech	hrachor širolistý
Dutch	brede lathyrus
English	broadleaf everlasting pea; broad-leaved everlasting-pea; everlasting pea; perennial pea; perennial peavine; perennial sweet pea
Esperanto	latiro larĝfolia; sovaĝa pizo larĝfolia
Estonian	laialehine seahernes
Finnish	ruusunätkelmä
French	gesse à larges feuilles; pois vivace
German	Breitblatt-Platterbse; Breitblättrige Platterbse; Bukettwicke
Hungarian	nagyvirágú lednek
Italian	cicerchia a foglie larghe; pisello a mazzetti; pisello vivace
Japanese	hiroha no renri sou
Lithuanian	plačialapis pelėžirnis
Occitan	gèissa de fuèlhas largas
Polish	groszek szerokolistny
Russian	china shirokolistnaia
Serbian	širokolisni grahor
Sorbian (Upper)	šěroki hróšik

(*Continued*)

TABLE 10.8 (*Continued*)
Popular Names Denoting *Lathyrus latifolius* L. in Some World Languages and Other Linguistic Taxa

Language/Taxon	Name
Spanish	albejana basta; alvejana bravía; arbejana borde; arveja silvestre; arvejana loca; arvejanas; arverjón; caravalle; crisantelmo; gallinicas; gallitos; guija de hoja ancha; zapatitos del Niño Jesús
Swedish	rosenvial
Turkish	her dam taze bizelya
Ukrainian	chyna shirokolysta; horoshok shirokolystyi; zillia od dannia
Welsh	ytbysen fythol lydanddail; ytbysen fythol lydanddail

TABLE 10.9
Popular Names Denoting *Lathyrus linifolius* (Reichard) Bassler in Some World Languages and Other Linguistic Taxa

Language/Taxon	Name
Azerbaijani	kətanyarpaq gülülcə
Belarusian	čyna gornaja; čyna l'nalistaja
Bulgarian	planinsko sekirče
Catalan	guixó muntanyenc; veça de muntanya
Czech	hrachor horský
Danish	krat-fladbælg
Dutch	knollathyrus
English	bitter vetch; earthnut pea, flaxlike-leafed vetchling; heath pea; wood pea
Esperanto	latiro ruĝblua
Estonian	mägi-seahernes
Finnish	syylälinnunherne
French	gesse à feuilles de lin; gesse aux tiges renflées; gesse des montagnes; orobe
German	Berg-Platterbse; Bergerbse; Knollige Bergerbse
Hungarian	hegyi lednek
Italian	cicerchia montana
Lithuanian	kalninis pelėžirnis
Norwegian (Bokmål)	knollerteknapp
Norwegian (Nynorsk)	knollerteknapp
Polish	groszek lnolistny; groszek skrzydlasty; groszek skrzydlaty
Russian	china gornaia; china lnolistnaia; sochevichnik lnolistnyi
Scots	mice pease; moose pease
Serbian	lanolisni grahor; planinski grahor
Sorbian (Upper)	hórski hróšik
Spanish	guija de monte
Swedish	gökärt; gökmat
Vietnamese	đậu tằm đắng
Welsh	pysen y coed gnapwreiddiog; pys y coed; pysen y coed; ytbysen y coed

TABLE 10.10
Popular Names Denoting *Lathyrus niger* (L.) Bernh. in Some World Languages and Other Linguistic Taxa

Language/Taxon	Name
Arabic	aljulban al'aswad
Belarusian	baravyja zajčyki; sačavičnik čorny
Catalan	guixó negre
Czech	hrachor černý
Danish	sort fladbælg
Dutch	zwarte lathyrus
English	black bitter vetch; black pea; blackening flat pea
Esperanto	latiro nigra
Estonian	must kurelääts; must seahernes
Finnish	mustalinnunherne
French	gesse noire
German	Schwarze Platterbse; Schwarzwerdende Platterbse
Hungarian	fekete lednek
Italian	cicerchia nera
Japanese	seiyouebira fuji
Lithuanian	saldžiašaknis pelėžirnis
Norwegian (Bokmål)	svarterteknapp
Norwegian (Nynorsk)	svarterteknapp
Polish	groszek czerniejący
Russian	china chornaia; sochevichnik chornyi
Serbian	crni grahor
Sorbian (Upper)	ćmowy hróšik
Spanish	orobo; yerbo de Panonia
Swedish	vippärt
Ukrainian	chornozillia; chyna chorna; horoshok chornyi
Welsh	pysen borfor; ytbysen ddu

TABLE 10.11
Popular Names Denoting *Lathyrus nissolia* L. in Some World Languages and Other Linguistic Taxa

Language/Taxon	Name
Arabic	aljulban alnyswly
Czech	hrachor trávolistý
Dutch	graslathyrus
English	grass pea; grass vetchling
Esperanto	latiro herbeca
Finnish	heinänätkelmä
French	gesse de nissole; gesse sans vrille

(*Continued*)

TABLE 10.11 (*Continued*)
Popular Names Denoting *Lathyrus nissolia* L. in Some World Languages and Other Linguistic Taxa

Language/Taxon	Name
German	Gras-Platterbse
Greek	láthyros I nissolía
Hebrew	tvfch dn
Hungarian	fűképű lednek; kacstalan lednek
Italian	cicerchia semplice
Persian	xlr tcmni
Polish	groszek liściakowy
Russian	china nissolia; china zlakolistnaia
Scots	grass vetchling
Serbian	travolisni grahor; žitolisni grahor
Sorbian (Upper)	trawowy hróšik
Spanish	alverja nisolia
Swedish	gräsvial
Turkish	otsu mürdümük
Ukrainian	chyna nissolia; chyna zlakolista; horoshok travolistyi; horoshok zlakolistyi
Welsh	ytbysen feinddail; ytbysen goch

TABLE 10.12
Popular Names Denoting *Lathyrus ochrus* (L.) DC. in Some World Languages and Other Linguistic Taxa

Language/Taxon	Name
Arabic	aljulban al'asfar
Catalan	tapissot; veça plana
Czech	hrachor žlutoplodý
Danish	småært
English	Cyprus vetch; ochrus vetch; winged vetchling
French	cicerole; gesse ochre; gesse ocre
German	Eselsohren; Flügel-Platterbse; Gelbe Erbse; Ockerplatterbse; Scheidige Platterbse
Hebrew	tvfch gdvl
Italian	cicerchia pisellina
Japanese	higerenrisou
Norwegian (Bokmål)	bleikflatbelg
Norwegian (Nynorsk)	bleikskolm
Polish	groszek bladożółty
Portuguese	chicharão preto; ervilhaca bastarda; ervilha dos campos; ervilhaca dos campos
Russian	china okhryanaia
Serbian	kiparski grahor; oker grahor
Spanish	alverjana; tapisote
Swedish	skaftvial
Turkish	Kıbrıs mürdümüğü

TABLE 10.13
Popular Names Denoting *Lathyrus odoratus* L. in Some World Languages and Other Linguistic Taxa

Language/Taxon	Name
Arabic	baslat alzzuhur; zahr el bisellah
Armenian	tap'volorr palaravor
Azerbaijani	ətirli gülülcə
Basque	ilar usainduna
Belarusian	čyna pachuchaja; pachučy harošak
Catalan	pèsol d'olor
Chinese	hua xiang wan dou; xiāng wāndòu
Czech	hrachor vonný
Danish	almindelig ærteblomst; ærteblomst; vellugtende ærteblomst
Dutch	pronkerwt; reukerwt; welriekende lathyrus
English	sweet pea
Erzya	idem vika; tantej čina
Esperanto	aroma latiro; odora latiro; odorpizo
Estonian	lillhernes
Finnish	hajuherne; tuoksuherne
French	gesse odorante; pois de senteur; pois fleur; pois musqué
French (Ontario)	frijol de olor
German	Bunte Wicke; Duftende Platterbse; Duftwicke; Gartenwicke; Wohlriechende Platterbse
Greek	glykó mpizéli; láthyros o eúosmos; moshompízelo
Hungarian	szagos bükköny
Ido	latiro
Irish	pis chumhra
Italian	cicerchia odorosa; pisello fiore; pisello odoroso
Japanese	jakou endou; suītopī
Jèrriais	pais-flieur
Latvian	puku zirnisi; smarzigas destina
Lithuanian	kvapusis pelėžirnis
Moldovan	linte odorate
Norwegian (Bokmål)	blomsterert
Norwegian (Nynorsk)	blomsterert
Occitan	pese de sentor
Persian	gl nxud
Polish	groszek pachnący
Portuguese	ervilha de cheiro
Romanian	orastica mirositoare; singele voiniculi
Russian	china dushistaia; dushistyi goroshek
Samogitian	žemčiūgs; žėrnėkielis; žėrnelē; žėrniokā; žėrnotē
Scots	sweet peas
Serbian	mirisni grahor; mirišljavi grašak; ukrasna grahorica; ukrasni grašak

(Continued)

TABLE 10.13 (*Continued*)
Popular Names Denoting *Lathyrus odoratus* L. in Some World Languages and Other Linguistic Taxa

Language/Taxon	Name
Sorbian (Upper)	wonjaty hróšik
Spanish	caracoles; caracolillo de olor; caracolillos; caracolillos de olor; chicharito de olor; chicharro de olor; chícharo de olor; chorreque; clarín; gayomba de Indias; guisante de olor; guisante dulce; guisantes de olor; látiro
Swedish	luktärt
Tagalog	gisante; gisantes
Turkish	itırşahi; kokulu bizelya
Ukrainian	chyna zapashna; horoshok dushistyi; horoshok pakhuchyi; horoshok zapashnyi
Vietnamese	hương đậu
Welsh	pys pêr

TABLE 10.14
Popular Names Denoting *Lathyrus palustris* L. in Some World Languages and Other Linguistic Taxa

Language/Taxon	Name
Bulgarian	blatno sekirche
Chinese	ōu shān lí dòu; shan li dou; wú chì shān lí dòu; xiàn yèshān lí dòu; zhao sheng xiang wan dou
Czech	hrachor bahenní
Danish	kær-fladbælg
Dutch	moeraslathyrus
English	blue marsh vetchling; marsh pea; marsh vetchling
English (U.S.)	marsh pea-vine
Erzya	çejädavksonj idem vika
Estonian	soo-seahernes
Finnish	rantanätkelmä; suonätkelmä
French	gesse des marais
German	Sumpf-Platterbse; Sumpfwicke
Hungarian	mocsári lednek
Italian	cicerchia palustre; veccia delle paludi
Japanese	seiyourenrisou
Lithuanian	pelkinis pelėžirnis
Norwegian (Bokmål)	myrflatbelg
Norwegian (Nynorsk)	myrskolm
Polish	groszek błotny
Russian	china bolotnaia
Scots	mice pease; moose pease

(*Continued*)

TABLE 10.14 (*Continued*)
Popular Names Denoting *Lathyrus palustris* L. in Some World Languages and Other Linguistic Taxa

Language/Taxon	Name
Serbian	barski grahor
Sorbian (Upper)	bahnowy hróšik
Swedish	kärrvial
Turkish	bataklık mürdümüğü
Ukrainian	chyna bolotna
Welsh	ydbysen las y morfa; ytbysen las y morfa; ytbysen y gors

TABLE 10.15
Popular Names Denoting *Lathyrus pannonicus* (Jacq.) Garcke in Some World Languages and Other Linguistic Taxa

Language/Taxon	Name
Bulgarian	panonsko sekirche
Czech	hrachor panonský
English	Pannonian vetchling
Esperanto	latiro kremkolora
French	gesse de Pannonie
German	Pannonische Platterbse
Hungarian	Magyar lednek
Italian	cicerchia pannonica
Polish	groszek pannoński
Russian	china vengerskaia
Serbian	panonski grahor
Sorbian (Upper)	wuherski hróšik
Ukrainian	chyna pannons'ka; horoshok pannons'kyi; horoshok pol'ovyi; horoshok uhors'kyi; zillia od pohaniuki

TABLE 10.16
Popular Names Denoting *Lathyrus pisiformis* L. in Some World Languages and Other Linguistic Taxa

Language/Taxon	Name
Chinese	dà tuō yèshān lí dòu
Czech	hrachor hrachovitý
English	pea-like vetchling

(*Continued*)

TABLE 10.16 (*Continued*)
Popular Names Denoting *Lathyrus pisiformis* L. in Some World Languages and Other Linguistic Taxa

Language/Taxon	Name
Erzya	ksnavonj kondämo idem vika
Estonian	hernelehine seahernes
German	Erbsenartige Platterbse
Hungarian	borsóképű lednek
Lithuanian	žirnialapis pelėžirnis
Mongolian	banduikhai temerdee
Polish	groszek wielkoprzylistkowy
Russian	china gorokhovidnaia
Serbian	graškoliki grahor
Sorbian (Upper)	hrochojty hróšik
Swedish	ärtvial
Ukrainian	chyna horokhovydna; horoshok horokhopodibnyi; horoshok horokhuvatyi

TABLE 10.17
Popular Names Denoting *Lathyrus pratensis* L. in Some World Languages and Other Linguistic Taxa

Language/Taxon	Name
Arabic	aljulban almarji
Armenian	tap'volorr margagetnayin
Azerbaijani	çəmən güllücəsi; çəmən lərgəsi
Belarusian	čyna luhavaja; haroshak zajačyj; lesavy ljadzej
Catalan	gerdell bord; guixa de prat; guixeta
Chinese	mù dì shān lí dòu; mu di xiang wan dou
Chuvash	uləh tərna pərçi
Czech	hrachor luční
Danish	gul fladbælg
Dutch	veldlathyrus
English	meadow pea; meadow peavine; meadow vetchling; yellow vetchling
Erzya	lugalangonj idem vika
Esperanto	latiro herbeja
Estonian	aas-seahernes
Finnish	niittynätkelmä
French	gesse des prés
Georgian	mdelos matkvarts'ana

(*Continued*)

TABLE 10.17 (*Continued*)
Popular Names Denoting *Lathyrus pratensis* L. in Some World Languages and Other Linguistic Taxa

Language/Taxon	Name
German	wiesen-Platterbse
Hungarian	réti lednek
Icelandic	fuglaertur
Italian	cicerchia dei prati; erba galletta
Japanese	kibananorenrisou
Kazakh	şalğındıq noğatıq
Latvian	pļavas dedestiņa; virklītes
Lithuanian	pievinis pelėžirnis
Moldovan	linte de pratur
Mongolian	nugyn temerdee
Norwegian (Bokmål)	gulflatbelg
Norwegian (Nynorsk)	gulskolm
Occitan	gèissa dels prats
Ossetian	qædurgærdæg
Persian	xlr zrd
Polish	groszek łąkowy; groszek żółty; lędźwian żółty
Portuguese	ervilha do prado; ervilhaca do campo
Portuguese (Brazil)	ervilha do prado; ervilhaca do campo
Romanian	orastica de finete
Russian	china lugovaia
Sami (Northern)	niitosáhpal
Scots	craw's pease; dug's-pease
Serbian	livadski grahor
Slovenian	travniški grahor
Sorbian (Upper)	łučny hróšik
Spanish	almorta de los prados; arbelleta; arveja de campo; arvejana; hierba guisantina; latiro de prado; latiro de prados; látiro de prado
Swedish	gulvial
Tajik	sirdirk
Turkish	çayır mürdümüğü
Ukrainian	chyna luchna; chyna lugova; chyna zhovta; horoshok kotliachyi; horoshok lisnyi; horoshok luhovyi; horoshok lukovyi; horoshok potiachyi; horoshok zhovtyi; horoshok zhuravlynyi
Welsh	ffacbysen y weirglodd; pupys y waun; ydbys y borfa; ydbysen y waun; ytbysen y ddôl; ytbysen y waun

TABLE 10.18
Popular Names Denoting *Lathyrus sativus* L. in Some World Languages and Other Linguistic Taxa

Language/Taxon	Name
Albanian	koçkulla
Albanian (Arvanitika)	lathurè
Amharic	gwaya
Arabic	aljulban almazrue; gulbân; khullar; turmos
Arabic (Egypt)	gelban
Aragonese	guicheras; guiheras; guiseras; guixeras
Armenian	tap'volorr sovorakany; tap'volorr ts'anovin
Asturian	almorta; alverjón; arbeya; arbeyera; arvejo cantudo; arvejote; bichas; cantuda; chichu; cicércula; diente de muertu; fríjol de yerba; garbanzu de yerba; muela; pedrete; pedruelo; pinsol; pito
Azerbaijani	ekin gülülcə
Bashkir	sərmalsəq
Basque	aixkol
Belarusian	čyna pasjaounaja
Bengali	khēsāri ḍāla
Bulgarian	obiknoveno sekirche
Burmese	pèsali
Catalan	guixa; guixera
Chinese	jiāshān lí dòu; ou zhou xiang wan dou; ou zhou shan li dou
Chuvash	tərna pərçi
Croatian	graholika; sijani grahor; sikirica
Czech	hrachor setý
Danish	agerfladbælg; fladbælg; dyrket fladbælg; græs-fladbælg; sædfladbælg
Dutch	wikkenlatyrus; zaailathyrus
English	blue vetchling; chickling pea; chickling vetch; dogtooth pea; giant lentil; grass peavine; grass pea; grasspea; Indian pea; khesari; riga pea; wedge peavine; white pea; white vetch
Erzya	idem vika
Esperanto	platgrajna latiro
Estonian	seahernes
Finnish	peltonätkelmä
French	dent de brebis; févette; gesse; gesse blanche; gesse commune; gesse cultivée; gesse domestique; jarosse; lentille d'Espagne; pois breton; pois carré; pois cornu
Georgian	ts'ulispira
German	Platterbse; Saat-Platterbse
Greek	láthyros
Greek (Ancient)	láthuros

(*Continued*)

TABLE 10.18 (*Continued*)
Popular Names Denoting *Lathyrus sativus* L. in Some World Languages and Other Linguistic Taxa

Language/Taxon	Name
Hindi	dholl kessari; khasari; khesari; khesari dhal; lataree
Hungarian	szegletes lednek
Italian	cicerchia commune
Japanese	gurasupī
Kashubian	séwny groszk
Kazakh	äyken; egistik noğatıq
Latin	lathyros
Latin (Medieval)	geissas caninas
Lithuanian	sėjamasis pelėžirnis
Macedonian	gravorika
Marathi	laakh
Moldovan	linte kultivate
Mongolian	tarimal temerdee
Nepali	khēsarī
Norwegian (Bokmål)	forflatbelg
Norwegian (Nynorsk)	forskolm
Occitan	gèissa comuna; pese carrat
Persian	xlr
Polish	groszek siewny; groszek zwyczajny; lędźwian siewny
Portuguese	chícharo
Romanian	latir; mazaroi; orastica
Russian	china aziatskaia; china posevnaia
Rusyn (Pannonian)	liednik
Serbian	gajeni grahor; obični grahor; sastrica; sekirče; sekirica; sikirica
Slovak	hrachor
Slovenian	grahor
Sorbian (Upper)	sywny hróšik
Spanish	almorta; alverjón; arveja; arvejo cantudo; arvejote; bichas; cantuda; cicércula; chícharo; diente de muerto; fríjol de yerba; garbanzo de yerba; guija; muela; pedruelo; pinsol; pito; tito
Swedish	akervial; plattvial
Telugu	lāthiras
Tigre	säbbära
Tigrinya	sebere
Turkish	ak burçak; mürdemeg; mürdümük; nazende; yası mercimek
Ukrainian	chyna posivna; horoshok nimetskyi; horoshok siinyi; horoshok uhluvatyi
Uzbek	burchak; burchoq
Vietnamese	chi đậu hoa; chi đậu thơm
Welsh	ytbysen India

TABLE 10.19
Popular Names Denoting *Lathyrus sphaericus* Retz. in Some World Languages and Other Linguistic Taxa

Language/Taxon	Name
Arabic	aljulban alkurawi
Bulgarian	sferichno sekirche
Catalan	guixó esfèric
English	grass pea; round-seeded vetchling; slender wild pea
French	gesse à fruits ronds; gesse à graines rondes
German	Kugelsamige Platterbse
Hebrew	tvfch chdvr
Hungarian	téglaszínű lednek
Italian	cicerchia bastarda
Persian	xlr du brgtcx'ai
Russian	china sharovidnaia
Serbian	okrugli grahor
Swedish	vårvial
Ukrainian	chyna kuliasta; horoshok kuliastyi

TABLE 10.20
Popular Names Denoting *Lathyrus sylvestris* L. in Some World Languages and Other Linguistic Taxa

Language/Taxon	Name
Armenian	tap'volorr antarrayin
Catalan	guixera de bosc; pèsol pirinenc
Czech	hrachor lesní
Danish	smalbladet skov-fladbælg; skov-fladbælg
Dutch	boslathyrus
English	flat pea; flat peavine; narrow-leaf everlasting pea
Erzya	virenj idem vika
Esperanto	latiro arbara
Estonian	mets-seahernes
Finnish	metsänätkelmä
French	gesse des bois; gesse sauvage
German	Wald-Platterbse
Hungarian	erdei lednek
Italian	cicerchia silvestre
Japanese	yanagibarenrisou
Lithuanian	miškinis pelėžirnis
Norwegian (Bokmål)	skogflatbelg

(Continued)

TABLE 10.20 (*Continued*)
Popular Names Denoting *Lathyrus sylvestris* L. in Some World Languages and Other Linguistic Taxa

Language/Taxon	Name
Norwegian (Nynorsk)	skogskolm
Occitan	gèissa salvatja
Piedmontese	pòis sarvaj
Polish	groszek leśny
Portuguese	chícharo selvagem
Russian	china lesnaia
Scots	flat pea; narrow-leaved everlastin-pea
Serbian	šumski grahor
Sorbian (Upper)	lěsny hróšik
Spanish	cicércula silvestre; guija silvestre
Swedish	backvial; skogsvial
Ukrainian	chyna lisova; dekokt pol'ovyi; liubi-mene-ne-pokin'; horokh zhuravlevyi; horoshok dykyi; horoshok lisovyi; horoshok zhuravlevyi; horoshok zhuravlynyi
Welsh	ydbys y coed; ydbysen; ydbysen barhaus gulddail; ydbysen culddail barhaus; ytbysen barhaus gulddail; ytbysen fythol gulddail

TABLE 10.21
Popular Names Denoting *Lathyrus tingitanus* L. in Some World Languages and Other Linguistic Taxa

Language/Taxon	Name
Arabic	aljulban alttanji
Armenian	tap'volorr tan tserin
Catalan	guixó de Tànger
Czech	hrachor tangerský
English	Tangier pea; Tangier scarlet pea
Finnish	etelännätkelmä
French	gesse de Tanger
German	Afrikanische Wicke; Tanger-Platterbse
Italian	cicerchia di Tangeri
Japanese	jittokumame
Portuguese	chícharo de Tanger
Portuguese (Brazil)	ervilha Tangier
Serbian	marokanski grahor; tangerski grahor
Spanish	slmorta tangerina
Swedish	spanskvial

TABLE 10.22
Popular Names Denoting *Lathyrus tuberosus* L. in Some World Languages and Other Linguistic Taxa

Language/Taxon	Name
Armenian	tap'volorr palarakir
Belarusian	čyna klubnevaja; čyna klubnjanosnaja; zemljanyja arekhi
Catalan	guixera borda; guixó tuberós
Chinese	méi hóng shān lí dòu
Chuvash	sysna paranki
Czech	hrachor hlíznatý
Danish	knold-fladbælg
Dutch	aardaker
English	earth chestnut; earthnut pea; groundnut peavine; tuberous pea; tuberous vetch
Esperanto	tubera latiro
Estonian	mugul-seahernes; seapähkel
Finnish	mukulanätkelmä
French	châtaigne de terre; gesse tubéreuse; gland de terre; macjonc; мacusson; pois tubéreux; souris de Hollande; souris de terre
Georgian	t'ero
German	Erdnuss Platterbse; Knollen-Platterbse
Hungarian	gumós lednek; mogyorós lednek
Italian	cicerchia tuberose
Japanese	kyuukon'endou
Kyrgyz	çımıldık
Lithuanian	gumbinis pelėžirnis
Norwegian (Bokmål)	jordflatbelg
Norwegian (Nynorsk)	jordskolm
Occitan	gèissa tuberose
Ossetian	tero
Persian	xlr ghdx'dar
Polish	groszek bulwiasty
Russian	china klubnenosnaia; china klubnevaia; lugovoi gorokh; myshki; rozovaia china; rozovyi goroshek; zemlianye orekhi
Sorbian (Upper)	pólny hróšik
Spanish	arveja tuberose
Swedish	knölvial
Turkish	koşkoz
Ukrainian	chechevychka; chyna bul'bysta; chyna rozheva; chyna rozova; horoshok bul'bystyi; horoshok bul'vystyi; horoshok dushistyi; horoshok krasnyi; horoshok rozovyi; horoshok stepnyi; horoshok zemlianyi; koroliv tsvit krasnyi; laskovitsia; orishki zemliani; zhola pidzemna
Welsh	ytbysen gnapiog; utbysen oddfynog

TABLE 10.23
Popular Names Denoting *Lathyrus vernus* (L.) Bernh. in Some World Languages and Other Linguistic Taxa

Language/Taxon	Name
Armenian	kisam garnanain
Belarusian	haroshak baravy; pjatushki; sachavichnik vjasennyj
Catalan	guixó primerenc
Chuvash	çimĕk kurăkĕ
Czech	hrachor jarní
Danish	vår-fladbælg
Dutch	voorjaarslathyrus
English	spring pea; spring vetch; spring vetchling
Erzya	tundonj idem vika
Esperanto	latiro printempa
Estonian	kevadine kurelääts; kevadine seahernes
Finnish	kevätlinnunherne
French	gesse printanière; orobe printanier; pois-de-pigeon
German	fruhlings-Platterbse
Hungarian	tavaszi lednek
Italian	cicerchia primaticcia
Japanese	tsurunashirenrisou
Lithuanian	pavasarinis pelėžirnis
Norwegian (Bokmål)	vårerteknapp
Norwegian (Nynorsk)	vårerteknapp
Occitan	gèissa primaverenca
Persian	xlr bx'argl
Polish	groszek wiosenny
Russian	china vesennyaya; sochevichnik vesennii
Serbian	jari grahor; prolećni grahor
Sorbian (Upper)	nalětni hróšik
Swedish	värärt
Ukrainian	chyna vesniana; horoshok dykyi; horoshok horobynyi; horoshok lisnyi; horoshok vesnianyi; horoshok zaiachyi; volivnyk iaryi; zaiachyi horokh vesnianyi

10.2 ORIGIN OF SCIENTIFIC AND POPULAR TAXA NAMES

The exact etymology of the Linnean name *Lathyrus* still remains insufficiently clarified (Linnaeus 1753, 1758). Certainly, it comes from the Latinized form of the original Ancient Greek word *láthyros*, but its meaning is still ambiguous. There are several opinions on this complex issue. The first one, which is rather widespread, tends to consider the word compound, consisting of the prefix *lá-*, meaning *very*, and the suffix *-thyros*, meaning *passionate* (Fernald 1950). In such a way, the word *láthyros* would refer to a plant, not necessarily belonging to the genus *Lathyrus*, believed to be an aphrodisiac (Gligić 1954), although its exact way of use, whether it was consumed individually or as a part of certain religious rituals, was not clarified. On the other hand, the Ancient Greek *láthyros* is also regarded as a descendant of the Proto-Indo-European root **lent-*, **lent-s-*, denoting lentil (*Lens culinaris* [Medik.], and thus being a cognate of the corresponding derivatives in other Indo-European languages (see Chapter 11), although, with a shift of meaning from lentil to the *Lathyrus* species (Chantraine 1968, Mikić-Vragolić et al. 2007). It is possible to contribute to this viewpoint by adding that the Ancient Greek *láthyros* is also considered one of the ancient names referring to pea (*Pisum sativum* L.), without any supplementary data on its etymology and its relation to other words that design primeval Eurasian grain legume crops (Sterndale-Bennett 2005, Wisconsin State Herbarium 2017).

Since the genus *Lathyrus* is remarkably rich in species and their vernacular names—as presented in more than twenty accompanying tables and additional data, that supplements individual listed taxa—we shall try to simplify its etymological and lexicological analysis to a reasonable extent, and we hope to do so without omitting anything pivotal for understanding the subject.

The Dutch language seems to be one of the very few with no popular name for the genus, but, instead, uses the Latin or scientific one, *lathyrus*, to denote its species. Similar examples may be found in Romanian, with *latir*, and Telugu, with *lāthiras*, as well as in the constructed languages, such as Esperanto and Ido, both with *latiro*.

Some of the rare, genuine popular names denoting the *Lathyrus* species in South Europe are the Catalan *guixa*, *guixera*, *guixó*, or *guixonera*; the French *garousse*, *gesse*, *jaisse*, and *jesse*; the Occitan *gèissa* and *gieissa;* and the Spanish *guija*. In the Medieval Latin, these terms denoted *geissas caninas*, "vetches of bad quality," while the ultimate origin of all could be the corruption of the second segment of the Latin name *faba Aegyptia* (Littré 1863–1872). There are two more recorded terms in French: *garoube*, denoting exclusively *Lathyrus sativus*, and *jarosse*, being a local name for the *Lathyrus* species in the ancient province of France, Poitou.

Another example of a name specifically designating the *Lathyrus* species may be found among some of the most widely spoken Indo-Aryan languages, such as Bengali, with *khēsāri*, Hindi, with *khesari*, and Nepali, with *khēsarī*. Highly hypothetically, these words, with their common source, could be related to the Proto-Indo-European root **kek-*, **k'ik'-*, referring to pea (Nikolayev 2012), although this potential connection should be elaborated with an utmost attention.

In the majority of modern Slavic languages, there are almost mutually identical words relating to the *Lathyrus* species, such as the Croatian *graholika* and *grahor*, the Czech *hrachor*, the Kashubian *groszk*, the Macedonian *gravorika*, the Polish *groszek*, the Serbian *grahor*, the Slovak *hrachor*, the Slovenian *grahor*, and the Upper Sorbian *hróšik* (Figure 10.1). All these terms derived from the Proto-Slavic root denoting pea, **górxŭ* (Vasmer 1953), which, in turn, developed from the Proto-Indo-European root **ghArs-*, referring to a leguminous plant (Pokorny 1959, Nikolayev 2012). These Proto-Indo-European and Proto-Slavic roots are also the source of the

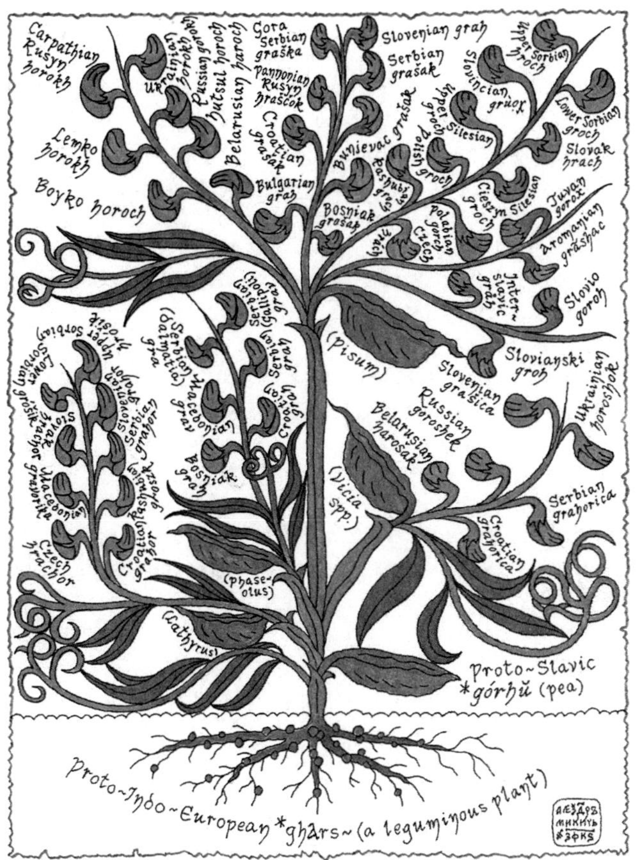

FIGURE 10.1 **(See color insert.)** One of the possible evolutions of the Proto-Indo-European root **ghArs-*, denoting a leguminous plant (Pokorny 1959, Nikolayev 2012), into Proto-Slavic and its contemporary descendants and some non-Slavic and constructed languages; the four basic meanings are written and drawn as pods, while their present words are rendered as flowers.

largest number of the names relating to pea and vetches (*Vicia* spp.), as described in more details in the Chapters 14 and 15.

It is noticeable that among the pan-Slavic names relating to the genus *Lathyrus*, as presented in the previous paragraph, the East Slavic languages are missing. There are two reasons for this. The first one is, as will be seen in the following passages, because the *Lathyrus* species in Belarusian, Russian, or Ukrainian are often regarded as sorts of pea or vetches. The second is the existence of another Proto-Slavic root, namely **lędo*, with a primeval meaning of *a wasteland plant* and, afterward, *birdsfoot trefoil* (*Lotus corniculatus* L.) (Vasmer 1955). While in Russian and Ukrainian the latter meaning was retained, this root produced the Belarusian *ljadzej*, the Polish *lędźwian*, and Pannonian Rusyn *liednek*, with a subsequent borrowing into a Uralic neighbor, Hungarian, as *lednek*, all of which denote the *Lathyrus* species.

Among the East Slavs, there is one more word specific for the *Lathyrus* species and with the unresolved etymology (Vasmer 1958): *čyna* in Belarusian, *china* in Russian, and *chyna* in Ukrainian, with a borrowing into the neighboring Uralic language, Erzya, as *čina*.

There is a very precious testimony, from both agronomic and linguistic viewpoints, that the species *Lathyrus japonicus* was used as a pulse crop. It is about the perishing Paleosiberian language isolate, Nivkh, and the Scmidt speech of its Northern Sakhalin dialect (Table 10.7), where this legume was used in various dishes (Shiraishi et al. 2016).

In a majority of the world's languages and in diverse ethnolinguistic families, the species of the genus *Lathyrus* are frequently associated with other grain legumes, such as with:

- Chickpea (*Cicer arietinum* L.), with the Arabic *alhumusy*, the Asturian *cicércula*, the English adjective *chickling*, the French *cicerole*, the Italian *cicerchia*, the Kazakh *noğatıq*, the Persian *nxud*, the Portuguese *chícharo,* and the Spanish *cicércula* and *garbanzo*;
- Lentil, with the Belarusian *sačavičnik*, English *lentil*, Moldovan *linte*, Russian *sochevichnik*, Spanish *lenteja* and Turkish *mercimek*, *mürdemeg*, *mürdümüğü*, or *mürdümük*, all being imports of the Persian *marcumak* or *mardumak* (Nişanyan 2017) and the Ukrainian *chechevychka*;
- White lupin (*Lupinus albus* L.), with the Arabic *turmos;*
- Common beans (*Phaseolus* spp.) or faba bean (*Vicia faba* L.), with the Asturian *fríjol*, the Chinese *dou*, the French *févette*, the Ontario French *frijol de olor*, the Icelandic *baunagrass,* and the Vietnamese *đậu*;
- Pea, with the Aragonese *guiseras*, the Armenian *tap'volorr*, the Asturian *albejana* or *almorta*, the Basque *ilar*, the Catalan *fesolet*, *pèsol,* or *pesolera*, the Chinese *wāndòu*, the Chuvash *pərçi*, the Danish *smååert*, the English *pea* and *peavine*, the Erzya *ksnavonj*, the Esperanto *pizo*, the Estonian *seahernes*, the Finnish *tuoksuherne*, the French *pois*, the German *Platterbse*, the Greek *mpizélo*, the Hindi *matter*, the Irish *pis*, the Istriot *bifèto*, the Japanese *endou*, the Jèrriais *pais*, the Latvian *zirnisi*, the Lithuanian *pelėžirnis*, the Occitan *pese*, the Ossetian *qædur*, the Piedmontese *pòis*, the Portuguese *ervilha* or *ervilhaca*, the Punjabi *mutter*, the Samogitian *žërnelē*, the Scots *pease*, the Serbian *grašak*, the Spanish *almorta*, *alverja*, *alverjón*, *arbejones*, *chícharo*, *guisante,* and many, many more, the Tagalog *gisante*, the Turkish *bizelya*, the Uzbek *burchoq,* and the Welsh *pysen* and *ytbysen*;
- To vetches, with the Belarusian *haroshak*, the Catalan *veça* and *veçó*, the English *vetch* and *vetchling*, the Erzya *vika*, the German *Sumpfwicke*, the Greek *víkos*, the Hungarian *bükköny*, the Italian *veccia*, the Romanian *mazaroi*, the Serbian *grahorica*, the Spanish *veza,* and the Ukrainian *horoshok*;

- To bitter vetch (*Vicia ervilia* [L.] Wildd.), with the English *black bitter vetch*, the French *orobe,* and the Turkish *burçak*;
- Both pea and vetch, with the Hungarian *borsóbükköny.*

It is noteworthy that, in some cases, the *Lathyrus* species are associated with animals, thus confirming their use as feed, such as with pigeons in French and horses in Spanish. In some South Slavic languages, *Lathyrus sativus.* brings a resemblance to an ax, due to the shape of its grain, and, along several other names, is called *sekirche* in Bulgarian, *sikirica* in Croatian, and *sekirče* in Serbian, all of which are, in fact, diminutive forms. Like many other pulses, the *Lathyrus* species are also imagined to have come from some other country or place, either faraway or quite close: Austria in English, Brittany in French, India in Welsh, Spain in Greek and French, and the city of Lisboa in Spanish.

FIGURE 1.1 Some of the most significant pulse crops, today and in the past: (from left to right and from above to below) *Arachis hypogaea* L., *Cajanus cajan* (L.) Huth, *Cicer arietinum* L., *Vicia ervilia* (L.) Willd., *Vicia faba* L., *Glycine max* (L.) Merr., *Lablab purpureus* (L.) Sweet, *Lathyrus odoratus* L., *Lens culinaris* Medik., *Lupinus albus* L., *Lupinus texensis* Hook., *Phaseolus lunatus* L., *Pisum sativum* L., *Vicia villosa* Roth, *Vigna angularis* (Willd.) Ohwi & H. Ohashi, *Vigna subterranea* (L.) Verdc.

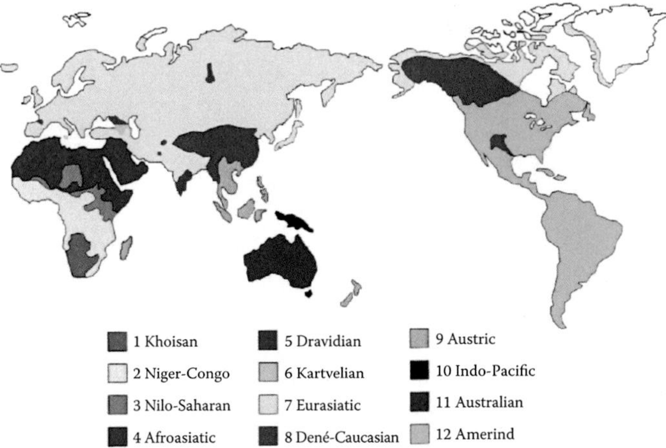

FIGURE 2.1 A simplified map of the major world ethnolinguistic families; *Eurasiatic* encompasses Indo-European, Altaic, Uralic, and Paleosiberian languages, *Dené-Caucasian* comprises Basque, Caucasian, Burushaski, Yenissenian, Sino-Tibetan, and Na-Dené languages, *Austric* denotes Austroasiatic and Austronesian languages and *Indo-Pacific* designates Andamanese, Trans-New Guinea, and Tasmanian languages. (Modified from Starostin, G., *The Tower of Babel, Evolution of Human Language Project*, http://starling.rinet.ru. With permission.)

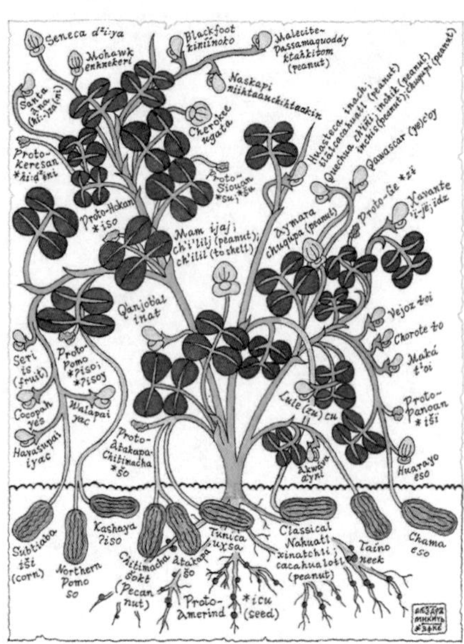

FIGURE 3.1 One of the possible evolutions of the root **icu* of the proposed Proto-Amerind language, denoting seed (Greenberg and Ruhlen 2007), into its extinct and modern descendants in North, Central, and South America; in all cases where the initial meaning was changed, the actual one is given within the brackets; the mediating protolanguages are shown as withered flowers, while the extinct languages are depicted as pods.

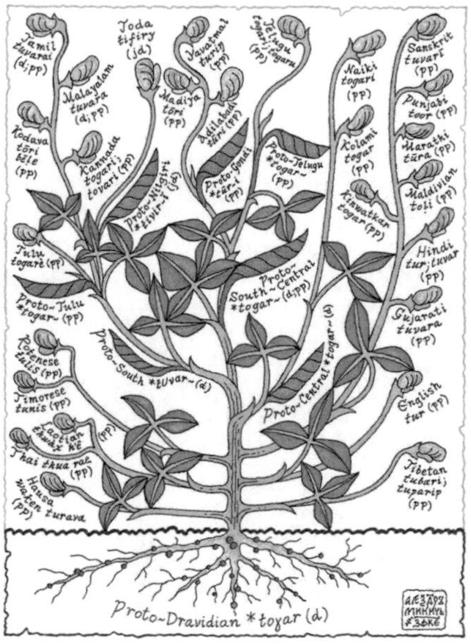

FIGURE 4.1 One of the possible evolutions of the Proto-Dravidian root **toɣar-*, denoting dal (Starostin 2006), into its direct derivatives, represented as pods, and contemporary descendants, portrayed as flowers, in the Central and Southern Dravidian languages; the meaning of each proto- and modern word is given within the brackets, namely (d) for *dal*, (jd) for *jungle pea,* and (pp) for *pigeon pea.*

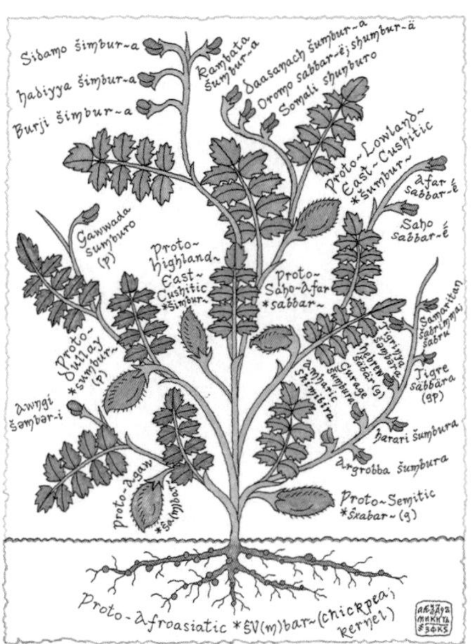

FIGURE 5.1 One of the possible evolutions of the Proto-Afroasiatic root **ŝV(m)bar-*, denoting both chickpea and kernel of corn (Militarev and Stolbova 2007), into its direct derivatives, interpreted as pods, and contemporary descendants, illustrated as flowers, in the Cushitic and Semitic languages; the meanings of each proto- and modern word are *chickpea*, if given without brackets, as well as *grain* (g), *grass pea* (gp), and *pea* (p).

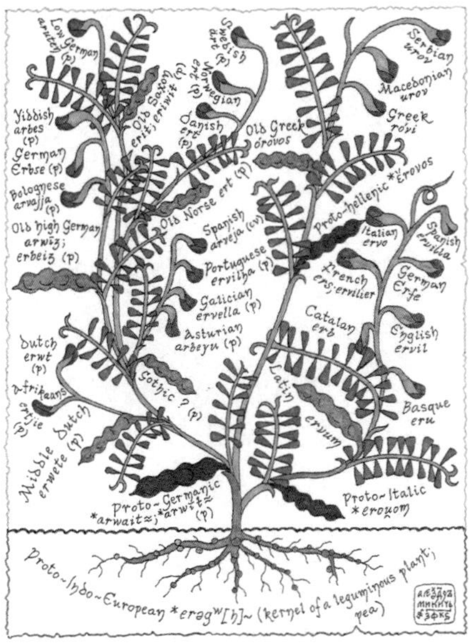

FIGURE 6.1 One of the possible evolutions of the Proto-Indo-European root **erəgʷ[h]-*, denoting both the kernel of a leguminous plant and pea (Pokorny 1959, Nikolayev 2012), into its direct derivatives, drawn as pods, and contemporary descendants, rendered as flowers, in the Indo-European languages; the meanings of the proto- and modern words are *bitter vetch*, if given without brackets, as well as *common vetch* (cv) and *pea* (p).

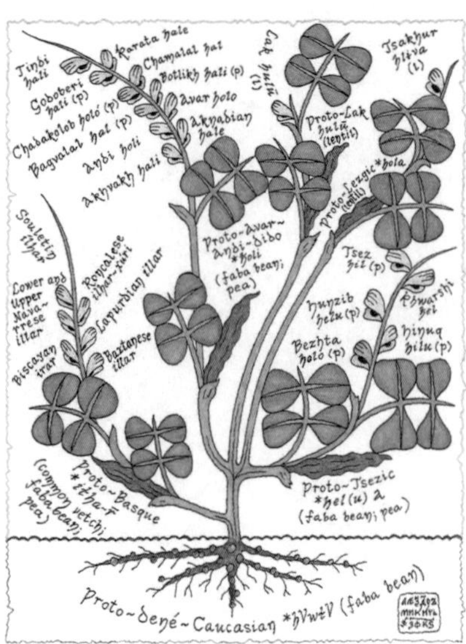

FIGURE 7.1 One of the possible evolutions of the root **hVwƛV* of the proposed Proto-Dené-Caucasian language, denoting faba bean (Starostin 2015), into its direct derivatives, shown as pods, and contemporary descendants, depicted as flowers, in the Basque and Caucasian languages; the meanings of the proto-words are given within brackets, while the meanings of the modern words are *faba bean*, if given without brackets, as well as *lentil* (l) and *pea* (p).

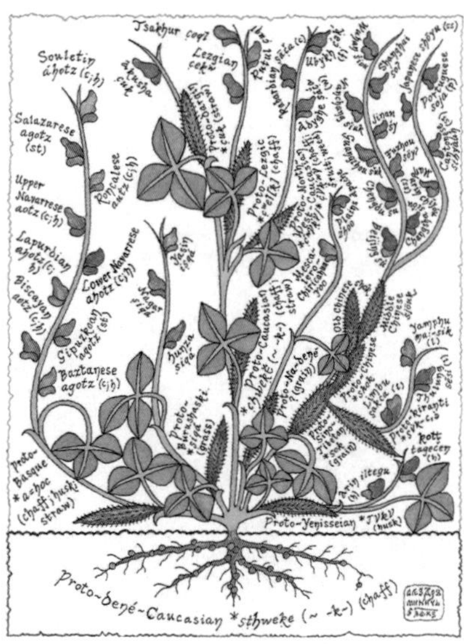

FIGURE 8.1 One of the possible evolutions of the root **sṭHwekĔ (~ -k-)* of the proposed Proto-Dené-Caucasian language, denoting chaff (Starostin 2015), into its direct derivatives, represented as pods, and contemporary descendants, portrayed as flowers, in the Basque, Burushaski, Caucasian, Na-Dené, Sino-Tibetan, and Yenisseian languages; the meanings of the proto-words are given within brackets and the meaning of each modern word, if given without brackets, is the same in its proto-word, while with the bracketed abbreviations, such as (c) for *chaff,* (f) for *fruit*; (g) for *grain*, (h) for *husk*, (l) for *lentil*, (p) for *soybean plant*, (ss) for *soybean sauce,* and (w) for *weed*, are given to mark the distinction where needed.

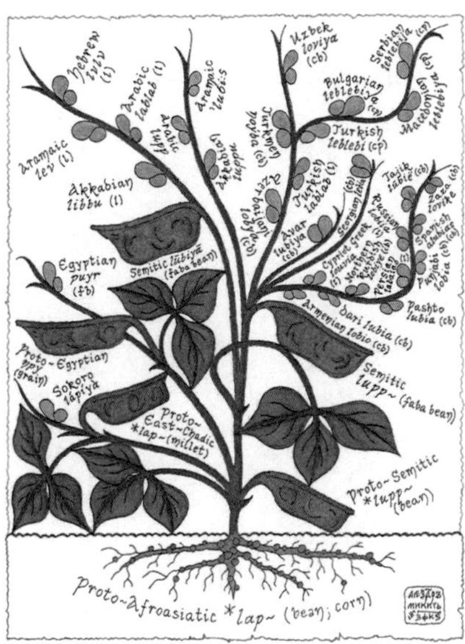

FIGURE 9.1 One of the possible evolutions of the Proto-Afroasiatic root **lap-* originally denoting a bean-like pulse and corn (Militarev and Stolbova 2007), into its direct derivatives, interpreted as pods, and contemporary descendants, illustrated as flowers, in the Chadic, Egyptian, and Semitic languages and with borrowings in other ethnolinguistic families; the meanings of the proto-words are given within brackets, while the meanings of the modern words are the same as in their own proto-words, if given without brackets, as well as *chickpea* (cp), *common bean* (cb), *faba bean* (fb), and *lablab* (l).

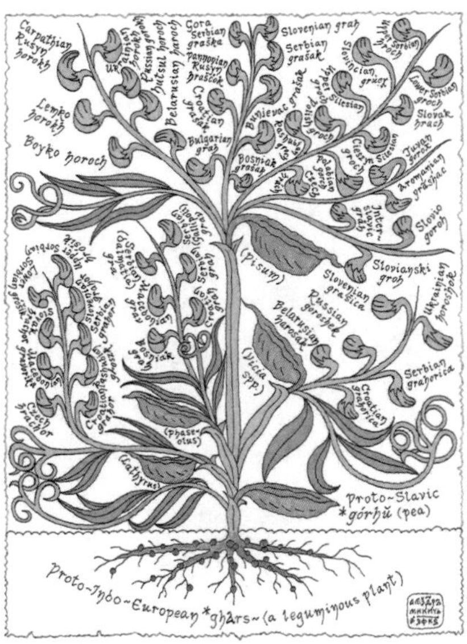

FIGURE 10.1 One of the possible evolutions of the Proto-Indo-European root **ghArs-*, denoting a leguminous plant (Pokorny 1959, Nikolayev 2012), into Proto-Slavic and its contemporary descendants and some non-Slavic and constructed languages; the four basic meanings are written and drawn as pods, while their present words are rendered as flowers.

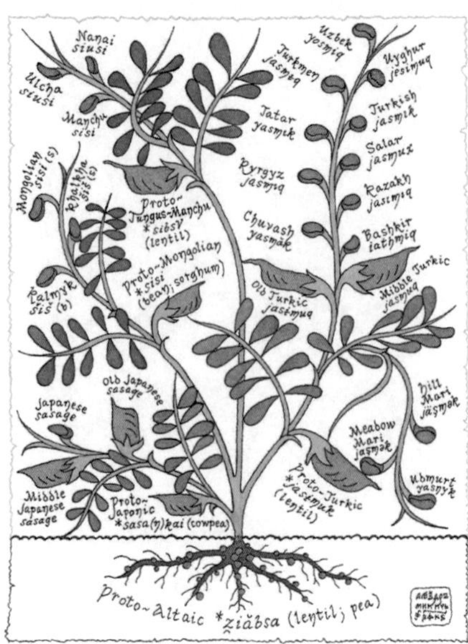

FIGURE 11.1 One of the possible evolutions of the root **zi̯ăbsa* of the proposed Proto-Altaic language, denoting lentil and pea (Starostin et al. 2003), into its direct derivatives, shown as pods, and contemporary descendants, depicted as flowers, in the Japonic, Mongolian, Tungus-Manchu, and Turkic languages; the meanings of the proto-words are given within brackets and the meaning of each modern word, if given without brackets, is the same in its proto-word, while with the bracketed abbreviations, such as (b) for *bean-like pulse* and (s) for *sorghum*, are given to mark the distinction where needed.

FIGURE 12.1 One of the possible evolutions of the root **tora* of the proposed Proto-Amerind language, denoting white color (Greenberg and Ruhlen 2007), into its modern descendants in North, Central and South America; the meanings of the modern words are *white*, if given without brackets, and *pearl lupine* (pl).

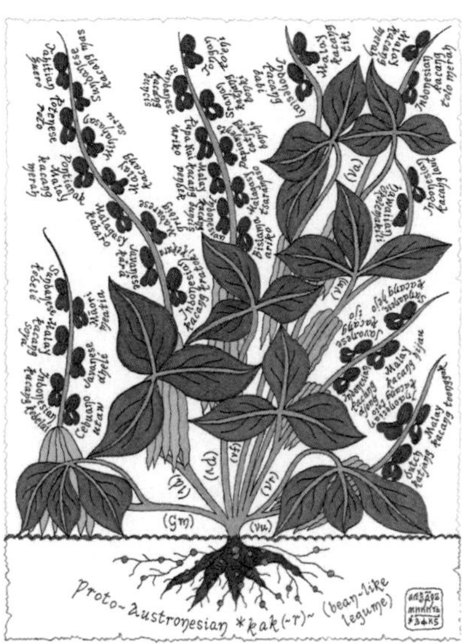

FIGURE 13.1 One of the possible evolutions of the hypothetical Proto-Austronesian morpheme denoting a bean-like legume into its contemporary descendants and some Indo-European languages; the six basic meanings are interpreted as branches with pods, with (Gm) for *Glycine max*, (Pl) for *Phaseolus lunatus*, (Pv) for *Phaseolus vulgaris*, (Vf) for *Vicia faba*, (Vr) for *Vigna radiata*, and (Vu) for *Vigna unguiculata*, while their present words are illustrated as flowers.

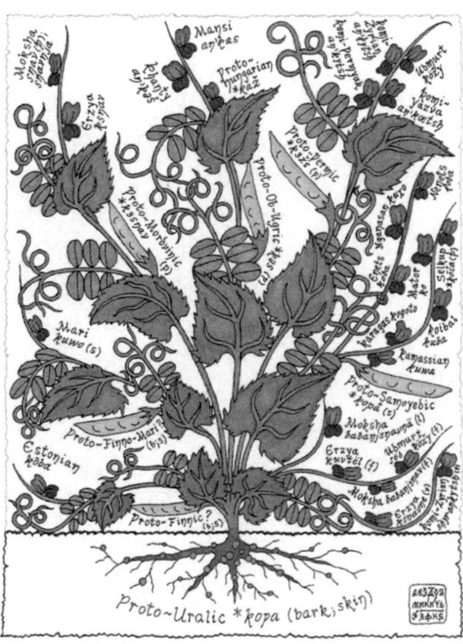

FIGURE 14.1 One of the possible evolutions of the Proto-Uralic root *kopa*, denoting bark and skin (Starostin 2006f), into its direct derivatives, drawn as pods, and contemporary descendants, rendered as flowers, in the Finnic, Finno-Mari, Mordivinic, Ob-Ugric, Permic, and Samoyedic languages; the meanings of the proto-words are given within brackets and the meaning of each modern word, if given without brackets, is the same in its proto-word, while with the bracketed abbreviations, such as (b) for *bark*, (f) for *faba bean*, (l) for lentil, (p) for *pea*, (s) for *skin*, and (v) for *vetches*, are given to mark the distinction where needed.

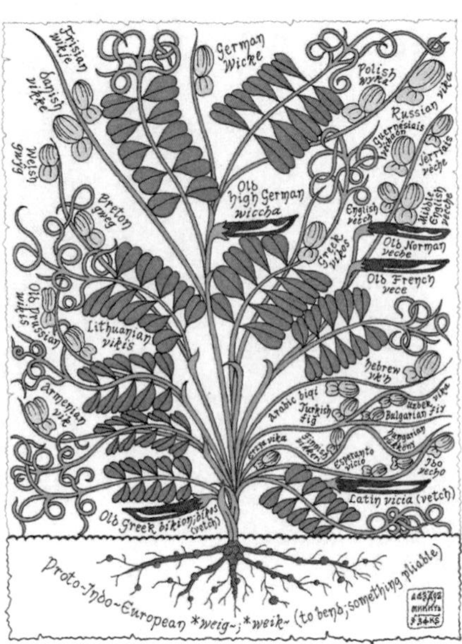

FIGURE 15.1 One of the possible evolutions of the Proto-Indo-European root **weig-*, **weik-*, denoting bending and something pliable (Militarev and Stolbova 2007), into its direct and indirect derivatives, shown as pods, and contemporary descendants, depicted as flowers, in the Indo-European, some non-Indo-European and constructed languages; the meanings of each proto-word and modern word is *vetch*.

FIGURE 16.1 One of the possible evolutions of the hypothetical Proto-Niger-Congo morpheme denoting something kidney-like and, possibly, grain of the *Vigna* species, into its contemporary descendants and with borrowing into French; the words associated with *Vigna subterranea* are represented as pods, while those referring to *Vigna unguiculata* are portrayed as flowers.

11 *Lens* Mill.

Synonyms: *Cicer* L.; *Ervum* L.; *Lathyrus* L.; *Lentilla* W. Wight ex D. Fairchild; *Vicia* L.

11.1 LIST OF TAXA SCIENTIFIC AND POPULAR NAMES

The first section of this chapter presents an overview of one of the less-abundant genera within the tribus *Fabeae* Rchb. and among the pulse crops. However, one of its species plays one of the most important economic roles in human diets throughout history and across the globe (Shahwar et al. 2017), as witnessed by the extraordinary richness in its vernacular names in the Old World languages and other linguistic categories (ISTA 1982, Rehm 1994, Erskine et al. 1998, Ferguson et al. 2000, Gledhill 2008, Porcher 2008, Reddy et al. 2010, The Plant List 2013, Ecocrop 2017, EPPO 2017, Ethnologue 2017, IBIS 2017, ILDIS 2017, Logos 2017, NPGS 2017, Wikipedia 2017, Wiktionary 2017). The other species of this g fakí; fakós o mageirikós enus may serve as gene pools of many desirable traits of the highest agronomic significance for introgressing into their cultivated relative (Gorim and Vandenberg 2017).

- *Lens culinaris* Medik.
English: lentil; common lentil; cultivated lentil
- *Lens culinaris* Medik. subsp. *culinaris* (Table 11.1)
Synonyms: *Cicer lens* (L.) Willd.; *Ervum lens* L.; *Lens esculenta* Moench; *Lens lens* Huth; *Lentilla lens* (L.) W. Wight ex D. Fairchild; *Vicia lens* (L.) Coss. & Germ.
- *Lens culinaris* Medik. subsp. *orientalis* (Boiss.) Ponert
Synonyms: *Ervum cyanea* Boiss. & Hohen.; *Ervum cyaneum* Boiss. & Hohen.; *Ervum orientale* Boiss.; *Ervum orientalis* Boiss.; *Lens cyanea* (Boiss. & Hohen.) Alef.; *Lens orientalis* (Boiss.) Hand.-Mazz.; *Lens orientalis* (Boiss.) Popov; *Lens orientalis* (Boiss.) Schmalh.; *Lens orientalis* (Boiss.) Schmalh. var. *cyaneum* (Boiss. & Hohen.) Popov; *Vicia orientalis* (Boiss.) Beg. & Diratz.
English: eastern lentil
Russian: chechevitsa golubaia; chechevitsa vostochnaia
Tajik: nask
Uzbek: burchak
- *Lens culinaris* Medik. subsp. *tomentosus* (Ladiz.) M. E. Ferguson et al.
Synonyms: *Lens tomentosus* Ladiz.
English: rough-haired lentil
Russian: chechevitsa linzoobraznaia

TABLE 11.1
Popular Names Denoting *Lens culinaris* Medik. Subsp. *culinaris* in Some World Languages and Other Linguistic Taxa

Language/Text	Name
Afrikaans	lensie; linz
Aka-Jeru	itbe:c'
Akkadian	mangu
Akusha	hulu-qara
Albanian	thjerrëz
Albanian (Arvanitika)	fierre; thierre
Amharic	misiri
Apulian	lendecchie
Arabic	adas; magg-
Aragonese	lentella; lentellera; lentilla; lentillera
Armenian	vosp; vosp utvokh
Aromanian	linti
Avar	çipalo
Azerbaijani	adi merchi; mərcimək
Bashkir	iathmiq
Basque	dilista
Belarusian	čačavica zvyčajnaja; sačavica; sačjouka
Bengali	masura ḍāla
Berber	taniltit
Bihari	dāli; ghaṁgharā
Bosniak	leća
Brescian	lentécia
Breton	pizenn rous
Budels	leenze
Bulgarian	leshta
Calabrian	lintìcchia
Caló	arités
Catalan	llentia; llentilla
Caterisani	lenticchja
Cebuano	balatong
Chechen	ẋozijn qo'š
Chewa	khobwe
Chinese (Cantonese)	xiăo biăndòu
Chinese (Mandarin)	biăndòu; bing dou; xiăo biăndòu
Chuvash	yasmăk
Colognian	Linze
Corsican	lintichja
Crimean Tatar	bercimek; mercimek
Croatian	leća

(Continued)

TABLE 11.1 (*Continued*)
Popular Names Denoting *Lens culinaris* Medik. Subsp. *culinaris* in Some World Languages and Other Linguistic Taxa

Language/Text	Name
Czech	čočka jedlá; čočka kuchyňská
Danish	linser
Dari	daal
Dutch	linze
Elfdalian	lins
English	common lentil; gram; lentil; lentille
Esperanto	lento; lentugo
Estonian	lääts
Finnish	linssi; kylvövirvilä; rokkalinssi
Flemish	lins
French	lentille; lentille comestible; lentille cultivée; lentillon
Friulian	lint
Gagauz	mercimek
Galician	lentella
Genoese	lentìggia
Georgian	ospis
German	Erve; Küchen-Linse; Linse
Greek	fakí; fakós o mageirikós
Greenlandic	eertaasaq
Griko	facì
Guarani	kumanda mirĩ
Gujarati	khādyānna; masura
Haitian Creole	lantiy
Hawaiian	papapa
Hebrew	adash; dshh trvvtt
Hindi	masur
Hungarian	főzeléklencse; lencse
Icelandic	linsubaun
Ilocano	lentéhas
Indonesian	kacang-kacangan; miju-miju
Irish	lintile
Istriot	lènto
Italian	lenticchia; lenticchia coltivata
Japanese	aoi mame; renzu mame
Javanese	kacang abang
Joratian	lintelye; neintelye
Kalmyk	nyet ulan burtsg
Kannada	masūra avare; thogare baylea

(*Continued*)

TABLE 11.1 (*Continued*)
Popular Names Denoting *Lens culinaris* Medik. Subsp. *culinaris* in Some World Languages and Other Linguistic Taxa

Language/Text	Name
Kapampangan	malobias
Kashubian	jôdnô soczewica
Kazakh	azıqtıq jasımıq; jasmıktan jasalgan; jasımıq
Khmer	sandekbay
Korean	len jeu kong; len se kong
Kurdish (Central)	nysk
Kurdish (Northern)	nîsk
Kyrgyz	jasmıq
Lak	hulū
Lao	fak
Latin	lēns; lenticula
Latvian	ēdamās lēcas; lēcas
Ligurian	lentìggia
Limbu	sakca
Limburgish	linze
Lithuanian	lęšis; paprastasis lęšis; valgomasis lęšis
Luxembourgish	lëns
Macedonian	kujnska lekja; lekja; obichna lekja
Malagasy	voa
Malay	kacang merah
Malayalam	payar̲
Maltese	għads
Manchu	sisi
Māori	pi
Mapundungun	jügi
Marathi	masoor; masūra
Mari (Hill)	jäşmək
Mari (Meadow)	jaşmək
Mazanderani	m'rd'ci
Meänkieli	linsi
Modenese (Eastern)	lèinta
Moksha	babanjsnavnä; tsetsavitsa
Mongolian	sevegzaram; khünsnii sevegzaram
Myanmar	pell hainn
Nahuatl	caxtillān pitzāhuac etl; ixtehuiloetl
Nanai	siusi
Nepali	musurō
Norwegian (Bokmål)	linse
Occitan	lentilha; lentilha cultivada

(*Continued*)

TABLE 11.1 (*Continued*)
Popular Names Denoting *Lens culinaris* Medik. Subsp. *culinaris* in Some World Languages and Other Linguistic Taxa

Language/Text	Name
Odia	masura
Otomi (Northwestern)	lentejas
Papiamento	lenteha
Pashto	nsk
Pennsylvania German	Linse
Persian	adas; marcumak; mardumak
Piedmontese	lentìa
Polish	soczewica jadalna
Portuguese	lentilha
Punjabi (Eastern)	dāḻa
Punjabi (Western)	masoor
Quechua	lantrihas; lintija
Romagnol	zizaròl
Romani	linta
Romanian	linte
Russian	chechevitsa kulturnaia; chechevitsa obyknovennaia; chechevitsa pishchevaia; liacha; liatsa
Rusyn (Pannonian)	lencha
Salar	jasmux
Sami (Inari)	linssä
Sami (Northern)	linsa
Sami (Skolt)	linsi
Sanskrit	dālaḥ; masūrakaḥ
Sardinian (Campidanese)	gentilla; gintilla
Scottish Gaelic	leantailean
Serbian	leća; sočivo
Serbian (Gallipoli)	sȍčivo
Sesotho	lensisi
Shona	nenyemba
Sindhi	dal
Sinhalese	parippu
Slovak	šošovica jedlá
Slovenian	leča; navadna leča
Somali	misir
Sorbian (Lower)	sok
Sorbian (Upper)	čóčk; čóčka; čok; sočk; sóčk; sok
Spanish	lenteja
Swahili	dengu; icho; mdengu
Swedish	åkerlins; lins
Tagalog	lentehas

(*Continued*)

TABLE 11.1 (*Continued*)
Popular Names Denoting *Lens culinaris* Medik. Subsp. *culinaris* in Some World Languages and Other Linguistic Taxa

Language/Text	Name
Tajik	nask
Tamil	maicūrp paruppu; misurupur
Tatar	kırlı borçak; yasmık
Telugu	pappu
Thai	mĕd ǐhåw
Thulung	sēsi
Tibetan	sran leb chung ngu
Tigrinya	birisini
Tsakhur	hlɨwa
Tswana	aditi chaddi
Turkic (Middle)	jasmuq
Turkic (Old)	jasɨmuq
Turkish	mercimek; jasmık
Turkmen	merjimek; jasmɨq
Udmurt	yasnyk
Ukrainian	sochevitsa kharchova; sochevitsa zvichaina
Ulch	siusi
Urdu	daal
Uyghur	jesimuq
Uzbek	burchak; yasmiq; yosmiq
Valencian	llentilla
Venetian	lentichia; lente
Vietnamese	thiết đậu
Walloon	lintile
Welsh	corbysen; corbysen corbys; ffacbysen; pysen felen
Xhosa	neentlumaya; nesityu
Yamphu	ma:-sik
Yiddish	lnzn
Yoruba	ẹwà
Yucatec	lentejas
Zaza	mercu
Zulu	amalentili; nesitshulu

- *Lens ervoides* (Brign.) Grande

Synonyms: *Cicer ervoides* Brign.; *Ervum ervoides* (Brign.) Hayek; *Ervum hohenakeri* Fisch. & C. A. Mey.; *Ervum lenticula* Schreb. ex Hoppe; *Ervum lenticula* Schreb. ex Sturm; *Lens lenticula* (Schreb.) Alef.; *Lens nigricans* (M. Bieb.) Godr. subsp. *ervoides* (Brign.) Ladiz.; *Vicia ervoides* (Brign.) Fiori; *Vicia lenticula* (Schreb.) Janka

English: ervum-like lentil
Russian: chechevitsa linzoobraznaia
- *Lens himalayensis* Alef.
English: Himalayan lentil
- *Lens lamottei* Czefr.
Synonyms: *Lens tenorei* auct.
English: Lamotte's lentil
- *Lens montbretii* (Fisch. & C. A. Mey.) P. H. Davis & Plitmann
English: Montbret's lentil
- *Lens nigricans* (M. Bieb.) Webb & Berthel.
Synonyms: *Ervum lentoides* Ten.; *Ervum nigricans* M. Bieb.; *Ervum sylvaticum* Fisch.; *Lathyrus nigricans* (M. Bieb.) Peterm.; *Lens culinaris* Medik. subsp. *nigricans* (M. Bieb.) Thell.; *Lens esculenta Medik.* subsp. *nigricans* (M. Bieb.) Thell.; *Lens tenorei* Lamotte; *Lens villosa* (Pomel) Batt.; *Vicia marschallii* Arcang.; *Vicia nigricans* (M. Bieb.) Janka
English: blackish lentil
Russian: chechevitsa chernovataia
- *Lens odemensis* Ladiz.
Synonyms: *Lens culinaris* subsp. *odemensis* (Ladiz.) M. E. Ferguson et al.
English: swollen lentil

11.2 ORIGIN OF SCIENTIFIC AND POPULAR TAXA NAMES

Cultivated lentil (*Lens culinaris* Medik.) is a crop with a considerable and reliable material for both etymological and lexicological research in diverse ethnolinguistic families. Several root words relating to lentil were assessed in Afroasiatic, Altaic, Caucasian, Indo-European, and Sino-Tibetan protolanguages.

One of the two attested Proto-Afroasiatic roots that are associated with lentil is **ʕadas-*, initially denoting faba bean (*Vicia faba* L.) (Militarev and Stolbova 2007, Mikić 2010). It shifted meaning to *lentil* in one of its derivatives, namely Proto-Semitic, with **ʕadaš-*, which gave the modern words in Arabic, Hebrew, and Maltese and was borrowed by the Indo-European Persian (Table 11.1). Another Proto-Afroasiatic root is **mang-*, originially designing lentil and millet (*Panicum miliaceum* L.). It brought forth the Proto-Semitic **mang-/*magg-*, referring to lentil and faba bean, with the first meaning retained in Akkadian and Arabic.

The Proto-Altaic root **zi̯ăbsa* denoted equally lentil and pea (*Pisum sativum* L.) and was a direct ancestor of two other ancient languages, which had a word relating to lentil (Figure 11.1). One of them is Proto-Turkic, with **jasi-muk*, that, through the Old Turkic *jasɨmuq* and Middle Turkic *jasmuq* (Starostin et al. 2003, 2006a), gave the modern words with the same meaning in Bashkir, Chuvash, Crimean Tatar, Kazakh, Kyrgyz, Salar, Tatar, Turkish, Turkmen, Uyghur, and Uzbek (Table 11.1) (Mikić and Perić 2011). Its basic form was also exported into some neighboring Uralic languages, such as Hill and Meadow Mari and Udmurt. Another is the Proto-Tungus-Manchu, with **sibsV*, which still persists, with unchanged meaning, in highly endangered languages Manchu, Nanai, and Ulcha (Mikić and Perić 2012).

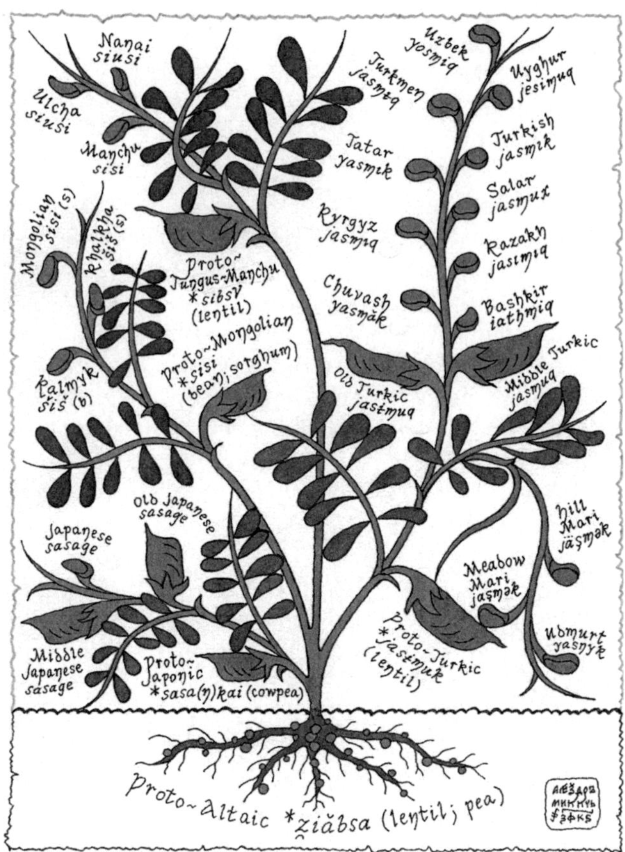

FIGURE 11.1 **(See color insert.)** One of the possible evolutions of the root **ziăbsa* of the proposed Proto-Altaic language, denoting lentil and pea (Starostin et al. 2003), into its direct derivatives, shown as pods, and contemporary descendants, depicted as flowers, in the Japonic, Mongolian, Tungus-Manchu, and Turkic languages; the meanings of the proto-words are given within brackets and the meaning of each modern word, if given without brackets, is the same in its proto-word, while with the bracketed abbreviations, such as (b) for *bean-like pulse* and (s) for *sorghum*, are given to mark the distinction where needed.

Although primarily referring to pea and faba bean in its modern representatives, the Proto-Caucasian root **hōwƚ[ā]* also denoted lentil (Starostin 2015). This meaning has been preserved in some of its direct descendants, such as the Proto-Lak *hulū* and the Proto-Lezgic **hola*, with the modern words in Lak and Tsakhur, respectively (Table 11.1) (Mikić and Vishnyakova 2012). The Dargwa dialect Akusha, most likely, borrowed its name relating to lentil from Lak (Starostin 2005d).

Lentil seems to have had a very prominent place in the everyday diets of the Proto-Indo-Europeans, as witnessed by the attested Proto-Indo-European root **lent-*, **lent-s-* (Pokorny 1959, Nikolayev 2012), defining lentil. Moreover, it produced several direct descendants with countless modern representatives, all in which the basic morpheme and the meaning have remained the same, despite the vast mutual spatial and temporal sundering that followed the Indo-European migrations, millennia ago. Among those are the following protolanguages of individual branches of the Indo-European ethnolinguistic family and their contemporary members (Table 11.1):

- The Proto-Baltic **leñš-ia-*, with Latvian and Lithuanian;
- The Proto-Germanic **lins-ī(n-)*, deriving into the Old High German *linsī* and *linsin* and the Middle High German *lins(e)*, with Afrikaans, Budels, Colognian, Danish, Dutch, Flemish, German, Icelandic, Limburgish, Luxembourgish, Norwegian, Pennsylvania German, Swedish, and Yiddish, and with exports to the Niger-Congo Sesotho and Zulu, and the Uralic Estonian, Finnish, Meänkieli and Inari, Northern and Skolt Sami;
- The Italic Latin *lēns*, which is the origin of the Linnean corresponding taxon name (Linnaeus 1753, 1758), with Apulian, Aragonese, Brescian, Calabrian, Catalan, Caterisani, Corsican, French, Friulian, Galician, Genoese, Istriot, Italian, Jorat, Ligurian, Eastern Modenese, Occitan, Piedmontese, Portuguese, Romanian, Sardinian, Spanish, Valencian, Venetian, and Walloon, and with borrowings into the Goidelic Celtic Irish and Scottish Gaelic, the Germanic English, the Indo-Aryan Romani, the Amerind Otomi, Quechua and Yucatec, the Austronesian Ilocano and Tagalog, the creoles Haitian Creoles and Papiamento, and the constructed Esperanto;
- The Proto-Slavic **lętjā* and the Old Church Slavonic *ljašta*, with Bulgarian, Croatian, Macedonian, Russian, Pannonian Rusyn, Serbian, and Slovenian, and with transferring into the Uralic Hungarian and Moksha.

Greek and its southern Italian offspring, Griko, have their names relating to lentil derived from the Proto-Indo-European root **bhabh-*, *bhabhā*, used to denote faba bean (Pokorny 1959, Nikolayev 2012) and thus is almost the only of its descendants without preserving the primeval meaning (see Chapter 7), but with a shift from faba bean to lentil (Table 11.1).

The Persian words for lentil, *marcumak* or *mardumak*, are present in some other Iranian languages, such as Mazanderani and Zaza (Table 11.1), and were borrowed by some Turkic neighbors, such as Gagauz, Turkish, and Turkmen (Nişanyan 2017).

Determining the etymology of the names designating lentil in some Indo-Aryan languages still seem unanswered. Such are those in Bengali, Gujarati, Hindi, Marathi, Nepali, Odia, Western Punjabi, and Sanskrit, as well as in the neighboring Dravidian languages Kannada and Tamil, with an approximate basic morpheme of *mas-* (Table 11.1). Could this be a derivation of the words referring to a pulse with an almost the same importance and way of use that is pea? In addition, could the crop or the word or both come from the Indian subcontinent via ocean to the Horn of Africa and leave the traces, as may be hinted from the words denoting lentil in the Afroasiatic Amharic and Somali? Some suggestions regarding this curious issue are presented in Chapter 14.

There is another Proto-Slavic root relating to lentil, **sočevica* (Vasmer 1955), with the modern descendants present in Belarusian, Czech, Kashubian, Polish, Russian, Serbian, Slovak, Lower and Upper Sorbian and Ukrainian. Its own origin is still unresolved, with the Proto-Indo-European root **s(w)ok^w^-*, meaning *juice* or *plant sap*, as one of the possible sources, although without any firm connection established so far (Nikolayev 2012).

The sole word referring to lentil in the proposed geographically scattered and linguistically abundant Indo-Pacific superstratum is the recently extinct Aka-Jeru, a member of the northern branch of the Great Andamanese language family (Table 11.1).

Being both geographically adjoining and belonging to different ethnolinguistic families, the Indo-European Armenian and the Kartvelian Georgian share the same word denoting lentil (Table 11.1), the origin of which remains insufficiently clarified.

Within Sino-Tibetan languages, the attested Proto-Kiranti root **sV̀k-c₁ə̀*, relating generally to seed, produced the word meaning exclusively lentil in one of its contemporary members, Limbu (Table 11.1) (Starostin 2005c).

Similarly to other pulses and various crops, the names for lentil in many languages are based upon the words denoting another plant, similar in grain shape or way of use (Table 11.1). In the case of lentil, we meet the names based upon those of:

- Pigeon pea (*Cajanus cajan* [L.] Huth.), such as in the Dravidian Kannada;
- Chickpea (*Cicer arietinum* L.), such as in the Romance Romagnol;
- Pea, such as in the Altaic Kalmyk, Tatar and Uzbek, the Caucasian Chechen, the Eskimo-Aleut Greenlandic, the Indo-European Brythonic Celtic Breton and Welsh, and the Uralic Moksha;
- Bean in a broad sense, such as in the Austroasiatic Khmer and Vietnamese, the Sino-Tibetan Cantonese and Mandarin Chinese, and the Tai-Kadai Thai language.

One of the rare examples where the name of the supposed country of origin constitutes the name for lentil is the Amerind Nahuatl (Table 11.1): it properly addresses Castile, one of the historical regions in Spain, as the place where lentil came from to Mesoamerica.

12 *Lupinus* L.

12.1 LIST OF TAXA SCIENTIFIC AND POPULAR NAMES

The first segment of this chapter provides the reader with an insight into the remarkably complex botanical classification of the genus *Lupinus,* encompassing the species from the Old and New Worlds, and listing the taxa that are well established and widely accepted and used, as well as those that are still a kind of unresolved issue (Ainouche and Bayer 1999). The collections of the popular names for several economically most significant species are given in details in the corresponding tables (ISTA 1982, Rehm 1994, Gledhill 2008, Porcher 2008, Drummond et al. 2012, The Plant List 2013, Atchison et al. 2016, Ecocrop 2017, EPPO 2017, Ethnologue 2017, IBIS 2017, ILDIS 2017, Logos 2017, NPGS 2017, Wikipedia 2017, Wiktionary 2017).

- *Lupinus aberrans* C. P. Sm.
- *Lupinus abramsii* C. P. Sm.
English: Abrams' lupine
- *Lupinus acaulis* Larrañaga
- *Lupinus achilleaphilus* C. P. Sm.
- *Lupinus acutilobus* A. Heller
- *Lupinus acopalcus* C. P. Sm.
- *Lupinus adinoanthus* C. P. Sm.
- *Lupinus adsurgens* Drew
Arabic: trms mrtf
English: Drew's silky lupine
- *Lupinus aegr-Aovium* C. P. Sm.
- *Lupinus affinis* J. Agardh
Arabic: trms mtqarb; trms mtqrb
English: fleshy lupine
- *Lupinus africanus* Lour.
- *Lupinus agardhianus* A. Heller
- *Lupinus agninus* Gand.
- *Lupinus agropyrophilus* C. P. Sm.
- *Lupinus alaimandus* C. P. Sm.
- *Lupinus alaristatus* C. P. Sm.
- *Lupinus albert-smithianus* C. P. Sm.
- *Lupinus albescens* Hook. & Arn.
English: hoary lupine
- *Lupinus albicaulis* Douglas
Synonyms: *Lupinus albicaulis* var. *shastensis* (A. Heller) C. P. Sm.; *Lupinus shastensis* A. Heller

English: pine lupine; sickle-keel lupine
Turkish: çam acı baklası
- *Lupinus albifrons* Benth.
Arabic: trms byd alwraq
English: evergreen lupine; foothill lupine; silver lupine; white-leaf bush lupine
Lithuanian: kalvinis lubinas
Turkish: gümüşi çalı acı baklası
- *Lupinus albifrons* Benth. var. *albifrons*
Synonyms: *Lupinus albifrons* var. *eminens* (Greene) C. P. Sm.; *Lupinus eminens* Greene; *Lupinus tricolor* Greene
- *Lupinus albifrons* Benth. var. *douglasii* (J. Agardh) C. P. Sm.
Synonyms: *Lupinus douglasii* J. Agardh
- *Lupinus albopilosus* A. Heller
- *Lupinus albosericeus* C. P. Sm.
- *Lupinus albus* L. (Table 12.1)
- *Lupinus albus* L. var. *albus*
Synonyms: *Lupinus termis* Forssk.; *Lupinus termis* subsp. *termis* (Forssk.) Ponert.
English: common white lupin
- *Lupinus albus* L. var. *graecus* (Boiss. & Spruner) Gladst.

TABLE 12.1
Popular Names Denoting *Lupinus albus* L. in Some World Languages and Other Linguistic Taxa

Language/Taxon	Name
Arabic	trms bid; trms byd; turmus
Arabic (Egypt)	termes
Arabic (Jordan)	termos
Arabic (Lebanon)	termos
Arabic (Palestine)	ttermos
Arabic (Syria)	termos
Armenian	lupin spitak
Bashkir	aq æse borsaq
Basque	eskuzuria
Belarusian	bely lubin
Bengali	sādā lupina
Catalan	llobí; llobí blanc; llobí ver; lupí blanc; tramús; tramús blanc
Caterisani	luppinu
Chinese (Cantonese)	bái yŭshàn dòu
Chinese (Mandarin)	bái yŭshàn dòu
Croatian	bijela lupina
Czech	lupina bílá
Dutch	witte lupine

(*Continued*)

TABLE 12.1 (*Continued*)
Popular Names Denoting *Lupinus albus* L. in Some World Languages and Other Linguistic Taxa

Language/Taxon	Name
English	European white lupin; Mediterranean white lupin; white lupin; white lupine
Esperanto	lupeno blanka
Finnish	valkolupiini
French	lupin blanc
Galician	chícharo bravo; chícharo de raposo; tremoceiro común
Galician-Asturian	chícharo bravo
Georgian	t'et'ri khanchkola
German	Weiße lupine
Greek	leukó loúpino; thermos
Greek (Ancient)	thérmos
Hebrew	tvrmvs trvvt
Icelandic	hvítur álfabaunur
Ido	lupino blanko
Irish	lúipíní bán
Italian	lupino bianco
Japanese	byakka rupinasu
Kazakh	aq böriburşaq; aq lyupïn
Latin	lupīnus
Latvian	baltā lupīna
Lithuanian	baltažiedis lubinas
Mari (Hill)	osh pivyrsa
Norwegian (Nynorsk)	siciliansk lupin
Polish	łubin biały
Portuguese	tremoçeiro-branco
Romanian	lupin alb
Russian	liupin belyi
Serbian	bela lupina; bijela lupina; bila lupina
Sicilian	luppina; luppinu
Slovak	lupina biela
Sorbian (Upper)	běła lupina
Spanish	almorta; altramuz; altramuz blanco; chocho; chorcho; entremozo; lupín blanco; lupino blanco
Swahili	mlupini mweupe
Swedish	vitlupin
Turkish	beyaz acı bakla
Ukrainian	liupin bilyi
Vietnamese	chi đậu cánh chim trắng
Welsh	bysedd-y-blaidd gwyn

Synonyms: *Lupinus graecus* Boiss. & Spruner; *Lupinus jugoslavicus* Kazim. et Now.; *Lupinus vavilovii* Atab. et Mais.
English: Greek white lupin; wild white lupin
- *Lupinus alcis-montis* C. P. Sm.
- *Lupinus aliamandus* C. P. Sm.
- *Lupinus aliattenuatus* C. P. Sm.
- *Lupinus alibicolor* C. P. Sm.
- *Lupinus alicanescens* C. P. Sm.
- *Lupinus aliceae* C. P. Sm.
- *Lupinus aliclementinus* C. P. Sm.
- *Lupinus alilatissimus* C. P. Sm.
- *Lupinus alinanus* C. P. Sm.
- *Lupinus alipatulus* C. P. Sm.
- *Lupinus alirevolutus* C. P. Sm.
- *Lupinus aliumbellatus* C. P. Sm.
- *Lupinus alivillosus* C. P. Sm.
- *Lupinus allargyreius* C. P. Sm.
- *Lupinus alopecuroides* Desr.
English: hairless lupine
Spanish (*Colombia*): chocho
- *Lupinus* × *alpestris* (A. Nelson) D. B. Dunn & J. M. Gillett
- *Lupinus altimontanus* C. P. Sm.
- *Lupinus altiplani* C. P. Sm.
- *Lupinus altissimus* Sessé & Moc.
- *Lupinus alturasensis* C. P. Sm.
- *Lupinus alveorum* C. P. Sm.
- *Lupinus amabayensis* C. P. Sm.
- *Lupinus amabilis* A. Heller
- *Lupinus amandus* C. P. Sm.
- *Lupinus amboensis* C. P. Sm.
- *Lupinus ammophilus* Greene
- *Lupinus ammophilus* Greene var. *ammophilus*
Synonyms: *Lupinus polyphyllus* var. *ammophilus* (Greene) Barneby
- *Lupinus ammophilus* Greene var. *crassus* (Payson) Isely
Synonyms: *Lupinus crassus* Payson
- *Lupinus amniculi-cervi* C. P. Sm.
- *Lupinus amniculi-salicis* C. P. Sm.
- *Lupinus amniculi-vulpum* C. P. Sm.
- *Lupinus amnis-otuni* C. P. Sm.
- *Lupinus ampaiensis* C. P. Sm.
- *Lupinus amphibius* Suksd.
- *Lupinus ananeanus* Ulbr.
- *Lupinus anatolicus* W. Swiecicki & W. K. Swiecicki
English: Anatolian lupin
Turkish: Anadolu acı baklası
- *Lupinus andersonianus* C. P. Sm.

- *Lupinus andersonii* S. Watson
English: Anderson's lupine
- *Lupinus andicola* Gillies
- *Lupinus andinus* Rose ex J. F. Macbr.
- *Lupinus anemophilus* Greene
- *Lupinus angustiflorus* Eastw.
Arabic: trms dyq alzhar
English: narrowflower lupine
- *Lupinus angustifolius* Blanco
- *Lupinus angustifolius* L. (Table 12.2)
Synonyms: *Lupinus linifolius* Roth; *Lupinus varius* L.
- *Lupinus angustifolius* L. subsp. *angustifolius*
English: common narrow-leaved lupin
- *Lupinus angustifolius* L. subsp. *reticulatus* (Desv.) Arcang.
Synonyms: *Lupinus reticulatus* Desv.
- *Lupinus antensis* C. P. Sm.
- *Lupinus antiplani* C. P. Sm.
- *Lupinus antoninus* Eastw.
Arabic: trms ntuny
English: Anthony Peak lupine
- *Lupinus apertus* A. Heller
English: summit lupine
- *Lupinus aphronorus* Blank.
- *Lupinus apodotropis* A. Heller
- *Lupinus appositus* C. P. Sm.
- *Lupinus aralloius* C. P. Sm.
- *Lupinus arborescens* Amabekova & Maisuran
- *Lupinus arboreus* Sims
Arabic: trms shgri
English (UK): tree lupin
English (U.S.): bush lupine; yellow bush lupine
Esperanto: lupeno arbusta; lupeno ĉiamverda
German: Baum-Lupine
Swedish: gul busklupin
Turkish: sarı çalı acı baklası
Welsh: coeden bysedd y blaidd
- *Lupinus arbustus* Douglas ex Lindl.
Synonyms: *Lupinus arbustus* subsp. *neolaxiflorus* D. B. Dunn; *Lupinus laxiflorus* auct.
Arabic: trms shjiri
English: longspur lupine; spur lupine
Turkish: mahmuzlu acı bakla
- *Lupinus arbutosocius* C. P. Sm.
- *Lupinus arceuthinus* Greene
- *Lupinus archeranus* C. P. Sm.
- *Lupinus arcticus* S. Watson

TABLE 12.2
Popular Names Denoting *Lupinus angustifolius* L. in Some World Languages and Other Linguistic Taxa

Language/Taxon	Name
Adyghe	bgedaxeʃhuʒ ʃhəxu
Arabic	trms dyq lwrq
Asturian	altramuz azul
Azerbaijani	ensizyarpaq acıpaxla
Bengali	nīla lupina
Catalan	llobí bord; llobí pilós; llobí salvatge
Chinese	xiá yè yǔshàn dòu
Czech	lupina úzkolistá; škrkavičnik
Danish	smalbladet lupin
Dutch	blauwe lupine
English	blue lupin; blue lupine; European blue lupine; narrow-leaf lupin; narrow-leaved lupin; narrowleaf lupin; New Zealand blue lupin; sweet lupinseed
Esperanto	lupeno mallarĝfolia
Finnish	sinilupiini; täplälupiini
French	lupin à feuilles étroites; lupin bleu; lupin petit bleu; lupin réticulé
German	Blaue Lupine; Schmalblättrige Lupine
Greek	kyanó loúpino
Hebrew	tvrmvs tzr lm
Italian	lupino azzurro
Kazakh	jiñişke japıraqtı lyupïn
Latvian	zilā lupīna
Lithuanian	siauralapis lubinas
Navajo	azeediilch'íīlii
Polish	łubin niebieski; łubin wąskolistny
Portuguese	tremoçeiro-azul
Romanian	lupin albastru
Russian	liupin uzkolistnyi
Serbian	plava lupina; uskolisna lupina
Slovak	lupina úzkolistá
Sorbian (Lower)	módry škrěkawnik
Sorbian (Upper)	wulkołopjenata lupina
Spanish	altramuz azul
Swahili	mlupini majani-membamba
Swedish	blålupin; fingerlupin
Turkish	mavi acı bakla
Ukrainian	liupin synii; liupin vuz'kolystyi
Welsh	bysedd-y-blaidd culddail

English: Arctic lupine; subalpine lupine
French: lupin arctique
Polish: łubin północny
Swedish: arktisk lupin
Turkish: Arktik acı bakla
- *Lupinus arcticus* S. Watson subsp. *arcticus*
English: Arctic lupine
- *Lupinus arcticus* S. Watson subsp. *subalpinus* (Piper & B. L. Rob.) D. B. Dunn
Synonyms: *Lupinus subalpinus* Piper & B. L. Rob.
English: subalpine lupine
- *Lupinus arenarius* Gardner
- *Lupinus arequipensis* C. P. Sm.
- *Lupinus argenteus* Pursh
Arabic: trms fdy
English: silvery lupine
Navajo: azee'bíni'í
Turkish: gümüşi acı bakla
- *Lupinus argenteus* Pursh var. *argentatus* (Rydb.) Barneby
Synonyms: *Lupinus argenteus* subsp. *spathulatus* (Rydb.) L. W. Hess & D. B. Dunn; *Lupinus argenteus* var. *laxiflorus* (Douglas ex Lindl.) Dorn; *Lupinus spathulatus* Rydb.
- *Lupinus argenteus* Pursh var. *argenteus*
Synonyms: *Lupinus argenteus* var. *tenellus* (Douglas ex D. Don) D. B. Dunn; *Lupinus laxiflorus* Douglas ex Lindl.; *Lupinus tenellus* Douglas ex G. Don
- *Lupinus argenteus* Pursh var. *argophyllus* (A. Gray) S. Watson
Synonyms: *Lupinus caudatus* subsp. *argophyllus* L. Ll. Phillips
- *Lupinus argenteus* Pursh var. *depressus* (Rydb.) C. L. Hitchc.
Synonyms: *Lupinus depressus* Rydb.
- *Lupinus argenteus* Pursh var. *fulvomaculatus* (Payson) Barneby
Synonyms: *Lupinus argenteus* subsp. *ingratus* (Greene) Harmon; *Lupinus ingratus* Greene
English: yellow-spot lupine
- *Lupinus argenteus* Pursh var. *heteranthus* (S. Watson) Barneby
Synonyms: *Lupinus caudatus* Kellogg; *Lupinus caudatus* subsp. *caudatus* Kellogg; *Lupinus greenei* A. Nelson; *Lupinus meionanthus* var. *heteranthus* S. Watson
English: Kellogg's spurred lupine; Lake Tahoe lupine
- *Lupinus argenteus* Pursh var. *hillii* (Greene) Barneby
- *Lupinus argenteus* Pursh var. *holosericeus* (Nutt.) Barneby
Synonyms: *Lupinus holosericeus* Nutt.
- *Lupinus argenteus* Pursh var. *montigenus* (A. Heller) Barneby
Synonyms: *Lupinus caudatus* subsp. *montigenus* (A. Heller) L. W. Hess & D. B. Dunn; *Lupinus montigenus* A. Heller
- *Lupinus argenteus* Pursh var. *palmeri* (S. Watson) Barneby
Synonyms: *Lupinus palmeri* S. Watson
English: Palmer's silvery lupine

- *Lupinus argenteus* Pursh var. *rubricaulis* (Greene) S. L. Welsh
Synonyms: *Lupinus alpestris* A. Nelson; *Lupinus argenteus* subsp. *rubricaulis* (Greene) L. W. Hess & D. B. Dunn; *Lupinus rubricaulis* Greene
English: silvery lupine
- *Lupinus argurocalyx* C. P. Sm.
- *Lupinus argyraeus* DC.
- *Lupinus ariste-josephii* C. P. Sm.
- *Lupinus arizelus* C. P. Sm.
- *Lupinus arizonicus* (S. Watson) S. Watson
Synonyms: *Lupinus arizonicus* (S. Watson) S. Watson subsp. *sonorensis* J. A. Christian & D. B. Dunn; *Lupinus concinnus* var. *arizonicus* S. Watson
Arabic: trms Aryzuni
English: Arizona lupine; Sonora lupine
Turkish: Arizona acı baklası
- *Lupinus arvensi-plasketti* C. P. Sm.
- *Lupinus arvensis* Benth.
- *Lupinus asa-grayanus* C. P. Sm.
- *Lupinus aschenbornii* S. Schauer
English: Aschenborn's lupin
- *Lupinus asplundianus* C. P. Sm.
- *Lupinus asymbepus* C. P. Sm.
- *Lupinus atacamicus* C. P. Sm.
English: Atacama lupine
- *Lupinus atlanticus* Gladst.
English: Atlas lupin
Swahili: mlupini wa kiatlantiki
- *Lupinus atropurpureus* C. P. Sm.
- *Lupinus aureonitens* Hook. & Arn.
- *Lupinus aureus* J. Agardh
- *Lupinus austrobicolor* C. P. Sm.
- *Lupinus austrohumifusus* C. P. Sm.
- *Lupinus austrorientalis* C. P. Sm.
- *Lupinus austrosericeus* C. P. Sm.
- *Lupinus axillaris* Blank.
- *Lupinus ballianus* C. P. Sm.
- *Lupinus bandelierae* C. P. Sm.
- *Lupinus bangii* Rusby
- *Lupinus barbatilabius* C. P. Sm.
- *Lupinus barkeri* Lindl.
- *Lupinus barkeriae* Knowles & Westc.
- *Lupinus bartlettianus* C. P. Sm.
- *Lupinus bartolomei* M. E. Jones
- *Lupinus bassett-maguirei* C. P. Sm.
- *Lupinus beaneanus* C. P. Sm.
- *Lupinus benthamii* A. Heller
English: spider lupine

- *Lupinus bi-inclinatus* C. P. Sm.
- *Lupinus bicolor* Lindl.

Synonyms: *Lupinus bicolor* subsp. *tridentatus* (Eastw. ex C. P. Sm.) D. B. Dunn; *Lupinus micranthus* Douglas; *Lupinus polycarpus* Greene
Arabic: trms thnai allun
English: bicolored lupine; Lindley's annual lupine; miniature lupine; pigmy-leaved lupine
Esperanto: lupeno brunhara
Istriot: ulìn; vulìn
Turkish: minyatür acı bakla

- *Lupinus bicolor* Lindl. subsp. *microphyllus* (S. Watson) D. B. Dunn
- *Lupinus bicolor* Lindl. subsp. *pipersmithii* (A. Heller) D. B. Dunn
- *Lupinus bicolor* Lindl. subsp. *umbellatus* (Greene) D. B. Dunn
- *Lupinus bimaculatus* Desr.
- *Lupinus bimaculatus* Hook. ex D. Don
- *Lupinus bingenensis* Suksd.

Synonyms: *Lupinus leucopsis* var. *bingenensis* (Suksd.) C. P. Sm.
English: bingen lupine; Suksdorf's lupine

- *Lupinus bivonii* C. Presl
- *Lupinus blaisdellii* Eastw.
- *Lupinus blankinshipii* A. Heller
- *Lupinus blaschkeanus* Fisch. & C. A. Mey.
- *Lupinus bogotensis* Benth.

English: Bogotan lupine
Spanish (Colombia): altramuz; chochitos

- *Lupinus bolivianus* C. P. Sm.

Synonyms: *Lupinus macrostachys* Rusby
English: Bolivian lupine

- *Lupinus bombycinocarpus* C. P. Sm.
- *Lupinus bonplandius* C. P. Sm.
- *Lupinus boyacensis* C. P. Sm.

English: Boyaca lupine

- *Lupinus brachypremnon* C. P. Sm.
- *Lupinus bracteolaris* Desr.
- *Lupinus brandegeei* Eastw.
- *Lupinus brevecuneus* C. P. Sm.
- *Lupinus brevicaulis* S. Watson

Arabic: trms qsir alsaq
English: sand lupine; shortstem lupine
Turkish: kum acı baklası

- *Lupinus brevior* (Jeps.) Christian & D. B. Dunn
- *Lupinus breviscapus* Ulbr.
- *Lupinus breweri* A. Gray

English: Brewer's lupine; matted lupine

- *Lupinus brittonii* Abrams
- *Lupinus bryoides* C. P. Sm.

- *Lupinus buchtienii* Rusby
- *Lupinus burkartianus* C. P. Sm.
- *Lupinus burkeri* Lindl.
- *Lupinus caballoanus* B. L. Turner
- *Lupinus cachupatensis* C. P. Sm.
- *Lupinus cacuminis* Standl.
- *Lupinus caeruleus* A. Heller
- *Lupinus caesius* Eastw.
- *Lupinus calcensis* C. P. Sm.
- *Lupinus caldasensis* C. P. Sm.
- *Lupinus camiloanus* C. P. Sm.
- *Lupinus campbelliae* Eastw.

English: Campbell's lupin

- *Lupinus campestris* Schltdl. & Cham.
- *Lupinus campestris-florum* C. P. Sm.
- *Lupinus candicans* Rydb.
- *Lupinus canus* Hemsl.
- *Lupinus capitatus* Greene
- *Lupinus capitis-amniculi* C. P. Sm.
- *Lupinus carazensis* Ulbr.
- *Lupinus carchiensis* C. P. Sm.
- *Lupinus cardenasianus* C. P. Sm.
- *Lupinus carhuamayus* C. P. Sm.
- *Lupinus carlos-ochoae* C. P. Sm.
- *Lupinus carolus-bucarii* C. P. Sm.
- *Lupinus carpapaticus* C. P. Sm.
- *Lupinus carrikeri* C. P. Sm.
- *Lupinus caucensis* C. P. Sm.
- *Lupinus caudatus* Kellogg subsp. *cutleri* (Eastw.) L. W. Hess & D. B. Dunn
- *Lupinus cavicaulis* C. P. Sm.
- *Lupinus ccorilazensis* Vargas ex C. P. Smith
- *Lupinus celsimontanus* C. P. Sm.
- *Lupinus cervinus* Kellogg

Arabic: trms yli
English: Santa Lucia lupine

- *Lupinus cesar-vargasii* C. P. Sm.
- *Lupinus cesaranus* C. P. Sm.
- *Lupinus chachas* Ochoa ex C. P. Sm.
- *Lupinus chamissonis* Eschsch.

English: Chamisso bush lupine

- *Lupinus chavanillensis* (J. F. Macbr.) C. P. Sm.
- *Lupinus chiapensis* Rose
- *Lupinus chihuahuensis* S. Watson

English: Chihuaha lupine

- *Lupinus chipaquensis* C. P. Sm.
- *Lupinus chlorolepis* C. P. Sm.

- *Lupinus chocontensis* C. P. Sm.
- *Lupinus chongos-bajous* C. P. Sm.
- *Lupinus christianus* C. P. Sm.
- *Lupinus christinae* A. Heller
- *Lupinus chrysanthus* Ulbr.
- *Lupinus chrysocalyx* C. P. Sm.
- *Lupinus chrysomelas* Casar.
- *Lupinus chumbivilcensis* C. P. Sm.
- *Lupinus citrinus* Kellogg

Arabic: trms brtqaly
English: fragrant lupine; orange lupine; orangeflower lupine
- *Lupinus clarkei* Oerst.

English: Clarke's lupin
- *Lupinus clementinus* Greene
- *Lupinus cochapatensis* C. P. Sm.
- *Lupinus colcabambensis* C. P. Sm.
- *Lupinus collinus* (Greene) A. Heller
- *Lupinus colombiensis* C. P. Sm.

English: Colombian lupine
- *Lupinus comatus* Rydb.
- *Lupinus compactiflorus* Rose
- *Lupinus comptus* Benth.
- *Lupinus condensiflorus* C. P. Sm.
- *Lupinus concinnus* J. Agardh

Adyghe: bgedaxeʃhuʒ ɭaχħe
Arabic: trms mrtb
English: Bajada lupine
Spanish: lupino bajada
- *Lupinus concinnus* J. Agardh subsp. *concinnus*
- *Lupinus concinnus* J. Agardh subsp. *orcuttii* (S. Watson) D. B. Dunn

Synonyms: *Lupinus orcuttii* S. Watson
- *Lupinus congdonii* (C. P. Sm.) D. B. Dunn
- *Lupinus conicus* C. P. Sm.
- *Lupinus consentinii* Walp.
- *Lupinus constancei* T. W. Nelson & J. P. Nelson

English: lassicus lupine; the Lassics lupine
- *Lupinus convencionensis* C. P. Sm.
- *Lupinus cookianus* C. P. Sm.
- *Lupinus coriaceus* Benth.

Synonyms: *Lupinus attenuatus* Gardner
- *Lupinus cosentinii* Guss.

English: hairy blue lupin; hairy lupin; sandplain lupin; West Australian blue lupin
Finnish: sisilianlupiini
German: bunte Lupine
- *Lupinus costaricensis* D. B. Dunn
- *Lupinus cotopaxiensis* C. P. Sm.

- *Lupinus couthouyanus* C. P. Sm.
- *Lupinus covillei* Greene
English: shaggy lupine
- *Lupinus crassulus* Greene
- *Lupinus croceus* Eastw.
English: saffron-flowered lupine
- *Lupinus crotalarioides* Benth.
- *Lupinus crucis-viridis* C. P. Sm.
- *Lupinus cuatrecasasii* C. P. Sm.
- *Lupinus culbertsonii* Greene
- *Lupinus culbertsonii* Greene subsp. *culbertsonii*
- *Lupinus culbertsonii* Greene subsp. *hypolasius* (Greene) B. J. Cox
- *Lupinus cumulicola* Small
- *Lupinus cusickii* S. Watson
- *Lupinus cusickii* S. Watson subsp. *abortivus* (Greene) B. J. Cox
- *Lupinus cusickii* S. Watson subsp. *brachypodus* (Piper) B. J. Cox
- *Lupinus cusickii* S. Watson subsp. *cusickii* S. Watson
- *Lupinus cuspidatus* Rusby
- *Lupinus cuzcensis* C. P. Sm.
- *Lupinus cymb-Aegressus* C. P. Sm.
- *Lupinus cymboides* C. P. Sm.
- *Lupinus czermakii* Briq. & Hochr.
- *Lupinus dalesiae* Eastw.
English: Quincy lupine
- *Lupinus dasyphyllus* Greene
- *Lupinus davisianus* C. P. Sm.
- *Lupinus debilis* Eastw.
- *Lupinus decaschistus* C. P. Sm.
- *Lupinus decemplex* C. P. Sm.
- *Lupinus decurrens* Gardner
- *Lupinus deflexus* Congdon
- *Lupinus delicatulus* Sprague & Riley
- *Lupinus diaboli-septem* C. P. Sm.
- *Lupinus diasemus* C. P. Sm.
- *Lupinus dichrous* Greene
- *Lupinus diehlii* M. E. Jones
- *Lupinus diffusus* Nutt.
Synonyms: *Lupinus cumulicola* Small
Arabic: trms mntshr
English: Oak Ridge lupine; sky-blue lupine; spreading lupine
- *Lupinus digitatus* Forssk.
English: Egyptian lupin
Finish: egyptinlupiini
- *Lupinus disjunctus* C. P. Sm.
- *Lupinus dispersus* A. Heller
- *Lupinus dissimulans* C. P. Sm.

- *Lupinus diversalpicola* C. P. Sm.
- *Lupinus dorae* C. P. Sm.
- *Lupinus dotatus* C. P. Sm.
- *Lupinus durangensis* C. P. Sm.
- *Lupinus duranii* Eastw.

English: Mono Lake lupine
- *Lupinus dusenianus* C. P. Sm.
- *Lupinus eanophyllus* C. P. Sm.
- *Lupinus eatonanus* C. P. Sm.
- *Lupinus edysomatus* C. P. Sm.
- *Lupinus egens* C. P. Sm.
- *Lupinus elaphoglossum* Barneby
- *Lupinus elatus* I. M. Johnst.

Arabic: trms twyl
English: tall silky lupine
- *Lupinus elegans* Kunth

Synonyms: *Lupinus californicus* hort. ex K. Koch
English: elegant lupine; Mexican lupine
Finnish: meksikonlupiini
- *Lupinus elegantulus* Eastw.
- *Lupinus ellsworthianus* C. P. Sm.
- *Lupinus elmeri* Greene

English: Elmer's lupine; South Fork Mountain lupine
- *Lupinus eramosus* C. P. Sm.
- *Lupinus erectifolius* C. P. Sm.
- *Lupinus eremonomus* C. P. Sm.
- *Lupinus equi-coeli* C. P. Sm.
- *Lupinus equi-collis* C. P. Sm.
- *Lupinus eriocalyx* (C. P. Sm.) C. P. Sm.
- *Lupinus eriocladus* Ulbr.
- *Lupinus erminens* S. Watson
- *Lupinus ermineus* S. Watson
- *Lupinus evermannii* Rydb.
- *Lupinus espinarensis* C. P. Sm.
- *Lupinus exaltatus* Zucc.
- *Lupinus excubitus* M. E. Jones

English: grape soda lupine
- *Lupinus exochus* C. P. Sm.
- *Lupinus expetendus* C. P. Sm.
- *Lupinus extrarius* C. P. Sm.
- *Lupinus falcifer* Nutt.
- *Lupinus falsoerectus* C. P. Sm.
- *Lupinus falsoformosus* C. P. Sm.
- *Lupinus falsograyi* C. P. Sm.
- *Lupinus falsomutabilis* C. P. Sm.
- *Lupinus falsoprostratus* C. P. Sm.

- *Lupinus falsorevolutus* C. P. Sm.
- *Lupinus famelicus* C. P. Sm.
- *Lupinus fiebrigianus* Ulbr.
- *Lupinus fieldii* Rose ex J. F. Macbr.
- *Lupinus filicaulis* C. P. Sm.
- *Lupinus finitus* C. P. Sm.
- *Lupinus fissicalyx* A. Heller
- *Lupinus flavescens* Rydb.
- *Lupinus flavoculatus* A. Heller

English: yellow-eyed lupine; yelloweyes
- *Lupinus foliolosus* Benth.
- *Lupinus formosus* Greene

Arabic: trms gmyl
English: summer lupine; western lupine
Turkish: yaz acı baklası
- *Lupinus formosus* Greene var. *bridgesii* (S. Watson) Greene
- *Lupinus formosus* Greene var. *formosus*
- *Lupinus forskahlei* Boiss.
- *Lupinus fragrans* A. Heller
- *Lupinus francis-whittieri* C. P. Sm.
- *Lupinus franciscanus* Greene
- *Lupinus fratrum* C. P. Sm.
- *Lupinus fraxinetorum* Greene
- *Lupinus fruticosus* Steud.
- *Lupinus fruticosus* Dum. Cours.
- *Lupinus fulcratus* Greene

English: greenstipule lupine
- *Lupinus gachetensis* C. P. Sm.
- *Lupinus garcianus* Bennett & Dunn
- *Lupinus garfieldensis* C. P. Sm.
- *Lupinus gaudichaudianus* C. P. Sm.
- *Lupinus gayanus* C. P. Sm.
- *Lupinus gentryanus* C. P. Sm.
- *Lupinus geophilus* Rose
- *Lupinus geraniophilus* C. P. Sm.
- *Lupinus gibertianus* C. P. Sm.
- *Lupinus giganteus* Rose
- *Lupinus glabellus* M. Martens & Galeotti
- *Lupinus glabratus* J. Agardh
- *Lupinus goodspeedii* J. F. Macbr.
- *Lupinus gormanii* Piper
- *Lupinus gracilentus* Greene
- *Lupinus graciliflorus* C. P. Sm.
- *Lupinus gratus* Greene
- *Lupinus grauensis* C. P. Sm.
- *Lupinus grayi* S. Watson

English: Sierra lupine
- *Lupinus gredensis* Gand.
Spanish: alberjón
- *Lupinus grisebachianus* C. P. Sm.
- *Lupinus guadalupensis* C. P. Sm.
Arabic: trms ghwadalubi
English: Guadalupe Island lupine
- *Lupinus guadalupensis* Greene
- *Lupinus guadiloupensis* Steud.
- *Lupinus guaraniticus* (Hassl.) C. P. Sm.
- *Lupinus guascensis* C. P. Sm.
- *Lupinus guatimalensis* auct.
- *Lupinus guggenheimianus* Rusby
- *Lupinus gussoneanus* J. Agardh
- *Lupinus habrocomus* Greene
- *Lupinus hamaticalyx* C. P. Sm.
- *Lupinus hartmannii* C. P. Sm.
- *Lupinus haudcytisoides* C. P. Sm.
- *Lupinus haughtianus* C. P. Sm.
- *Lupinus hautcarazensis* C. P. Sm.
- *Lupinus havardii* S. Watson
English: Big Bend bluebonnet; Chisos bluebonnet
- *Lupinus hazenanus* C. P. Sm.
- *Lupinus helleri* Greene
- *Lupinus hendersonii* Eastw.
English: Henderson's lupin
- *Lupinus heptaphyllus* (Vell.) Hassl.
- *Lupinus herreranus* C. P. Sm.
- *Lupinus herzogii* Ulbr.
- *Lupinus hexaedrus* E. Fourn.
- *Lupinus hieronymii* C. P. Sm.
- *Lupinus hilarianus* Benth.
- *Lupinus hinkleyorum* C. P. Sm.
- *Lupinus hintonii* C. P. Sm.
- *Lupinus hintoniorum* B. L. Turner
- *Lupinus hirsutissimus* Benth.
English: stinging annual lupine; stinging lupine
- *Lupinus* × *hispanicoluteus* W. Święcicki & W. K. Święcicki
- *Lupinus hispanicus* Boiss. & Reut.
English: Hispanian lupin
Hebrew: tvrmvs sfrd
Serbian: španska lupina
Spanish: alberjón; haba de lagarto; haba de lobo; titones
Swahili: mlupini wa Hispania
- *Lupinus hispanicus* Boiss. & Reut. subsp. *bicolor* (Merino) Gladst.
Synonyms: *Lupinus luteus* var. *bicolor* Merino; *Lupinus rothmaleri* Klink.

English: two-colored Hispanian lupin
- *Lupinus hispanicus* Boiss. & Reut. subsp. *hispanicus*
English: common Hispanian lupin
- *Lupinus holmgrenianus* C. P. Sm.
English: Holmgren's lupine
- *Lupinus honoratus* C. P. Sm.
- *Lupinus horizontalis* A. Heller
- *Lupinus hornemannii* J. Agardh
- *Lupinus hortonianus* C. P. Sm.
- *Lupinus hortorum* C. P. Sm.
- *Lupinus howard-scottii* C. P. Sm.
- *Lupinus howardii* M. E. Jones
English: Howard's lupin
- *Lupinus huachucanus* M. E. Jones
- *Lupinus huancayoensis* C. P. Sm.
- *Lupinus huariacus* C. P. Sm.
- *Lupinus huaronensis* J. F. Macbr.
- *Lupinus huigrensis* Rose ex C. P. Sm.
- *Lupinus humicolus* A. Nelson
- *Lupinus humilis* Rose ex Pittier
- *Lupinus humifusus* Sessé & Moc. ex G. Don
- *Lupinus hyacinthinus* C. F. Baker
English: San Jacinto lupine
Navajo: atsá ch'il
- *Lupinus hyacinthinus* Greene
- *Lupinus* × *hybridus* Lem.
- *Lupinus idoneus* C. P. Sm.
- *Lupinus ignobilis* C. P. Sm.
- *Lupinus imminutus* C. P. Sm.
- *Lupinus inamoenus* Greene ex C. F. Baker
- *Lupinus indigoticus* Eastw.
- *Lupinus indutus* Greene ex C. F. Baker
- *Lupinus inflatus* C. P. Sm.
- *Lupinus insignis* Glaz. ex C. P. Sm.
- *Lupinus* × *insignis* Lem.
- *Lupinus insulae* C. P. Sm.
- *Lupinus interruptus* Benth.
- *Lupinus intortus* C. P. Sm.
- *Lupinus inusitatus* C. P. Sm.
- *Lupinus involutus* C. P. Sm.
- *Lupinus inyoensis* A. Heller
- *Lupinus ione-grisetae* C. P. Sm.
- *Lupinus ione-walkerae* C. P. Sm.
- *Lupinus isabelianus* Eastw.
- *Lupinus jahnii* Rose ex Pittier
- *Lupinus jaimehintoniana* B. L. Turner

- *Lupinus james-westii* C. P. Sm.
- *Lupinus jamesonianus* C. P. Sm.
- *Lupinus javanicus* Burm.f.
- *Lupinus jean-julesii* C. P. Sm.
- *Lupinus jelskianus* C. P. Sm.
- *Lupinus johannis-howellii* C. P. Sm.
- *Lupinus jonesii* Rydb.
- *Lupinus jorgensenanus* C. P. Sm.
- *Lupinus jucundus* Greene
- *Lupinus jujuyensis* C. P. Sm.
- *Lupinus juninensis* C. P. Sm.
- *Lupinus kalenbornorum* C. P. Sm.
- *Lupinus kellermanianus* C. P. Sm.
- *Lupinus kerrii* Eastw.
- *Lupinus killipianus* C. P. Sm.
- *Lupinus kingii* S. Watson

English: King's lupine
- *Lupinus klamathensis* Eastw.
- *Lupinus kunthii* J. Agardh
- *Lupinus kuschei* Eastw.

English: Yukon lupin
- *Lupinus kunthii* J. Agardh
- *Lupinus kyleanus* C. P. Sm.
- *Lupinus labiatus* Nutt.
- *Lupinus lacticolor* Tamayo
- *Lupinus lacus-huntingtonii* C. P. Sm.
- *Lupinus lacuum-trinitatum* C. P. Sm.
- *Lupinus lanatus* Benth.
- *Lupinus lapidicola* A. Heller

English: Mt. Eddy lupine
- *Lupinus larsonanus* C. P. Sm.
- *Lupinus lassenensis* Eastw.
- *Lupinus latifolius* J. Agardh

Arabic: trms ryd aluraq
English: broadleaf lupine
- *Lupinus latifolius* J. Agardh subsp. *dudleyi* (C. P. Sm.) P. Kenney & D. B. Dunn
- *Lupinus latifolius* J. Agardh subsp. *latifolius*

Synonyms: *Lupinus rivularis* var. *latifolius* (J. Agardh) S. Watson
- *Lupinus latifolius* J. Agardh subsp. *latifolius* var. *barbatus*

English: Klamath lupine, bearded lupine
- *Lupinus latifolius* J. Agardh subsp. *latifolius* var. *latifolius* J. Agardh

English: broadleaf lupine
- *Lupinus latifolius* J. Agardh subsp. *leucanthus* (Rydb.) P. Kenney & D. B. Dunn
- *Lupinus latifolius* J. Agardh subsp. *longipes* (Greene) P. Kenney & D. B. Dunn
- *Lupinus latifolius* J. Agardh subsp. *parishii* (C. P. Sm.) P. Kenney & D. B. Dunn
- *Lupinus latifolius* J. Agardh subsp. *viridifolius* (A. Heller) P. Kenney & D. B. Dunn

- *Lupinus latissimus* Greene
- *Lupinus laudandrus* C. P. Sm.
- *Lupinus laxifolius* A. Gray
- *Lupinus lechlerianus* C. P. Sm.
- *Lupinus ledigianus* C. P. Sm.
- *Lupinus lelandsmithii* Eastw.
- *Lupinus lemmonii* C. P. Sm.
- *Lupinus lepidus* Douglas ex Lindl.
English: dwarf lupine; Pacific lupine; prairie lupine
- *Lupinus lepidus* Douglas ex Lindl. var. *aridus* (Douglas) Jeps.
Synonyms: *Lupinus aridus* Douglas
English: arid dwarf kupine
Finnish: aavikkolupiini
- *Lupinus lepidus* Douglas ex Lindl. var. *confertus* (Kellogg) C. P. Sm.
Synonyms: *Lupinus confertus* Kellogg
Swedish: nevadalupin
- *Lupinus lepidus* Douglas ex Lindl. var. *lepidus*
Synonyms: *Lupinus minimus* Douglas
English: common dwarf lupine
- *Lupinus lepidus* Douglas ex Lindl. var. *lobbii* (A. Gray ex S. Watson) C. L. Hitchc.
Synonyms: *Lupinus aridus* var. *lobbii* A. Gray ex S. Watson; *Lupinus lyallii* A. Gray; *Lupinus sellulus* var. *lobbii* (A. Gray ex S. Watson) B. J. Cox
English: dwarf mountain lupine; Lobb's lupine; Lobb's tidy lupine; Lyall's lupine
French: lupin de Lyall
Navajo: yiłch'ozh'azee'
- *Lupinus lepidus* Douglas ex Lindl. var. *sellulus* (Kellogg) Barneby
Synonyms: *Lupinus sellulus* Kellogg
- *Lupinus lepidus* Douglas ex Lindl. var. *utahensis* (S. Watson) C. L. Hitchc.
Synonyms: *Lupinus aridus* var. *utahensis* S. Watson; *Lupinus caespitosus* Nutt.; *Lupinus lepidus* subsp. *caespitosus* (Nutt.) Detling
English: stemless dwarf lupine
- *Lupinus leptocarpus* Benth.
- *Lupinus leptophyllus* Cham. & Schltdl.
- *Lupinus leptostachyus* Greene
- *Lupinus lespedezoides* C. P. Sm.
- *Lupinus lesueurii* Standl.
- *Lupinus leucophyllus* Douglas ex Lindl.
Synonyms: *Lupinus canescens* Howell; *Lupinus cyaneus* Rydb.; *Lupinus erectus* L. F. Hend.; *Lupinus leucophyllus* subsp. *erectus* (L. F. Hend.) Harmon; *Lupinus leucophyllus* subsp. *leucophyllus* Douglas ex Lindl.; *Lupinus leucophyllus* var. *belliae* C. P. Sm.; *Lupinus leucophyllus* var. *canescens* (Howell) C. P. Sm.
Arabic: trms swfi aluraq
English: velvet lupine; woolly-leaf lupine
- *Lupinus lilacinus* A. Heller
- *Lupinus lindenianus* C. P. Sm.
- *Lupinus lindleyanus* J. Agardh

- *Lupinus linearifolius* Larrañaga
- *Lupinus linearis* Desr.
- *Lupinus lingulae* C. P. Sm.
- *Lupinus littoralis* Douglas
Arabic: trms shaty
English: chinook-licorice; seashore lupine
Swedish: strandlupin
- *Lupinus lobbianus* C. P. Sm.
- *Lupinus longifolius* (S. Watson) Abrams
Arabic: trms twil aluraq
English: longleaf bush lupine
Turkish: uzun yapraklı acı bakla
- *Lupinus longilabrum* C. P. Sm.
- *Lupinus lorenzensis* C. P. Sm.
- *Lupinus louise-bucariae* C. P. Sm.
- *Lupinus louise-grisetae* C. P. Sm.
- *Lupinus lucidus* Benth. ex Loudon
- *Lupinus ludovicianus* Greene
- *Lupinus luetzelburgianus* C. P. Sm.
- *Lupinus luteolus* Kellogg
Arabic: trms shhb
English: butter lupine; pale yellow lupine
- *Lupinus lutescens* C. P. Sm.
- *Lupinus luteus* L. (Table 12.3)
- *Lupinus luteus* L. var. *albicans* Kurl. et Stankev.
- *Lupinus luteus* L. var. *arcellus* Kurl. et Stankev.
- *Lupinus luteus* L. var. *aurantiacus* Kurl. et Stankev.
- *Lupinus luteus* L. var. *aureus* Kurl. et Stankev.
- *Lupinus luteus* L. var. *cremeus* Kurl. et Stankev.
- *Lupinus luteus* L. var. *croceus* Kurl. et Stankev.
- *Lupinus luteus* L. var. *kazimierskii* Kurl. et Stankev.
- *Lupinus luteus* L. var. *leucospermus* Kurl. et Stankev.
- *Lupinus luteus* L. var. *luteus*
- *Lupinus luteus* L. var. *maculosus* Kurl. et Stankev.
- *Lupinus luteus* L. var. *melanospermus* Kurl. et Stankev.
- *Lupinus luteus* L. var. *niger* Kurl. et Stankev.
- *Lupinus luteus* L. var. *ochroleucus* Kurl. et Stankev.
- *Lupinus luteus* L. var. *sempolovskii* (Atab) Kurl. et Stankev.
- *Lupinus luteus* L. var. *sinskayae* Kurl. et Stankev.
- *Lupinus luteus* L. var. *stepanovae* Kurl. et Stankev.
- *Lupinus luteus* L. var. *sulphureus* (Atab.) Kurl. et Stankev.
- *Lupinus lutosus* A. Heller
- *Lupinus lyman-bensonii* C. P. Sm.
- *Lupinus lysichitophilus* C. P. Sm.
- *Lupinus macbrideanus* C. P. Sm.
English: Macbride's lupine

TABLE 12.3
Popular Names Denoting *Lupinus luteus* L. in Some World Languages and Other Linguistic Taxa

Language/Taxon	Name
Adyghe	bgedaxeʃhuʒ gueɩ
Arabic	trms sfr
Armenian	lupin deghin
Asturian	altramuz mariellu
Azerbaijani	sarı acıpaxla
Basque	eskuhori
Bengali	haluda lupina
Catalan	llobí groc; tramús groc
Chinese	huáng yǔshàn dòu
Czech	lupina žlutá
Danish	gul lupin
Dutch	gele lupine
English	annual yellow lupin; European yellow lupine; yellow lupine
Esperanto	lupeno flava
Finnish	keltalupiini
French	lupin jaune
German	Gelbe Lupine; Hasenklee
Greek	kítrino loúpino
Hebrew	tvrmvs tzhvv
Italian	lupino giallo
Japanese	kibana no wa uchiwa mame
Kashubian	żôłti lëpin
Kazakh	sarı böri burşaq; sarı lyupïn
Latvian	dzeltenā lupīna; Eiropas dzeltenā lupīna
Lithuanian	geltonasis lubinas
Polish	łubin żółty
Portuguese	lupino-amarelo; tremoçeiro-amarelo; tremoço-de-cheiro; tremoço-de-flor-amarela
Romanian	lupin galben
Russian	liupin zheltyi
Serbian	žuta lupina
Slovak	lupina žltá
Sorbian (Upper)	žołta lupina
Spanish	altramuz amarillo
Swahili	mlupini njano
Swedish	gul-lupin; gullupin
Turkish	sarı acı bakla
Ukrainian	liupin zhovtyi

- *Lupinus macranthus* Rose
- *Lupinus macrocarpus* Hook. & Arn.
- *Lupinus macrocarpus* Torr.
- *Lupinus macrophyllus* Benth.
- *Lupinus macrorhizos* Georgi
- *Lupinus maculatus* Rydb.
- *Lupinus madrensis* Seem.
- *Lupinus magdalenensis* C. P. Sm.
- *Lupinus magnificus* M. E. Jones
English: Panamint Mountain lupine
- *Lupinus magniflorus* C. P. Sm.
- *Lupinus magnistipulatus* Planchuelo & D. B. Dunn
- *Lupinus maissurianii* Atabek. & Polukhina
- *Lupinus malacophyllus* Greene
- *Lupinus malacotrichus* C. P. Sm.
- *Lupinus maleopinatus* C. P. Sm.
- *Lupinus mandonanus* C. P. Sm.
- *Lupinus mantaroensis* C. P. Sm.
- *Lupinus marcusianus* C. P. Sm.
- *Lupinus mariae-josephae* H. Pascual
Catalan: tramussera valenciana
English: Valencian lupin
- *Lupinus marinensis* Eastw.
- *Lupinus mariposanus* Eastw.
- *Lupinus markleanus* C. P. Sm.
- *Lupinus marschallianus* Sweet
- *Lupinus martensis* C. P. Sm.
- *Lupinus martinetianus* (C. P. Sm.) C. P. Sm.
- *Lupinus mathewsianus* C. P. Sm.
- *Lupinus matucanicus* Ulbr.
- *Lupinus mearnsii* C. P. Sm.
- *Lupinus meionanthus* A. Gray
English: Lake Tahoe lupine
- *Lupinus melaphyllus* C. P. Sm.
- *Lupinus meli-campestris* C. P. Sm.
- *Lupinus meridanus* Moritz ex C. P. Sm.
- *Lupinus metensis* C. P. Sm.
- *Lupinus mexiae* C. P. Sm.
- *Lupinus mexicanus* Cerv. ex Lag.
Synonyms: *Lupinus ehrenbergii* Schltdl.; *Lupinus hartwegii* Lindl.
English: Mexican lupine
Polish: łubin Hartwega
- *Lupinus micensis* M. E. Jones
- *Lupinus michelianus* C. P. Sm.
English: Michel's lupine
- *Lupinus micheneri* Greene

- *Lupinus micranthus* Guss.
Catalan: llobí hirsut
English: bitter blue lupin; hairy lupin
Finnish: rusokarvalupiini
French: lupin hérissé
German: Zottige Lupine
Portuguese: tremoçeiro-hirsuto
Spanish: altramuz peludo
Swahili: mlupini maua-madogo
Swedish: kranslupin
Turkish: tüylü acı bakla
- *Lupinus microcarpus* Sims
English: chick lupine; wide-bannered lupine
- *Lupinus microcarpus* Sims var. *densiflorus* (Benth.) Jeps.
Synonyms: *Lupinus densiflorus* Benth.; *Lupinus densiflorus* var. *aureus* (Kellogg) Munz; *Lupinus menziesii* var. *aureus* Kellogg
English: dense-flowered lupine; white-whorl lupin; whitewhorl lupine
German: Quirl-Lupine
Swedish: båglupin
- *Lupinus microcarpus* Sims var. *microcarpus*
Synonyms: *Lupinus ruber* A. Heller; *Lupinus subvexus* C. P. Sm.
English: chick lupin; wide-bannered lupin
- *Lupinus microphyllus* Desr.
- *Lupinus milleri* J. Agardh
- *Lupinus minearanus* C. P. Sm.
- *Lupinus minutissimus* Tamayo
- *Lupinus mirabilis* C. P. Sm.
- *Lupinus misticola* Ulbr.
- *Lupinus molle* A. Heller
- *Lupinus mollendoensis* Ulbr.
- *Lupinus mollis* A. Heller
- *Lupinus mollissifolius* Davidson
- *Lupinus monensis* Eastw.
- *Lupinus monettianus* C. P. Sm.
- *Lupinus monserratensis* C. P. Sm.
- *Lupinus montanus* Kunth
- *Lupinus montanus* Kunth subsp. *glabrior* (S. Watson) D. B. Dunn & Harmon
- *Lupinus montanus* Kunth subsp. *montanus* Kunth
- *Lupinus montanus* Kunth subsp. *montesii* (C. P. Sm.) D. B. Dunn & Harmon
- *Lupinus monticola* Rydb.
- *Lupinus moritzianus* Kunth
- *Lupinus mucronulatus* Howell
- *Lupinus muelleri* Standl.
- *Lupinus muellerianus* C. P. Sm.
- *Lupinus multicincinnis* C. P. Sm.
- *Lupinus multiflorus* Desr.

- *Lupinus munzianus* C. P. Sm.
- *Lupinus munzii* Eastw.
- *Lupinus mutabilis* Sweet (Table 12.4)

Synonyms: *Lupinus cruckshanksii* Hook.

- *Lupinus nanus* Douglas ex Benth.

Arabic: trms qzm
English: douglas' annual lupin; dwarf lupin; field lupin; ocean-blue lupin; sky lupin
Finnish: kääpiölupiini
German: Zwerg-Lupine
Polish: łubin karłowy
Turkish: bodur acı baklası

- *Lupinus neglectus* Rose
- *Lupinus nehmadae* C. P. Sm.
- *Lupinus nemoralis* Greene
- *Lupinus neocotus* C. P. Sm.
- *Lupinus neomexicanus* Greene
- *Lupinus nepubescens* C. P. Sm.
- *Lupinus nevadensis* A. Heller

Arabic: trms nyfadi

TABLE 12.4
Popular Names Denoting *Lupinus mutabilis* Sweet in Some World Languages and Other Linguistic Taxa

Language/Taxon	Name
Arabic	trms tfry
Aymara	tarwi
English	altramuz; Andean lupine; pearl lupin; pearl lupine; Peruvian field lupin; South American lupin
Finnish	tuoksulupiini
French	lupin des Andes
German	Anden-Lupine; Andenlupine
Italian	chocho; lupino; tarwi
Japanese	zashoku novoli fuji
Lithuanian	Andinis lubinas
Polish	łubin andyjski; łubin zmienny
Portuguese	lupino-mutável; tremoço
Quechua (Bolivia)	chuchu; tarwi; tawri
Quechua (Peru)	taazsi; tarwi
Russian	liupin izmenchivyi
Serbian	andska lupina; južnoamerička lupina
Spanish	chocho; tarwi
Spanish (Peru)	altramuz; chocho; lupino; tarhui; tarwi
Swedish	doftlupin
Turkish	and acı baklası

English: Nevada lupine
- *Lupinus niederleinianus* C. P. Sm.
- *Lupinus niger* Wehmer
- *Lupinus nipomensis* Eastw.
English: Nipomo Mesa lupine
- *Lupinus niveus* S. Watson
- *Lupinus noldekae* Eastw.
- *Lupinus nonoensis* C. P. Sm.
- *Lupinus nootkatensis* Donn ex Sims (Table 12.5)
- *Lupinus notabilis* C. P. Sm.
- *Lupinus nubigenus* Kunth
- *Lupinus nubilorum* C. P. Sm.
- *Lupinus nutcanus* Spreng.
- *Lupinus nutkatensis* J. G. Cooper
- *Lupinus obscurus* C. P. Sm.
- *Lupinus obtunsus* C. P. Sm.
- *Lupinus obtusilobus* A. Heller
English: bluntlobe lupine
- *Lupinus ochoanus* C. P. Sm.
- *Lupinus ochroleucus* Eastw.
- *Lupinus octablomus* C. P. Sm.
- *Lupinus odoratus* A. Heller
Arabic: trms try
English: Mojave lupine; royal Mojave lupin
- *Lupinus onustus* S. Watson
Arabic: trms mmtl

TABLE 12.5
Popular Names Denoting *Lupinus nootkatensis* Donn ex Sims in Some World Languages and Other Linguistic Taxa

Language/Taxon	Name
English	Nootka lupin; Nootka lupine
Esperanto	lupeno malmultfolia
Finish	alaskanlupiini
French	lupin d'Alaska; lupin d'Écosse
German	Alaska-Lupine
Icelandic	Alaskalúpína
Italian	lupino nootka
Norwegian (Bokmål)	alaskalupin; sandlupin
Norwegian (Nynorsk)	sandlupin
Serbian	aljaska lupina
Swedish	sandlupin
Turkish	Nutka acı baklası
Welsh	bysedd-y-blaidd gwyllt

English: Plumas lupine; yellow pine lupine
- *Lupinus opertospicus* C. P. Sm.
- *Lupinus opsianthus* Amabekova & Maisuran
- *Lupinus oquendoanus* C. P. Sm.
- *Lupinus oreophilus* Phil.
- *Lupinus oscar-haughtii* C. P. Sm.
- *Lupinus ostiofluminis* C. P. Sm.
- *Lupinus otto-buchtienii* C. P. Sm.
- *Lupinus otto-kuntzeanus* C. P. Sm.
- *Lupinus otuzcoensis* C. P. Sm.
- *Lupinus ovalifolius* Benth.
- *Lupinus pachanoanus* C. P. Sm.
- *Lupinus pachitensis* C. P. Sm.
- *Lupinus pachylobus* Greene

English: big-pod lupine
- *Lupinus padre-crowleyi* C. P. Sm.

English: DeDecker's lupine; Father Crowley's lupine
- *Lupinus palaestinus* Boiss.

English: white-grey lupin
Hebrew: tvrmvs rtzshrl
Swahili: mlupini kijivucheupe
- *Lupinus pallidus* Brandegee
- *Lupinus paniculatus* Desr.
- *Lupinus paraguariensis* Chodat & Hassl.
- *Lupinus paranensis* C. P. Sm.
- *Lupinus paruroensis* C. P. Sm.
- *Lupinus parvifolius* Gardner
- *Lupinus parviflorus* Hook. & Arn.

English: lodgepole lupin
- *Lupinus parviflorus* subsp. *myrianthus* (Greene) Harmon var. *myrianthus*
- *Lupinus parviflorus* Hook. & Arn. subsp. *parviflorus*
- *Lupinus pasachoensis* C. P. Sm.
- *Lupinus pasadenensis* Eastw.
- *Lupinus patulus* C. P. Sm.
- *Lupinus paucartambensis* C. P. Sm.
- *Lupinus paucovillosus* C. P. Sm.
- *Lupinus pavonum* C. P. Sm.
- *Lupinus paynei* Davidson
- *Lupinus pearceanus* C. P. Sm.
- *Lupinus peirsonii* H. Mason

English: Peirson's lupine; long lupine
- *Lupinus penlandianus* C. P. Sm.
- *Lupinus pendeltonii* A. Heller
- *Lupinus pendentiflorus* C. P. Sm.
- *Lupinus pendletonii* A. Heller
- *Lupinus perblandus* C. P. Sm.

TABLE 12.6
Popular Names Denoting *Lupinus perennis* L. in Some World Languages and Other Linguistic Taxa

Language/Taxon	Name
Arabic	trms mmr
Chinese	sùgēn yǔshàn dòu
English	blue-bean; Indian beet; old maid's bonnets; sundial lupine; wild lupine; wild perennial lupine
Finnish	pulskalupiini
German	Ausdauernde Lupine
Hungarian	évelő csillagfürt
Italian	lupino perenne
Kazakh	köpjıldıq böriburşaq
Norwegian (Bokmål)	jærlupin
Norwegian (Nynorsk)	jærlupin
Ossetian	æhsædæntæ
Portuguese	tremoço-perene
Portuguese (Brazil)	tremoço-perene
Romanian	lupinul peren
Russian	liupin mnogoletntyi
Sami (Northern)	lupiidnat
Serbian	višegodišnja lupina
Swedish	gruslupin
Thai	thùng lūphin p̀ā
Turkish	yabani acı bakla
Ukrainian	liupin bahatorichnyi

- *Lupinus perbonus* C. P. Sm.
- *Lupinus perconfertus* C. P. Sm.
- *Lupinus perennis* L. (Table 12.6)
- *Lupinus perennis* L. subsp. *gracilis* (Nutt.) D. B. Dunn
- *Lupinus perennis* L. subsp. *occidentalis* S. Watson
- *Lupinus perennis* L. subsp. *perennis*
- *Lupinus perglaber* Eastw.
- *Lupinus perissophytus* C. P. Sm.
- *Lupinus perplexus* C. P. Sm.
- *Lupinus persistens* Rose
- *Lupinus peruvianus* Ulbr.
- *Lupinus philippianus* C. P. Sm.
- *Lupinus philistaeus* Boiss.
- *Lupinus physodes* Douglas
- *Lupinus pickeringii* A. Gray
- *Lupinus pilosus* L.

Synonyms: *Lupinus hirsutus* L.

Arabic: trms shry
English: blue lupine
Hebrew: tvrmvs hhrm
Polish: łubin kosmaty
Swahili: mlupini buluu
- *Lupinus pilosissimus* M. Martens & Galeotti
- *Lupinus pinguis* Ulbr.
- *Lupinus pinus-contortae* C. P. Sm.
- *Lupinus piperi* B. L. Rob. ex Piper
- *Lupinus piperitus* Davidson
- *Lupinus pipersmithianus* J. F. Macbr.
- *Lupinus pisacensis* C. P. Sm.
- *Lupinus piurensis* C. P. Sm.
- *Lupinus platamodes* C. P. Sm.
- *Lupinus platanophilus* M. E. Jones
- *Lupinus plattensis* S. Watson
English: Platte lupine
- *Lupinus platyptenus* C. P. Sm.
- *Lupinus plebeius* Greene ex C. F. Baker
- *Lupinus polyphyllus* Lindl. (Table 12.7)
- *Lupinus polyphyllus* Lindl. var. *burkei* (S. Watson) C. L. Hitchc.
Synonyms: *Lupinus burkei* S. Watson; *Lupinus polyphyllus* subsp. *bernardinus* (Abrams ex C. P. Sm.) Munz
English: Burke's lupine
- *Lupinus polyphyllus* Lindl. var. *humicola* (A. Nelson) Barneby
Synonyms: *Lupinus humicola* A. Nelson; *Lupinus wyethii* S. Watson
- *Lupinus polyphyllus* Lindl. var. *pallidipes* (A. Heller) C. P. Sm.
- *Lupinus polyphyllus* Lindl. var. *polyphyllus*
Synonyms: *Lupinus polyphyllus* var. *grandifolius* Lindl. ex J. Agardh
- *Lupinus polyphyllus* Lindl. var. *prunophilus* (M. E. Jones) L. Ll. Phillips
Synonyms: *Lupinus biddlei* L. F. Hend. ex C. P. Sm.; *Lupinus prunophilus* M. E. Jones
English: hairy bigleaf lupin
- *Lupinus poopoensis* C. P. Sm.
- *Lupinus popayanensis* C. P. Sm.
- *Lupinus potosinus* Rose
- *Lupinus praealtus* C. P. Sm.
- *Lupinus praestabilis* C. P. Sm.
- *Lupinus praetermissus* C. P. Sm.
- *Lupinus pratensis* A. Heller
Arabic: trms mrjy
English: Inyo Meadow lupine
- *Lupinus prato-lacuum* C. P. Sm.
- *Lupinus princei* Harms
- *Lupinus pringlei* Rose
- *Lupinus proculaustrinus* C. P. Sm.
- *Lupinus prolifer* Desr.

TABLE 12.7
Popular Names Denoting *Lupinus polyphyllus* Lindl. in Some World Languages and Other Linguistic Taxa

Language/Taxon	Name
Arabic	trms dyd aluraq
Armenian	darnarvuit
Chinese	duō yè yǔshàn dòu
Czech	lupina mnoholistá; vlčí bob mnoholistý
Danish	mangebladet lupin; staude-lupin
Dutch	vaste lupine
English	big-leaf lupin; garden lupin; large-leaf lupin; Washington lupin
Esperanto	lupeno multfolia
Finnish	komealupiini; lupiini
French	lupin des jardins; lupin pérenne; lupin polyphylle; lupin vivace
German	Staudenlupine; Vielblättrige Lupine
Greek	margaritódes loúpino
Hebrew	tvrmvs rv-'lm
Hungarian	erdei csillagfürt
Italian	lupino fogliuto
Kashubian	wiôlgòlëstny lëpin
Lithuanian	gausialapis lubinas
Norwegian (Bokmål)	hagelupin
Norwegian (Nynorsk)	hagelupin
Persian	lun
Polish	łubin trwały
Portuguese	tremoçeiro-de-jardim
Russian	liupin mnogolistnyi
Serbian	mnogolisna lupina; ukrasna lupina; višegodišnja lupina; višelisna lupina
Slovak	lupina mnoholistá; lupína mnoholistá; vlčí bôb mnoholistý
Sorbian (Upper)	wjelełopjenkata lupina
Spanish	altramuz perenne; lupino
Swedish	blomsterlupin
Turkish	acıbakla-lüpen; büyük yapraklı acıbakla
Ukrainian	liupin bahatolystyi
Welsh	bysedd-y-blaidd y gerddi

- *Lupinus propinquus* Greene
- *Lupinus prostratus* J. Agardh
- *Lupinus proteanus* Eastw.
- *Lupinus protrusus* C. P. Sm.
- *Lupinus prouvensalanus* C. P. Sm.
- *Lupinus pseudopolyphyllus* C. P. Sm.
- *Lupinus pseudotsugoides* C. P. Sm.
- *Lupinus psoraleoides* Pollard

- *Lupinus pubescens* Benth.
Chinese: máo yŭshàn dòu
Finnish: nukkalupiini
Spanish (Peru): chocho del páramo
- *Lupinus pucapucensis* C. P. Sm.
- *Lupinus pulloviridus* C. P. Sm.
- *Lupinus pulvinaris* Ulbr.
- *Lupinus pumviridis* C. P. Sm.
- *Lupinus punto-reyesensis* C. P. Sm.
- *Lupinus puracensis* C. P. Sm.
- *Lupinus purdieanus* C. P. Sm.
- *Lupinus pureriae* C. P. Sm.
- *Lupinus purosericeus* C. P. Sm.
- *Lupinus puroviridis* C. P. Sm.
- *Lupinus purpurascens* A. Heller
- *Lupinus pusillus* Pursh
Bengali: chōṭa lupina
English: dwarf lupine; rusty lupine; small lupine
- *Lupinus pusillus* Pursh subsp. *intermontanus* (A. Heller) D. B. Dunn
- *Lupinus pusillus* Pursh subsp. *pusillus*
- *Lupinus puyupatensis* C. P. Sm.
- *Lupinus pycnostachys* C. P. Sm.
- *Lupinus pygmaeus* Tamayo
- *Lupinus quellomayus* C. P. Sm.
- *Lupinus quercus-jugi* C. P. Sm.
- *Lupinus quercuum* C. P. Sm.
- *Lupinus quitensis* C. P. Sm.
- *Lupinus radiatus* C. P. Sm.
- *Lupinus rainierensis* Eastw.
- *Lupinus ramosissimus* Benth.
- *Lupinus reflexus* Rose
- *Lupinus* × *regalis* (auct.) Bergmans (*Lupinus arboreus* × *Lupinus polyphyllus*)
Danish: regnbuelupin
English: rainbow lupin; Russell lupin
Finnish: kirjolupiini
French: lupin des jardins; lupin hybride de Russell
Japanese: Rasseru rupinasu
Polish: łubin królewski
Swedish: regnbågslupin
- *Lupinus regius* Rudolph ex Torr. & A. Gray
- *Lupinus regnellianus* C. P. Sm.
- *Lupinus reineckianus* C. P. Sm.
- *Lupinus reitzii* Burkart ex M. Pinheiro & Miotto
- *Lupinus retrorsus* L. F. Hend.
- *Lupinus revolutus* C. P. Sm.
- *Lupinus rhodanthus* C. P. Sm.

- *Lupinus richardianus* C. P. Sm.
- *Lupinus rickeri* C. P. Sm.
- *Lupinus rimae* Eastw.
- *Lupinus rivetianus* C. P. Sm.
- *Lupinus rivularis* Douglas ex Lindl.
Arabic: trms nhry
English: riverbank lupine
Polish: łubin łąkowy
- *Lupinus romasanus* Ulbr.
- *Lupinus roseolus* Rydb.
- *Lupinus roseorum* C. P. Sm.
- *Lupinus rotundiflorus* M. E. Jones
- *Lupinus rowleeanus* C. P. Sm.
- *Lupinus rubriflorus* Planchuelo
- *Lupinus ruizensis* C. P. Sm.
- *Lupinus rupestris* Kunth
Arabic: trms skhry
- *Lupinus rusbyanus* C. P. Sm.
- *Lupinus russellianus* C. P. Sm.
- *Lupinus rydbergii* Blank.
- *Lupinus sabinianus* Lindl.
- *Lupinus sabinii* Hook.
- *Lupinus sabuli* C. P. Sm.
- *Lupinus sabulosus* A. Heller
- *Lupinus salicisocius* C. P. Sm.
- *Lupinus salinensis* C. P. Sm.
- *Lupinus salticola* Eastw.
- *Lupinus sandiensis* C. P. Sm.
- *Lupinus santanderensis* C. P. Sm.
- *Lupinus sarmentosus* Desr.
- *Lupinus sativus* Gaterau
- *Lupinus saxatilis* Ulbr.
- *Lupinus saxosus* Howell
English: rock lupine
- *Lupinus scaposus* Rydb.
- *Lupinus scheuberae* Rydb.
- *Lupinus schickendantzii* C. P. Sm.
- *Lupinus schiedeanus* Steud.
- *Lupinus schumannii* C. P. Sm.
- *Lupinus schwackeanus* C. P. Sm.
- *Lupinus seclusus* C. P. Sm.
- *Lupinus seifrizianus* (C. P. Sm.) C. P. Sm.
- *Lupinus sellowianus* Harms
- *Lupinus semiaequus* C. P. Sm.
- *Lupinus semiprostratus* C. P. Sm.
- *Lupinus semiverticillatus* Desr.

- *Lupinus semperflorens* Hartw. ex Benth.
- *Lupinus sergenti* Tamayo ex Pittier
- *Lupinus sergentii* Tamayo
- *Lupinus sericatus* Kellogg

English: Cobb Mountain lupine

- *Lupinus sericeus* Pursh

Synonyms: *Lupinus flexuosus* Lindl. ex J. Agardh; *Lupinus leucopsis* J. Agardh; *Lupinus ornatus* Douglas; *Lupinus sericeus* var. *egglestonianus* auct.; *Lupinus sericeus* var. *flexuosus* (Lindl. ex J. Agardh) C. P. Sm.; *Lupinus sericeus* var. *maximus* Fleak & D. B. Dunn; *Lupinus sericeus* var. *sericeus* Pursh

English: Pursh's silky lupine; silky lupine

French: lupin soyeux

- *Lupinus sericeus* Pursh var. *barbiger* (S. Watson) S. L. Welsh
- *Lupinus sericeus* Pursh var. *sericeus*
- *Lupinus serradentum* C. P. Sm.
- *Lupinus setifolius* Planchuelo & D. B. Dunn
- *Lupinus shockleyi* S. Watson

English: purple desert lupine

- *Lupinus shrevei* C. P. Sm.
- *Lupinus sierrae-blancae* Wooton & Standl.
- *Lupinus sierrae-blancae* Wooton & Standl. subsp. *aquilinus* (Wooton & Standl.) L. S. Fleak & D. B. Dunn
- *Lupinus sierrae-blancae* Wooton & Standl. subsp. *sierrae-blancae*
- *Lupinus sierrae-zentae* C. P. Sm.
- *Lupinus sileri* S. Watson
- *Lupinus simonsianus* C. P. Sm.
- *Lupinus simulans* Rose
- *Lupinus sinaloensis* C. P. Sm.
- *Lupinus sinus-meyersii* C. P. Sm.
- *Lupinus sitgreavesii* S. Watson
- *Lupinus smithianus* Kunth

English: Smith's lupine

- *Lupinus solanagrorum* C. P. Sm.
- *Lupinus somaliensis* Baker f.

English: Somali lupin

Swahili: mlupini somali

- *Lupinus sonomensis* A. Heller
- *Lupinus soratensis* Rusby
- *Lupinus soukupianus* C. P. Smith ex J. F. Macbr.
- *Lupinus sparhawkianus* C. P. Sm.
- *Lupinus sparsiflorus* Benth.

Synonyms: *Lupinus sparsiflorus* subsp. *mohavensis* Dziekanowski & D. B. Dunn

Arabic: trms mtnathr alzhar

English: Coulter's lupine; desert lupine; Mojave lupine

Turkish: çöl acı baklası

- *Lupinus spatulata* Larrañaga
- *Lupinus speciosus* Voss
- *Lupinus spectabilis* Hoover
Arabic: trms mbhrj
English: shaggyhair lupine
- *Lupinus splendens* Rose
- *Lupinus spragueanus* C. P. Sm.
- *Lupinus spruceanus* C. P. Sm.
- *Lupinus staffordiae* C. P. Sm.
- *Lupinus standleyensis* C. P. Sm.
- *Lupinus stationis* C. P. Sm.
- *Lupinus stipulatus* J. Agardh
- *Lupinus stiveri* Kellogg
- *Lupinus stiversii* Kellogg
English: harlequin annual lupine; harlequin lupine
- *Lupinus stoloniferus* L.
- *Lupinus strigulosus* Gand.
- *Lupinus storkianus* C. P. Sm.
- *Lupinus subacaulis* Griseb.
Synonyms: *Lupinus chilensis* C. P. Sm.
- *Lupinus subcarnosus* Hook.
English: buffalo lupine; Texas bluebonnet
- *Lupinus subcuneatus* C. P. Sm.
- *Lupinus subhamatus* C. P. Sm.
- *Lupinus subhirsutus* Davidson
- *Lupinus subinflatus* C. P. Sm.
- *Lupinus sublanatus* Eastw.
- *Lupinus submontanus* Rose
- *Lupinus subsessilis* Benth.
- *Lupinus subtomentosus* C. P. Sm.
- *Lupinus subvolutus* C. P. Sm.
- *Lupinus succulentus* Douglas ex K. Koch
Arabic: trms sary
English: arroyo lupine; hollowleaf annual lupin; succulent lupin
- *Lupinus sufferrugineus* Rusby
- *Lupinus suksdorfii* B.L. Rob. ex Piper
- *Lupinus suksdorfii* Robinson
- *Lupinus sulphureus* Douglas
Arabic: trms kbryti
English: sulphur lupine; sulphur-flower lupine
Turkish: sülfür acı baklası
- *Lupinus sulphureus* Douglas subsp. *kincaidii* (C. P. Sm.) L. Ll. Phillips
Synonyms: *Lupinus oreganus* A. Heller; *Lupinus oreganus* var. *kincaidii* C. P. Sm.
English: Kincaid's lupin
- *Lupinus sulphureus* Douglas subsp. *subsaccatus* (Suksd.) L. Ll. Phillips
Synonyms: *Lupinus bingenensis* var. *subsaccatus* Suksd.

- *Lupinus sulphureus* Douglas subsp. *sulphureus*
English: sulphur lupin; sulphur-flowered lupin
- *Lupinus summersianus* C. P. Sm.
- *Lupinus surcoensis* C. P. Sm.
- *Lupinus sylvaticus* Hemsl.
- *Lupinus syriggedes* C. P. Sm.
- *Lupinus tacitus* C. P. Sm.
- *Lupinus tafiensis* C. P. Sm.
- *Lupinus talahuensis* C. P. Sm.
- *Lupinus tamayoanus* C. P. Sm.
- *Lupinus tarapacensis* C. P. Sm.
- *Lupinus tarijensis* Ulbr.
English: Tarija lupine
- *Lupinus tarmaensis* C. P. Sm.
- *Lupinus tatei* Rusby
- *Lupinus taurimortuus* C. P. Sm.
- *Lupinus tauris* Benth.
- *Lupinus tayacajensis* C. P. Sm.
- *Lupinus tegeticulatus* Eastw.
- *Lupinus thermis* Gasp.
- *Lupinus thermus* St.-Lag.
- *Lupinus tetracercophorus* C. P. Sm.
- *Lupinus texanus* Hook.
- *Lupinus texensis* Hook.
Arabic: trms tksasy
English: Texas bluebonnet; Texas lupine
Finnish: Teksasinlupiini
Japanese: Tekisasu rupinasu
Spanish: altramuz de Texas
Swedish: texaslupin
Turkish: Teksas acı baklası
- *Lupinus thompsonianus* C. P. Sm.
- *Lupinus tidestromii* Greene
English: clover lupine; Tideström's lupine
Turkish: Tidestrom acı baklası
- *Lupinus tidestromii* Greene var. *layneae* (Eastw.) Munz
- *Lupinus tidestromii* Greene var. *tidestromii*
- *Lupinus tilcaricus* C. P. Sm.
- *Lupinus timotensis* Tamayo
- *Lupinus tolimensis* C. P. Sm.
- *Lupinus tomentosus* DC.
- *Lupinus tominensis* Wedd.
- *Lupinus toratensis* C. P. Sm.
English: lito; warwanzo
- *Lupinus tracyi* Eastw.
English: Tracy's lupine

- *Lupinus triananus* C. P. Sm.
- *Lupinus tricolor* G. Nicholson
- *Lupinus trifidus* Torr. ex S. Watson
- *Lupinus tristis* Sweet
- *Lupinus trochophyllus* Hoffmanns.
- *Lupinus truncatus* Hook. & Arn.
English: collared annual lupine
- *Lupinus tuckeranus* C. P. Sm.
- *Lupinus tucumanensis* C. P. Sm.
- *Lupinus ulbrichianus* C. P. Sm.
- *Lupinus uleanus* C. P. Sm.
- *Lupinus ultramontanus* C. P. Sm.
- *Lupinus umidicola* C. P. Sm.
- *Lupinus uncialis* S. Watson
- *Lupinus uncinatus* Schltdl.
- *Lupinus urcoensis* C. P. Sm.
- *Lupinus urubambensis* C. P. Sm.
- *Lupinus vaginans* Benth.
- *Lupinus valdepallidus* C. P. Sm.
- *Lupinus valerioi* Standl.
- *Lupinus vallicola* A. Heller
English: open lupin
- *Lupinus vallicola* A. Heller subsp. *apricus* (Greene) D. B. Dunn
- *Lupinus vallicola* subsp. *vallicola* A. Heller
- *Lupinus vandykeae* Eastw.
English: Van Duke's lupine
- *Lupinus vargasianus* C. P. Sm.
- *Lupinus varicaulis* C. P. Sm.
- *Lupinus variegatus* A. Heller
- *Lupinus variegatus* Poir.
- *Lupinus variicolor* Steud.
Arabic: trms mlwn
English: Lindley's varied lupine; manycolored lupine; varicolored lupine; varied lupin
- *Lupinus varneranus* C. P. Sm.
- *Lupinus vavilovii* Atabekova & Maissurjan
English: Vavilov's lupin
Russian: liupin Vavilova
- *Lupinus velillensis* C. P. Sm.
- *Lupinus velutinus* Benth.
- *Lupinus venezuelensis* C. P. Sm.
English: Venezuelan lupin
- *Lupinus ventosus* C. P. Sm.
- *Lupinus venustus* Bailly
- *Lupinus verbasciformis* Sandwith
- *Lupinus verjonensis* C. P. Sm.
- *Lupinus vernicius* Rose

- *Lupinus* × *versicolor* Caball.
- *Lupinus viduus* C. P. Sm.
- *Lupinus vilcabambensis* C. P. Sm.
- *Lupinus villosus* Willd.
- *Lupinus violaceus* A.Heller
- *Lupinus viridicalyx* C. P. Sm.
- *Lupinus visoensis* J. F. Macbr.
- *Lupinus volcanicus* Greene

English: volcano lupine

- *Lupinus volubilis* C. P. Sm.
- *Lupinus watsonii* A. Heller

English: Watson's lupin

- *Lupinus weberbaueri* Ulbr.
- *Lupinus werdermannianus* C. P. Sm.
- *Lupinus westiana* Small
- *Lupinus westianus* Small
- *Lupinus westianus* Small var. *aridorum* (McFarlin ex Beckner) Isely

Synonyms: *Lupinus aridorum* McFarlin ex Beckner
Arabic: trms dhabl
English: scrub lupine

- *Lupinus westianus* Small var. *westianus*
- *Lupinus whiltoniae* Eastw.
- *Lupinus wilkesianus* C. P. Sm.
- *Lupinus williamlobbii* C. P. Sm.
- *Lupinus williamsianus* C. P. Sm.
- *Lupinus wolfianus* C. P. Sm.
- *Lupinus xanthophyllus* C. P. Sm.
- *Lupinus xenophytus* C. P. Sm.
- *Lupinus yanahuancensis* C. P. Sm.
- *Lupinus yanlyensis* C. P. Sm.
- *Lupinus yaruahensis* C. P. Sm.
- *Lupinus yarushensis* C. P. Sm.
- *Lupinus ynesiae* C. P. Sm.

12.2 ORIGIN OF SCIENTIFIC AND POPULAR TAXA NAMES

Pearl lupin (*Lupinus mutabilis*) has certainly been the most important of all the *Lupinus* species of the New World, which are by far more numerous than those originating in the Old World. This pulse crop had already played a significant role in the Andean highlands and adjacent regions in pre-Incan civilizations (Morris 1999) and later became a widely cultivated grain legume for human consumption in the Incan Empire. At the same time, various *Lupinus* species were used by many Native American peoples, such as the Amerind Yavapai and the Na-Dené Navajo in North America (Jacobsen and Mujica 2008). As a consequence of complex and often unhappy historical circumstances, only a few names related to the native *Lupinus* species in the American Aboriginal languages survived until today. Among these rare

FIGURE 12.1 **(See color insert.)** One of the possible evolutions of the root **tora* of the proposed Proto-Amerind language, denoting white color (Greenberg and Ruhlen 2007), into its modern descendants in North, Central and South America; the meanings of the modern words are *white*, if given without brackets, and *pearl lupine* (pl).

examples are the words denoting pearl lupin in the Amerind Aymara and Quechua languages (Table 12.4). With great caution, we dare to suggest that the first segment of these words, the morpheme *tar*-, could be derived from the Proto-Amerind root **tora,* describing something white (Greenberg and Ruhlen 2007), which, in our case, may be the grains and what, interestingly enough, is equivalent to the English adjective *pearl* in the name of this crop (Figure 12.1). The second segment of the word, morpheme *-awi*, may simply refer to something bean-like (Castetter and Underhill 1935), with a possible identical meaning in Aymara with *chuwi*, Quechua with *chuwi*, and Xavante with *uhi*. The Aymara and Quechua words denoting pearl lupin were borrowed by a few European languages, retaining their basic meaning, such as in Italian and both European and South American Spanish.

It has been well known that white lupin (*Lupinus albus*) was widely cultivated in ancient times in Egypt, Greece, and Rome (van der Mey 1996). Certain pioneering studies with seminal significance defined ancient Egypt as its center of domestication (Zhukovsky 1929), estimating that other lupins except white were grown in the Nile Valley as far back as 2000 BC (Gladstones 1976). The opinion that white lupin originated in Greece is today the most widely accepted (Kurlovich 2002), mostly due to the existence of wild botanical varieties in the south of the Balkan Peninsula. This viewpoint is supported by the recent results of complex archaeobotanical analyses, concluding that white lupin could be imported into ancient Egypt much later (Cappers and Hamdy 2007), possibly after the conquests of Alexander the Great in the fourth century BC and already in its cultivated form (Gladstones 1974). From Greece, white lupin gradually spread both westwards, to the Appenine Peninsula, and eastwards, along the East Mediterranean coast, to Egypt. Thus, we may assume that it was primarily grown by farmers in ancient Greece, that it was adopted by various Semitic and other peoples, especially ancient Egyptians, and that it became well established in the Roman Empire (Annicchiarico et al. 2010), as both a natively cultivated and imported crop, curiously enough, from Egypt, as testified by the name of *Aegyptian bean* (du Cange 1883).

The travel of the white lupin crop and its name eastbound was thousands of miles long and lasted a few millennia, starting in Europe, passing through Asia Minor with the Near East and North Africa, and finishing on the steep slopes of the Pyrenees: a kind of ancient *Tour de Méditerranée*. The only attested Ancient Greek name denoting white lupin was *thérmos*, which, according to many, meant *hot* (Kurlovich 2002). On the other hand, it may be possible that the Proto-Indo-European root **(s)ter(ə)p-*, meaning something terminal or point-like (Pokorny 1959, Nikolayev 2012), produced not only the Latin *terminus* or the Old Indo-Aryan *tarman*, but also the Ancient Greek *térmi-s*, a fundamentally descriptive term depicting the crop's inflorescence shape (Mikić 2011a). As such, the word came in contact with some of the Near East Semitic languages, such as Hebrew, and with a later transfer into Arabic. Via their military conquests and cultural impacts, the Arabs spread the modified word across the Iberian Peninsula, as witnessed by the names in the Romance Catalan, Galician, or Spanish, as well as to the newly arrived Turkish tribes, also confirmed by one of the terms for white and some other lupin species.

Regarding the road that white lupin took westward toward the shores of ancient Italy, the fact that we have no linguistic trace of the Greek *thermos* in Latin or in the languages of some other neighboring Italic peoples may represent a veritable curiosity for centuries. All we know about a historical linguistic side of white and other lupins in the Roman Republic and Empire is the remarkably known Latin adjective *lupīnus*. This word became a designation for the Linnean genus (Linnaeus 1753, 1758) and, as presented in all the lists of the vernacular names and the accompanying tables in the first segment of this chapter, gave a basis for the words denoting all the lupin (spelling more common for UK English) or lupine (typical spelling for North American English) species in a vast number of ethnolinguistic families. Among them are:

- The Altaic, such as the Japonic with Japanese and the Turkic with Kazakh;
- The Indo-European, such as the Armenian, the Baltic with Latvian and Lithuanian, the Celtic with Irish, the Germanic, where Old High German *luvina* was borrowed from Latin and with Danish, Dutch, English,

German, Norwegian, and Swedish, the Hellenic with Greek, the Indo-Aryan with Bengali, the Romance with Catalan, Caterisani, French, Istriot, Italian, Portuguese, Romanian, Sicilian, and Spanish, and the Slavic with Belarusian, Croatian, Czech, Kashubian, Polish, Russian, Serbian, Slovak, Sorbian, and Ukrainian;
- The Niger-Congo with Swahili;
- The Tai-Kadai with Thai;
- The Uralic with Finnish and Northern Sami;
- The constructed languages, such as Esperanto and Ido.

It is usually considered that the names *lupīnus* and *Lupinus* are derived from the Latin noun *lupus*, referring to wolf, and, being an adjective, meaning *of a wolf* and *wolf's* (Lewis and Short 1879, Lewis 1890). One of the reasons for naming it in this way was an opinion that it preferred growing on *wolfish*—that is, poor and deserted, soils—where the wolves abide (Gligić 1954, Marin and Tatić 2004). The term *wolfish* is used in a metaphorical sense, such as an ancient Roman misconception that white lupin ravenously exhausts the soil (Collins Dictionaries 2014, Kew Science 2017). What was also considered a crucial similarity between lupins and wolves is that both bring death to animals, where the former did it with its naturally high alkaloid content in its grains (Lee et al. 2006).

Although it may cause more confusion in this already complex etymological and lexicological account, we will mention two words that, at least morphologically and semantically, could have some possible links with the white lupin crop. One of them is the Afroasiatic root **lap-*, meaning both faba bean (*Vicia faba* L.) and corn, and producing the Proto-Semitic **lupp-*, denoting only faba bean (Militarev and Stolbova 2007). Its only descendant is attested in the extinct Akkadian language, as *luppu*, spoken in ancient Mesopotamia, geographically not so close to the Mediterranean coast. Thus, the probability that this word had some connection with the Latin *lupīnus* seems verily small. Another root is the Proto-Indo-European root **leb-*, primarily referring to something blade-like, elongated, concave, and sharp (Pokorny 1959, Nikolayev 2012). It produced the Ancient Greek *lobó-s*, which denotes legume beans in general and which only traces in other languages are found eastward, in Armenian, as *lobi*, and Georgian, as *lobio*, both meaning faba bean (see Chapters 7 and 9). If this Ancient Greek word had anything to do with the Latin *lupīnus*, it has never drawn attention of comparative linguistics and, frankly, seems to leave us puzzled at the moment and with a hope that this complex riddle could be solved sometime in the future. It is probable that the Ancient Greek *thérmos*, through the Arabic modification *turmus*, and the Latin *lupīnus* met each other somewhere in the northernmost parts of the Iberian Peninsula, as seen in many Catalan names.

In some languages belonging to various ethnolinguistic families, lupins bear resemblance to various other legumes, such as to licorice (*Glycyrrhiza glabra* L.), with an English name for *Lupinus littoralis* Douglas; to pea (*Pisum sativum* L.), with the names in Bashkir, Galician, Galician-Asturian, Kazakh, Hill Mari, Spanish, and Welsh; and to an unspecified kind of legume bean, with the names in Adyghe, Azerbaijani, Cantonese and Mandarin Chinese, Icelandic, Japanese, Navajo, Turkish, and Vietnamese. One of the English names for *Lupinus perennis* associates it with beet (*Beta vulgaris* L.), most likely because of an abundant growth habit (Table 12.6).

The references to wolf are present in an extraordinarily large number of languages of all ethnolinguistic families and it would not make too much sense to provide the reader with their list. Much more interesting may be that the words of Slavic origin in Czech and Lower Sorbian, denoting *Lupinus angustifolius* (Table 12.2), link it with other beautiful beings, namely butterflies, or chicks in one of the English names for *Lupinus microcarpus.* The petals in some lupin species, native to the southern regions of the United States, especially *Lupinus texensis*, resemble the shape of the bonnets worn by the American pioneer women, and it is no wonder that it became a beloved state flower of Texas (Elliott 2008). Many American lupins are very popular and highly esteemed because of their ornamental value (Mikić 2015a).

There are some properties of the lupin flowers and the lupin grains composing the names denoting them in various languages: for instance, the scent of flowers in Finnish or the bitter taste of grains in Azerbaijani, Georgian, or Turkish. Almost sole references to the medicinal properties of lupins are found in Navajo, where the morpheme *yiłch'ozh* associates *Lupinus lepidus* with leaves used for chewing, probably as a kind of healing treatment, as well as the word *azee'bíni'í*, denoting *Lupinus argenteus,* that points out its role as a medicament. These examples confirm a genuine recognition of lupins as health-beneficial plants from time immemorial to this very day (Arnoldi et al. 2015).

13 *Phaseolus* L.

Synonyms: *Alepidocalyx* Piper; *Cadelium* Medik.; *Candelium* Medik.; *Caracalla* Tod.; *Lipusa* Alef.; *Minkelersia* M. Martens & Galeotti; *Phasellus* Medik.; *Phaseolos* St. Lag.; *Rudua* F. Maek.

13.1 LIST OF TAXA SCIENTIFIC AND POPULAR NAMES

The chapter on the genus *Phaseolus* is one of the extensive ones in the book. Unlike the genera *Lathyrus* L., *Lupinus* L. and *Vicia* L. that are presented across all the continents, or *Vigna* Savi, typical almost only for the Old World, the genus *Phaseolus* is, according to the present level of our knowledge, characteristic solely for the Americas (Debouck 1999, Delgado-Salinas et al. 1999, Freytag and Debouck 2002, Delgado-Salinas et al. 2006, Spataro et al. 2011, Bitocchi et al. 2017). The first section of this chapter brings a combined overview of its species and other taxa and the popular names for those that are most widely cultivated. These names resemble bridges among mutually distant regions of the world, with a constant and immensely rich traffic of word meanings and agricultural practices (ISTA 1982, Rehm 1994, Gledhill 2008, Porcher 2008, The Plant List 2013, Ecocrop 2017, EPPO 2017, Ethnologue 2017, IBIS 2017, ILDIS 2017, Logos 2017, NPGS 2017, Wikipedia 2017, Wiktionary 2017).

- *Phaseolus acinaciformis* Freytag & Debouck
English: spinal-ginger-like bean
- *Phaseolus acutifolius* A. Gray (Table 13.1)
- *Phaseolus acutifolius* A. Gray var. *acutifolius*
Synonyms: *Phaseolus acutifolius* var. *latifolius* G. F. Freeman
English: common tepary bean
- *Phaseolus acutifolius* A. Gray var. *tenuifolius* A. Gray
Synonyms: *Phaseolus tenuifolius* (A. Gray) Wooton & Standl.
English: fine-leaved tepary bean
Spanish: frejolillo
Spanish (Mexico): ejotillo
- *Phaseolus albescens* McVaugh ex R. Ramírez & A. Delgado
English: whitening bean
- *Phaseolus albiflorus* Freytag & Debouck
English: white-flowered bean
- *Phaseolus albinervus* Freytag & Debouck
English: white-veined bean
- *Phaseolus albiviolaceus* Freytag & Debouck
English: white-violet bean
- *Phaseolus altimontanus* Freytag & Debouck
English: high-mountainous bean

- *Phaseolus amabilis* Standl.
English: lovely bean
- *Phaseolus amblyosepalus* (Piper) C. V. Morton
Synonyms: *Alepidocalyx amblyosepalus* Piper
English: blunt-sepal bean
- *Phaseolus angustissimus* A. Gray
English: slimleaf bean
- *Phaseolus anisophyllus* (Piper) Freytag & Debouck
Synonyms: *Alepidocalyx anisophyllus* Piper
English: unequal-leaved bean
- *Phaseolus augusti* Harms
Synonyms: *Phaseolus bolivianus* Piper
English: majestic bean
Quechua: huillka
- *Phaseolus campanulatus* Freytag & Debouck
English: bell-shaped bean
- *Phaseolus carteri* Freytag & Debouck
English: Carter's bean
- *Phaseolus chiapasanus* Piper
English: Chiapas bean
Spanish (Mexico): frijol del monte; frijolillo
- *Phaseolus coccineus* L. (Table 13.2)
Synonyms: *Lipusa multiflora* Alef.; *Phaseolus griseus* Piper; *Phaseolus leiosepalus* Piper; *Phaseolus multiflorus* Lam.; *Phaseolus multiflorus* Willd.; *Phaseolus striatus* Brandegee; *Phaseolus strigillosus* Piper; *Phaseolus superbus* A. DC.; *Phaseolus vulgaris* var. *coccineus* L.
- *Phaseolus coccineus* L. subsp. *coccineus*
English: runner bean
- *Phaseolus coccineus* L. subsp. *coccineus* var. *argenteus* Freytag
English: silvery runner bean
- *Phaseolus coccineus* L. subsp. *coccineus* var. *coccineus*
Synonyms: *Phaseolus coccineus* subsp. *formosus* (Kunth) Maréchal et al.; *Phaseolus coccineus* subsp. *obvallatus* (Schltdl.) Maréchal et al.; *Phaseolus formosus* Kunth; *Phaseolus leiosepalus* Piper; *Phaseolus multiflorus* Lam.; *Phaseolus obvallatus* Schltdl.
English: true runner bean
- *Phaseolus coccineus* L. subsp. *coccineus* var. *condensatus* Freytag
English: dense runner bean
- *Phaseolus coccineus* L. subsp. *coccineus* var. *griseus* (Piper) Freytag
Synonyms: *Phaseolus coccineus* subsp. *griseus* (Piper) A. Delgado, nom. inval.; *Phaseolus griseus* Piper
English: gray runner bean
- *Phaseolus coccineus* L. subsp. *coccineus* var. *lineatibracteolatus* Freytag
English: linear-bracteole runner bean

- *Phaseolus coccineus* L. subsp. *coccineus* var. *parvibracteolatus* Freytag
English: small-bracteole runner bean
- *Phaseolus coccineus* L. subsp. *coccineus* var. *pubescens* Freytag
English: hairy runner bean
- *Phaseolus coccineus* L. subsp. *coccineus* var. *semperbracteolatus* Freytag
English: ever-bracteole runner bean
- *Phaseolus coccineus* L. subsp. *coccineus* var. *splendens* Freytag
English: splendid runner bean
- *Phaseolus coccineus* L. subsp. *coccineus* var. *strigillosus* (Piper) Freytag
Synonyms: *Phaseolus strigillosus* Piper
English: short-bristled runner bean
- *Phaseolus coccineus* L. subsp. *coccineus* var. *tridentatus* Freytag
English: three-toothed runner bean
- *Phaseolus coccineus* L. subsp. *coccineus* var. *zongolicensis* Freytag
English: Zongolica runner bean
- *Phaseolus coccineus* L. subsp. *striatus* (Brandegee) Freytag
English: striped runner bean
- *Phaseolus coccineus* L. subsp. *striatus* (Brandegee) Freytag var. *guatemalensis* Freytag
English: Guatemalan runner bean
- *Phaseolus coccineus* L. subsp. *striatus* (Brandegee) Freytag var. *minuticicatricatus* Freytag
English: fine-scar runner bean
- *Phaseolus coccineus* L. subsp. *striatus* (Brandegee) Freytag var. *pringlei* Rose ex Freytag
English: Pringle's runner bean
- *Phaseolus coccineus* L. subsp. *striatus* (Brandegee) Freytag var. *purpurascens* Freytag
English: purplish runner bean
- *Phaseolus coccineus* L. subsp. *striatus* (Brandegee) Freytag var. *rigidicaulis* Freytag
English: rigid-stem runner bean
- *Phaseolus coccineus* L. subsp. *striatus* (Brandegee) Freytag var. *striatus* (Brandegee) Freytag
Synonyms: *Phaseolus striatus* Brandegee; *Phaseolus striatus* var. *purpurascens* Freytag, nom. inval.
English: true striped runner bean
- *Phaseolus coccineus* L. subsp. *striatus* (Brandegee) Freytag var. *timilpanensis* Freytag
English: Timilpan runner bean
- *Phaseolus costaricensis* Freytag & Debouck
English: Costa Rican bean
- *Phaseolus dasycarpus* Freytag & Debouck
English: hairy-podded bean

- *Phaseolus dumosus* Macfad.
Synonyms: *Phaseolus coccineus* subsp. *darwinianus* Hern.-Xol. & Miranda-Colín; *Phaseolus coccineus* subsp. *polyanthus* (Greenm.) Maréchal et al.; *Phaseolus leucanthus* Piper; *Phaseolus polyanthus* Greenm.
English: bushy bean
French: haricot botil
Spanish (Colombia): cacha; matatropa; metatropa; petaco
Spanish (Ecuador): toda la vida
Spanish (Guatemala): dzich; ixich; juruna; piligüe; piloy; piloya
Spanish (Mexico): acaletl; botil; cenek; ibis; patlaxte; xuyumel
Spanish (Venezuela): frijol gallinazo; murutungo
- *Phaseolus esperanzae* Seaton
English: Esperanza bean
- *Phaseolus esquincensis* Freytag
English: twisted bean
- *Phaseolus filiformis* Benth.
Synonyms: *Phaseolus wrightii* A. Gray
Arabic: fasulia khiti
English: slimjim bean; slender-stem bean; Wright's phaseolus
Indonesian: kacang slimjim
- *Phaseolus glabellus* Piper
Synonyms: *Phaseolus coccineus* subsp. *glabellus* (Piper) A. Delgado, nom. inval.
English: glabrous-like bean
- *Phaseolus gladiolatus* Freytag & Debouck
English: sword-like bean
- *Phaseolus grayanus* Wooton & Standl.
Synonyms: *Phaseolus palmeri* Piper; *Phaseolus pedicellatus* var. *grayanus* (Wooton & Standl.) A. Delgado ex Isely; *Phaseolus pyramidalis* Freytag
English: Gray's bean
- *Phaseolus harmsianus* Diels
English: Harms' bean
- *Phaseolus hintonii* A. Delgado
English: Hinton's bean
- *Phaseolus jaliscanus* Piper
Synonyms: *Phaseolus sempervirens* Piper
English: Jalisciense bean
- *Phaseolus juquilensis* A. Delgado
English: Juquila bean
- *Phaseolus laxiflorus* Piper
English: loose-flowered bean
- *Phaseolus leptophyllus* G. Don
English: thin-leaved bean
- *Phaseolus leptostachyus* Benth.
Synonyms: *Phaseolus leptostachyus* f. *purpureus* Freytag, nom. inval.; *Phaseolus leptostachyus* var. *incisus* (Piper) Freytag, nom. inval.; *Phaseolus leptostachyus* var. *pinnatifolius* Freytag, nom. inval.

English: thin-stemmed bean
- *Phaseolus leptostachyus* Benth. var. *intonsus* (Piper) Freytag
English: leafy thin-stemmed bean
- *Phaseolus leptostachyus* Benth. var. *leptostachyus*
Synonyms: *Phaseolus anisotrichus* Schltdl.
English: true thin-stemmed bean
- *Phaseolus leptostachyus* Benth. var. *lobatifolius* Freytag
English: lobe-leaved thin-stemmed bean
- *Phaseolus leptostachyus* Benth. var. *nanus* Freytag
English: dwarfish thin-stemmed bean
- *Phaseolus lignosus* Britton
English: woody bean
- *Phaseolus longiplacentifer* Freytag
English: long-placenta bean
- *Phaseolus lunatus* L. (Table 13.3)
Synonyms: *Dolichos tonkinensis* Bui-Quang-Chieu; *Phaseolus bipunctatus* Jacq.; *Phaseolus falcatus* Benth. ex Hemsl., nom. nud.; *Phaseolus ilocanus* Blanco; *Phaseolus inamoenus* L.; *Phaseolus limensis* Macfad.; *Phaseolus lunatus* var. *lunatus* L.; *Phaseolus lunatus* var. *macrocarpus* (Moench) Benth.; *Phaseolus lunatus* var. *silvester* Baudet; *Phaseolus macrocarpus* Moench; *Phaseolus portoricensis* Spreng.; *Phaseolus puberulus* Kunth; *Phaseolus rosei* Piper; *Phaseolus saccharatus* Macfad.; *Phaseolus tunkinensis* Lour.; *Phaseolus viridis* Piper; *Phaseolus xuaresii* Zuccagni
- *Phaseolus macrolepis* Piper
English: large-scale bean
- *Phaseolus maculatifolius* Freytag & Debouck
English: spot-leaf bean
- *Phaseolus maculatus* Scheele
Arabic: fasulia mbq
Catalan: mongeta tacada
English: Metcalfe bean; prairie bean; spotted bean
Hebrew: sh'v't mnvkdt
Indonesian: kacang belang
- *Phaseolus maculatus* Scheele subsp. *maculatus*
Synonyms: *Phaseolus metcalfei* Wooton & Standl.; *Phaseolus ovatifolius* Piper; *Phaseolus retusus* Benth.
English: true spotted bean
- *Phaseolus maculatus* Scheele subsp. *ritensis* (M. E. Jones) Freytag
Synonyms: *Phaseolus ritensis* M. E. Jones
English: Santa Rita mountain bean
Spanish (Mexico): cocolmeca
- *Phaseolus macvaughii* A. Delgado
English: McVaugh's bean
Spanish: frijolillo
- *Phaseolus magnilobatus* Freytag & Debouck

English: large-lobed bean
- *Phaseolus marechalii* A. Delgado
English: Maréchal's bean
- *Phaseolus micranthus* Hook. & Arn.
Synonyms: *Phaseolus brevicalyx* Micheli
English: small-flowered bean
- *Phaseolus microcarpus* Mart.
English: small-fruited bean
- *Phaseolus mollis* Hook. f.
English: graceful bean
- *Phaseolus neglectus* F. J. Herm.
English: overlooked bean
- *Phaseolus nelsonii* Maréchal et al.
English: Nelson's bean
- *Phaseolus nodosus* Freytag & Debouck
English: conspicuous-nodded bean
- *Phaseolus novoleonensis* Debouck
English: Nuevo Léon bean
- *Phaseolus oaxacanus* Rose
Synonyms: *Phaseolus pedicellatus* var. *oaxacanus* (Rose) A. Delgado & nom. inval.
English: Oaxacan bean
- *Phaseolus oligospermus* Piper
English: few-seeded bean
- *Phaseolus opacus* Piper
English: dark-shaded bean
- *Phaseolus pachyrrhizoides* Harms
English: thick-rooted bean
- *Phaseolus parvifolius* Freytag
English: small-flowered bean
- *Phaseolus parvulus* Greene
English: very small bean
- *Phaseolus pauciflorus* Sessé & Moc. ex G. Don
Synonyms: *Minkelersia galactioides* M. Martens & Galeotti
English: few-flowered bean
- *Phaseolus pedicellatus* Benth.
Synonyms: *Phaseolus floribundus* Piper; *Phaseolus pedicellatus* var. *pedicellatus* Benth.
English: pallet-stalked bean
- *Phaseolus persistentus* Freytag & Debouck
English: persisting bean
- *Phaseolus plagiocylix* Harms
English: oblique-cup bean
- *Phaseolus pluriflorus* Maréchal et al.

Synonyms: *Phaseolus anisotrichus* subsp. *incisus* Piper
English: multiple-flowered bean
- *Phaseolus polymorphus* S. Watson
English: many-formed bean
- *Phaseolus polymorphus* S. Watson var. *albus* Freytag
English: white many-formed bean
- *Phaseolus polymorphus* S. Watson var. *polymorphus*
Synonyms: *Phaseolus pedicellatus* var. *polymorphus* (S. Watson) A. Delgado & nom. inval.; *Phaseolus schaffneri* Piper
English: true many-formed bean
- *Phaseolus polystachios* (L.) Britton et al.
English: beanvine; thicket bean
- *Phaseolus polystachios* (L.) Britton et al. subsp. *polystachios*
Synonyms: *Dolichos polystachios* L.
English: true beanvine
- *Phaseolus polystachios* (L.) Britton et al. subsp. *sinuatus* (Nutt. ex Torr. & A. Gray) Freytag
Synonyms: *Phaseolus polystachios* var. *sinuatus* (Nutt. ex Torr. & A. Gray) Maréchal et al.; *Phaseolus sinuatus* Nutt. ex Torr. & A. Gray
English: wavy-edged bean
- *Phaseolus polystachios* (L.) Britton et al. subsp. *smilacifolius* (Pollard) Freytag
Synonyms: *Phaseolus smilacifolius* Pollard
English: catbrier-leafed bean
- *Phaseolus purpusii* Brandegee
Synonyms: *Phaseolus pedicellatus* var. *purpusii* (Brandegee) A. Delgado & nom. inval.
English: Purpus' bean
- *Phaseolus reticulatus* Freytag & Debouck
English: netted bean
- *Phaseolus rimbachii* Standl.
English: Rimbach's bean
- *Phaseolus rosei* Piper
Arabic: fasulia urdi
English: Rose's bean
- *Phaseolus rotundatus* Freytag & Debouck
English: roundish bean
- *Phaseolus salicifolius* Piper
English: willow-leafed bean
- *Phaseolus scabrellus* Benth. ex S. Watson
English: rough bean
- *Phaseolus scrobiculatifolius* Freytag
English: valve wide-leafed bean
- *Phaseolus sonorensis* Standl.

English: Sonora bean
- *Phaseolus talamancensis* Debouck & Torres González
English: Talamanca bean
- *Phaseolus tenellus* Piper
English: tender bean
- *Phaseolus teulensis* Freytag
English: tiled bean
- *Phaseolus texensis* A. Delgado & W. R. Carr
English: Texas bean
- *Phaseolus trifidus* Freytag
English: three-segment bean
- *Phaseolus tuerckheimii* Donn. Sm.
English: Turckheim bean
- *Phaseolus venosus* Piper
English: veined bean
- *Phaseolus vulgaris* L. (Table 13.4)
Synonyms: *Phaseolus aborigineus* Burkart; *Phaseolus communis* Pritz.; *Phaseolus compressus* DC.; *Phaseolus esculentus* Salisb.; *Phaseolus nanus* L.; *Phaseolus vulgaris* var. *mexicanus* Freytag, nom. inval.
- *Phaseolus vulgaris* L. var. *aborigineus* (Burkart) Baudet
Synonyms: *Phaseolus aborigineus* Burkart
English: original common bean
- *Phaseolus vulgaris* L. var. *vulgaris*
Synonyms: *Phaseolus compressus* DC.; *Phaseolus compressus* var. *carneus* G. Martens; *Phaseolus compressus* var. *cervinus* G. Martens; *Phaseolus compressus* var. *ferrugineus* G. Martens; *Phaseolus ellipticus* var. *albus* G. Martens; *Phaseolus ellipticus* var. *aureolus* G. Martens; *Phaseolus ellipticus* var. *helvolus* Savi; *Phaseolus ellipticus* var. *mesomelos* Haberle; *Phaseolus ellipticus* var. *pictus* Caval.; *Phaseolus ellipticus* var. *spadiceus* G. Martens; *Phaseolus gonospermus* var. *oryzoides* G. Martens; *Phaseolus gonospermus* var. *variegatus* Savi; *Phaseolus oblongus* var. *albus* G. Martens; *Phaseolus oblongus* var. *spadiceus* Savi; *Phaseolus oblongus* var. *zebrinus* G. Martens; *Phaseolus sphaericus* var. *atropurpureus* G. Martens; *Phaseolus sphaericus* var. *minor* G. Martens; *Phaseolus vulgaris* var. *albus* Haberle; *Phaseolus vulgaris* var. *nanus* G. Martens; *Phaseolus vulgaris* var. *niger* G. Martens; *Phaseolus vulgaris* var. *ochraceus* Savi; *Phaseolus vulgaris* var. *variegatus* DC.; *Phaseolus zebra* var. *carneus* G. Martens; *Phaseolus zebra* var. *purpurascens* G. Martens
English: common bean; wild bean
- *Phaseolus xanthotrichus* Piper
English: yellow-haired bean
- *Phaseolus xolocotzii* A. Delgado
English: Xolocotzia bean
- *Phaseolus zimapanensis* A. Delgado
Synonyms: *Phaseolus xanthotrichus* var. *zimapanensis* ined.
English: Zimapán bean

13.2 ORIGIN OF SCIENTIFIC AND POPULAR TAXA NAMES

As already mentioned in the first paragraphs of this chapter, the genus *Phaseolus* is, today, considered endemic to the Americas. Although we may expect that extensive historical linguistic evidence relating to *Phaseolus* survived in the languages of the Native American peoples, it is extremely far from etymological abundance existing in other ethnolinguistic families of the world, especially the Indo-European. However, recent advances in a novel research discipline, paleobiolinguistics, bring hope that we may learn more about the history of terms related to *Phaseolus* crops in the Americas (Brown et al. 2014).

One of the very few examples referring to *Phaseolus* is the name *pawi*, used in the Northern Amerind Tohono O'odham language (Table 13.1) and

TABLE 13.1
Popular Names Denoting *Phaseolus acutifolius* A. Gray in Some World Languages and Other Linguistic Taxa

Language/Taxon	Name
Arabic	fasulia tibari
Catalan	mongetera tepary
Czech	fazol ostrolistý
Danish	tepary-bønne
Dutch	tepary-boon
English	escomite; pawi; pavi; tepari; tepary; yori mui; yori muni
Esperanto	pintfolia fazeolo; tepari-fazeolo
Estonian	tepariuba; teravalehine aeduba
French	haricot tépari
German	Teparybohne
Greek	fasóli tepari; meksikániko fasóli
Italian	fagiolo tepari; tepary
Japanese	teparī bīn
Malay	kacang tepary
Myanmar	taw-pe-nauk
Opata	tépar
Polish	fasola ostrolistna
Portuguese	feijão-tepari
Russian	fasol' ostrolistnaia; tepari
Serbian	oštrolisni grah; oštrolisni pasulj; tepari
Slovak	fazuľa končistolistá; fazuľa ostrolistá
Spanish	escomite; frijol tépari
Spanish (Costa Rica)	frijol piñuelero
Spanish (Mexico)	escomite; escumite; judía tépari; tépari
Tohono O'odham	pawi
Ukrainian	kvasolia hostrolista; tepari
Vietnamese	đậu tepary
Yucatec	xmayum

spoken by the people of the same name, living in the Sonoran Desert in Arizona and New Mexico. This was recorded as an answer by the Tohono O'odham farmers, renown for cultivating tepary bean (*Phaseolus acutifolius*), to the European visitors asking them what they were sowing: 'T'pawi!', meaning 'It's a bean!' (Castetter and Underhill 1935). The word denoting common bean (*Phaseolus vulgaris*) in two Southern Amerind languages Quechua, with *chuwi*, and Xavante, with *uhi*, could be in a distant relationship with the aforementioned Tohono O'odham name.

There are several more isolated examples of the names associated with *Phaseolus acutifolius* in the Amerind languages. An account from the seventeenth century mentions the word *tépar* in another Northern Amerind language, Opata, used to be spoken in north-central Sonora in Mexico and now officially extinct, also designating tepary bean (Pennington 1981). Among the English names referring to tepary bean used mainly in the United States, there is also an import of Amerind origin, namely *yori*, which is believed to denote non-native *Phaseolus* species, with a meaning of *non-Indian person's bean* (Felger and Moser 1985); however, this may not denote solely the European crops, such as faba bean (*Vicia faba* L.), but also the *Phaseolus* species from other regions of the vast American continent and, thus, not known to the native peoples living in the above-mentioned regions of Mexico and the United States.

It may be worth mentioning a possible connection among the names relating to tepary bean and common bean in three Northern Amerind languages, namely Cheyenne, Cree, and Yucatec (Tables 13.1 and 13.4), where the associative morpheme in all three words may refer to a flower color. The Nahuatl name referring to common bean was first borrowed and modified by the Spanish and then transferred into French and Jèrriais (Table 13.1), in which form it became in the international cuisine worldwide (Chamoux 1997, Polese 2006).

There exist more Amerind words relating to *Phaseolus* species, which were embraced together with a specific crop and modified by the West European overlords, with the examples like some Spanish names for *Phaseolus coccineus* and *Phaseolus vulgaris*, such as *ejote* from Nahuatl or *poroto* from Quechua (Tables 13.2 and 13.4).

The rich Austronesian language family readily welcomed the *Phaseolus* crops and easily found the words to denote them (Tables 13.1 through 13.4), since its members already had names to identify bean-like grain legumes, such as soybean (*Glycine max* [L.] Merr.), faba bean, and various *Vigna* species. All these designations could develop from a common ancestor, some still unattested Proto-Austronesian root, which may contain a hypothetical proto-morpheme **kac-, (k)ar-*. Such assumption may explain the variety of names, seen in the Borneo-Philippine with Malagasy and Minahasan, the Nuclear Malayo-Polinesian with

TABLE 13.2
Popular Names Denoting *Phaseolus coccineus* L. in Some World Languages and Other Linguistic Taxa

Language/Taxon	Name
Arabic	fasulia qrmzi
Basque	babarrun loregorria
Belarusian	fasolja ahnista-čyrvonaja; tureckija baby
Catalan	fesol fava de Sóller; fesolera vermella; mongetera vermella
Chinese (Cantonese)	hùhng fà choi dáu
Chinese (Mandarin)	hébāo dòu; hong hua cai dou; long zhao dou
Czech	fazol šarlatový
Danish	pralbønne
Dutch	pronkboon
English	multiflora bean; Oregon lima bean; perennial bean; red flowered runner bean; red flowered vegetable bean; runner bean; scarlet runner; scarlet runner bean; seven year bean
Esperanto	fajrofazeolo; skarlata fazeolo
Estonian	õisuba
Finnish	ruusupapu
French	haricot d'Espagne; haricot écarlate; haricot-fleur
German	Feuerbohne; Prunkbohne; Schminkbohne
German (Austria)	Käferbohne
Greek	fasóli ispanías
Hebrew	sh'v't skrlt
Indonesian	kacang runner
Irish	pónaire reatha
Italian	fagiolo a fiori rossi; fagiolo americano; fagiolo della regina; fagiolo di Spagna; fagiolo rampicante di Spagna; fagiolo scarlatto; fagiolone
Japanese	benibanaingen; hana mame; hana-sasage
Jèrriais	pais brantcheurs
Kashubian	wielekwiatowy bónk
Kazakh	türik burşağı
Korean	bulgeungangnamkong
Lithuanian	raudonžiedė pupelė
Nahuatl	ayecohtli; ayocohtli
Norwegian (Bokmål)	pralbønne
Norwegian (Nynorsk)	blomsterbønne; pralbønne; prydbønne
Persian	lubiai qrmz
Polish	fasola wielokwiatowa
Portuguese	feijão-da-Espanha; feijão-de-sete-anos; feijão-flor; feijão-trepador; feijoca; feijoeiro-escarlate
Russian	fasol' iarkokrasnaia; fasol' mnogotsvetkovaia; fasol' ognenno-krasnaia; fasol' turetskaia; turetskie boby
Serbian	mnogocvetni pasulj
Slovak	fazuľa šarlátová

(*Continued*)

TABLE 13.2 (*Continued*)
Popular Names Denoting Phaseolus coccineus L. in Some World Languages and Other Linguistic Taxa

Language/Taxon	Name
Slovenian	laški fižol; turški fižol
Sorbian (Upper)	turkowski bob
Spanish	ayecote; ayocote; judía encarnada; judía pinta; pilay
Spanish (Argentina)	chamborote; chilipuca; chomborote; judía encarnada; judia escarlata; judía pinta; pallar; poroto de espagna; poroto pallar
Spanish (Colombia)	frijol calentano
Spanish (Costa Rica)	cubá
Spanish (Cuba)	frijol angolano
Spanish (Ecuador)	popayán
Spanish (Guatemala)	chomborote; frijol chamborote; ixtapacal; piloy
Spanish (Mexico)	ayocote; botil; patol; tukamulil
Swedish	rosenböna
Ukrainian	kvasolia bahatokvitkova
Vietnamese	đậu sọn
Welsh	ffeuen ddringo

TABLE 13.3
Popular Names Denoting *Phaseolus lunatus* L. in Some World Languages and Other Linguistic Taxa

Language/Taxon	Name
Arabic	fasulia hlali
Basque	limako babarruna
Bulgarian	bob Lima; madagaskarski bob; maslen bob
Catalan	garrofó; mongeta de Lima
Chinese (Cantonese)	choi dáu
Chinese (Mandarin)	ai sheng xue dou; cai dou; jin dou; li ma dou; ling dou; long ya dou; mián dòu; xi mian dou; xue dou; yu dou
Czech	fazol barmský; fazol měsíční
Danish	limabønne; månebønne; sukkerbønne
Dutch	indische maanboon; lima-boon
English	Burma bean; butter bean; Duffin bean; large Lima bean; large white bean; lima bean; Madagascar bean; Rangoon bean; sieva bean; sugar bean
English (U.S.)	butter bean; Carolina bean
Esperanto	limofazeolo; lunfazeolo
Estonian	liima aeduba; liimauba
Finnish	limanpapu; voipapu
French	fève créole; fève de Java; haricot de Lima; haricot de Madagascar; haricot du Cap; haricot Lima à gros grains; pois du Cap

(*Continued*)

TABLE 13.3 (*Continued*)
Popular Names Denoting *Phaseolus lunatus* L. in Some World Languages and Other Linguistic Taxa

Language/Taxon	Name
French (Antilles)	pois chouche; pois doux
French (Canada)	fève créole; haricot du Cap; haricot de Madagascar; haricot de Siéva; haricot du Tchad; pois de 7 ans; pois de Java; pois du Cap; pois savon; pois souche
French (Haiti)	pois souche
French (Réunion)	gros pois
Galician	faba de Lima; feixón de Lima
German	Indische Mondbohne; Limabohne; Mondbohne; Sievabohne
Greek	fasólia tis Límas
Haitian Creole	pwachouk; pwadchous
Hebrew	sh'v't lmh
Hungarian	Limabab
Ilocano	patani
Indonesian	kacang jawa; kacang kratok; kacang lima; kacang mentega; kekara; koro legi; kratok
Irish	pónaire mhór; pónaire lima
Italian	fagiolo del Capo; fagiolo di Giava; fagiolo detto di Lima; fagiolo di Lima; fagiolo lunato
Japanese	aoi mame; rai mame
Javanese	kårå; kårå bethik; kårå legi; kårå manis; kratok
Korean	chaedu
Madurese	gribig; kratok
Malagasy	kabaro
Malay	kacang cina; kacang jawa; kacang kara; kacang s'ringing; kekara; kratok
Malay (Pontianak)	kacang merah
Malayalam	baṭṭar bīns
Minahasan	saru
Myanmar	htawbat pe; kal beir kan; kawl be; pe bra; pe byu gyi; pe gya; santagu pe; tim sin; tunoran
Polish	fasola limeńska; fasola półksiężycowata
Portuguese	bonje; fava-Belém; fava-de-lima; fava-terra; feijão-bonje; feijão-de-lima; feijão-espadinho; feijão-farinha; feijão-favona; feijão-fígado-de-galinha; feijão-manteiga; feijão-verde; feijoal; mangalô-amargo
Portuguese (Brazil)	feijão-fava
Russian	fasol' lima; fasol' limskaia; fasol' lunnaia; fasol' lunoobraznaia; fasol' lunovidnaia
Serbian	grah Lima; pasulj Lima
Slovak	fazuľa mesiacovitá
Spanish	frijol de luna; frijol Lima; frijol mantequilla; garrofó; garrofón; guaracaro; haba de Lima; haba lima; habones; judía de Lima; judía de manteca; Lima; pallar; poroto pallar
Spanish (Argentina)	frijol manteca; poroto manteca

(*Continued*)

TABLE 13.3 (*Continued*)
Popular Names Denoting *Phaseolus lunatus* L. in Some World Languages and Other Linguistic Taxa

Language/Taxon	Name
Spanish (Bolivia)	palato
Spanish (Colombia)	carauta; comba; frijol de año; torta
Spanish (Costa Rica)	kedeba
Spanish (Cuba)	frijol caballero
Spanish (Ecuador)	haba pallar
Spanish (El Salvador)	chilipuca
Spanish (Honduras)	chilipuca; chilipuco; fríjol chilipuca; fríjol reina; frijol viterra; judía limeña
Spanish (Mexico)	furuna
Spanish (Peru)	layo; pallar
Spanish (Puerto Rico)	haba
Sundanese	kacang jawa; kacang mas; kekara; roay
Swahili	mfiwi
Swedish	limaböna
Tagalog	patani
Tahitian	huero
Tetum	koto fuik; koto moruk; koto tisi
Thai	thua rachamat
Tongan	pe
Ukrainian	limska kvasolia
Uzbek	lima loviya
Vietnamese	đậu bơ; đậu Lima; đậu ngự; đậu sieva
Yucatec	ixtapacal; patashete

TABLE 13.4
Popular Names Denoting *Phaseolus vulgaris* L. in Some World Languages and Other Linguistic Taxa

Language/Taxon	Name
Abruzzan	fascinale
Afrikaans	boontjie
Ainu	mame
Albanian	fasule; grosha
Albanian (Arbëresh)	fasolla
Albanian (Arvanitika)	fasùlé
Amharic	adenigwarē
Andalusian	habichuelas; vainas verdes
Apulian	fasùle; pasulu; pasuli
Arabic	fasulia sha
Aragonese	alubia; chudía; chudiera; chodiga; fesol; fresol; fresuelo
Armenian	lobi; lobio

(*Continued*)

TABLE 13.4 (*Continued*)
Popular Names Denoting *Phaseolus vulgaris* L. in Some World Languages and Other Linguistic Taxa

Language/Taxon	Name
Aromanian	fâsuľilu
Asturian	fabas; fabes
Australian Kriol	bin
Avar	lubiya
Aymara	chuwi; phurut'i
Azerbaijani	adi lobya; bağla
Bashkir	fasol; noqat borsaσȷь
Basque	babarruna; banarra; indaba; leka; mailar
Bavarian	Buanschoan; Fisoin
Belarusian	fasolja zvyčajnaja
Berg en Terblijt	bwoan
Bergamasque	fasöl
Bilzen	baun
Bislama	ariko
Bolognese	fasôl
Bosniak	baštenski grah; boranija; grah; obični grah; pasulj; poljski grah; pop grah; vrtni grah; zelena boranija; zeleni grah
Brakel (Gelderland)	boèn
Bree	buu-uun
Brescian	fasól
Breton	fav-brezil; fav-glas; fav-sec'h
Budels	boén
Bulgarian	fasul
Bunjevac	gra; grava
Calabrian	faggiòla; fasciola; fasòla; fasòlu; fasúali; fasualu; pòsa; suriàca; vasuli
Caló	quindia
Carinthian German	Strankerl
Carinthian Slovenian	strok
Catalan	bajoquera; fesol; fesolera; mongeta; mongetera
Catanian	ttriáca
Caterisani	faggiola; posa
Chechen	qö
Cherokee	tuya
Cheyenne	ma'emonêškeho
Chinese (Cantonese)	sìjì dòu
Chinese (Mandarin)	bai fan dou; ban wen dou; cài dòu; qing dou; shi jia cai dou; si ji dou; yun dou
Chinese (Taiwan)	pan wên tou
Chuvash	šalśa părśi
Cree	miyicimin; pimiciacis
Crimean Tatar	baqla; paqla

(*Continued*)

TABLE 13.4 (*Continued*)
Popular Names Denoting *Phaseolus vulgaris* L. in Some World Languages and Other Linguistic Taxa

Language/Taxon	Name
Croatian	grah
Czech	fazol obecný
Danish	almindelig bønne; bønne; buskbønne
Dari	lubia
Dené	jígay
Dhao	kabui ae
Dutch	boon; gewone boon
Dyula	sɔsɔ
Elfdalian	byöna
Emilian	fasol; fasulein
Emilian-Romagnol	faṡö
English	bean; beans; black bean; common bean; field bean; flageolet bean; French bean; garden bean; green bean; haricot bean; kidney bean; navy bean; pinto bean; pop bean; snap; snap bean; string bean; wax bean
Esperanto	fazeolo
Estonian	harilik aeduba; türgi uba
Extremaduran	alubias; frijones
Eys	boeën
Ferraresi	fasòl
Fijian	bini
Finnish	tarhapapu
Finnish (Helsinki)	bōōna
Finnish (South-West)	pōōnā
French	haricot; haricot commun; haricot vert; mange-tout; mogette
French (Canada)	fèves; haricot
Frisian (North)	buan; green buan; guardbuan
Friulian	fasûl
Galician	chicho; fabeira; faba; feixó; feixón; feixóns verdes; habas
Gawwada	älälo
Geffen	bôn
Genk	boen
Genoese	faxeu
Georgian	ch'veulebrivi lobio; lebia; lobio
German	Buschbohne; Gartenbohne; Grüne Bohne; Kidney-Bohne; Nierenbohne; Perlbohne; Pintobohne; Schwarze Bohne; Stangenbohne; Wachtelbohne; Weiße Bohne
German (Austria)	Fisole
Greek	fasíolos o koinós; fasoláki annirihetikó
Greenladic	amitsukujooq eertaq
Griko	pasùli
Guarani	kumanda; saporo

(*Continued*)

TABLE 13.4 (*Continued*)
Popular Names Denoting *Phaseolus vulgaris* L. in Some World Languages and Other Linguistic Taxa

Language/Taxon	Name
Gujarati	vāla
Haitian Creole	pwa nouris
Hasselt	boeën
Hawaiian	pāpapa
Hebrew	sh'v't mtzvh
Heusden (Belgium)	boen
Hindi	bakala; biins
Hmong	taum
Horst	boën
Hungarian	fuszulyka; paszuly; paszulyka; veteménybab
Hunsel	boeën
Icelandic	matbaun
Ido	fazeolo
Ilocano	pardá
Indonesian	boncis; kacang; kacang biasa; kacang buah pinggang; kacang buncis; kacang hijau; kacang hitam; kacang liar; kacang pinto; kacang umum
Innu	shaieu
Interlingua	phaseolo
Irish	pónaire dhuánach; pónaire Fhrancach
Istriot	faʃól
Italian	fagiolo; fagiolo comune; fagiolo nano; fagiolo rampicante; fagioli da sgranare; fagiolini o cornetti; fagiuolo; fragiolo; mangiatutto; piattone
Japanese	ingen mame
Javanese	kacang boncis; kacang buncis
Jèrriais	haricot; pais d'mai
Joratian	favioûla
Kabyle	ibawen
Kampen	bone
Kannada	huruḷi
Kanne	boen
Kashubian	zwëczajny bónk
Kazakh	kädimgi ürme burşaq
Kempenlands (Eersel)	bón
Khmer	sandek
Kinyarwanda	igishyimbo
Kirundi	ibiharage; igiharage
Koersel	Boen
Kongo	dideso
Korean	deonggulgangnamkong; gangnangkong; juldangkong
Kupang Malay	boncis; kacang
Kurdish (Central)	fasuljha

(*Continued*)

TABLE 13.4 (*Continued*)
Popular Names Denoting *Phaseolus vulgaris* L. in Some World Languages and Other Linguistic Taxa

Language/Taxon	Name
Kurdish (Northern)	fasûlî; lobî; lobiye; lovî
Kyrgyz	buurçak
Ladin	fajöl
Ladino	frejol
Lak	şaαηnal qjuruv
Lao	makthov
Latvian	dārza pupiņas; parastās pupiņas
Lebbeke	boeën
Ligurian	faisoe; faxeu; faxoe
Lingala	lidɛ́só; lidɛ́su; madɛ́su; nkunde
Lithuanian	daržinė pupelė
Lombard	cornett; fasöl; fasöö
Lombard (Eastern)	fasoeu
Low German	breekbonen
Lunteren	boan
Luxembourgish	gréng boun
Maasai	engamuriki
Macedonian	grav
Malagasy	tsaramaso
Malay	kacang buncis; kacang mérah; kacang pendek
Malayalam	payar
Maltese	fażola; fulla
Mantovani	fasoeul
Manx	poanrey
Māori	pī nunui
Mapundungun	denüll; zegüj
Marathi	ghēvaḍā
Marchigiano	fasciólu; fagiòlu; fagiolettu; faciuole
Maxakalí	pëyõg; püyõg
Meänkieli	pööna
Mirandolesi	faśól
Modenese	fasól
Mofu-Gudur	aŋgar
Moravian	buny
Nahuatl	etl; exotl
Navarrese	calbotes; pochas
Neapolitan	fasulo
Nederasselt	bôn
Nepali	raajama
Nijswiller	bòn
Nivkh	tur

(*Continued*)

TABLE 13.4 (*Continued*)
Popular Names Denoting *Phaseolus vulgaris* L. in Some World Languages and Other Linguistic Taxa

Language/Taxon	Name
Norwegian (Bokmål)	hagebønne
Norwegian (Nynorsk)	hagebønne
Occitan	baneta; cotelet; faiòu; fasòls; favòl; mongeta; tecon
Odia	sima; śimẇa
Onze-Lieve-Vrouw-Waver	Boeën
Opglabbeeks	buun
Ossetian	qædoræ; qædur
Papiamento	bonchi; catjang bonchi
Parmesan	fasó
Pashto	lubia
Persian	lubia
Picard	po d'chuc
Piedmontese	faseul; fasoel
Polish	fasola zwykła
Portuguese	feijão; feijão-comum; feijão-vagem; feijoeiro
Punjabi	lobia
Quechua	chuwi; poroto; purutu; wayruru
Rapa Nui	'ariko; arikō
Reggiano	fasol
Reggiano Arsàve	gioiei
Riojan	caparrones; pochas
Romagnol	fasòl
Romanesco	faciolo
Romani	fasoj; graho
Romani (Serbia)	grasulj
Romanian	faseolea
Russian	fasol' obyknovennaia; fasol' ovoshchnaia; fasol' zelionaia
Salentino	fasule
Sami (Inari)	páápu
Sami (Northern)	bábut; báhpu
Sami (Skolt)	kåårak
Samogitian	popalė
Sanskrit	śimbī
Sardinian (Campidanese)	fasolu
Sardinian (Logudorese)	basolu
Sardinian (Unified)	basolu; fasoleddu; fasolu; fasou; pisu
Serbian	boranija; grah; običan pasulj; pasulj
Serbian (Belgrade)	suljpa
Serbian (Bosnian Frontier)	gra'; grah
Serbian (Dalmatia)	gra'
Serbian (Gallipoli)	grav; pasûj
Serbian (Gora)	grâ

(*Continued*)

TABLE 13.4 (*Continued*)
Popular Names Denoting *Phaseolus vulgaris* L. in Some World Languages and Other Linguistic Taxa

Language/Taxon	Name
Serbian (Montenegro)	gra'
Serbian (Great Morava Valley)	boranја; pasúlj; pasuljévina
Serbian (Užice)	grâ; grȁ; grȁorovina; maȕnljika; pasulj
Shona	bhinzi
Sicilian	faciola; fasòlu; fasulina; trujaca
Sinhalese	mǣkarala
Slovak	fazuľa obyčajná; fazuľa záhradná
Slovenian	navadni fižol
Somali	cambuulo; digir; digir guud
Soqotri	dengo
Sorbian (Upper)	niska buna; niski bob
Sotho	nawa
Spanish	alubias; caraotas; chícharos; fabas; fréjoles; frijoles; frijones; granos; habichuelas; judía común; judías; nuña; ñuñas; pochas; porotos; vainita
Spanish (Argentina)	chaucha; porotos
Spanish (Bolivia)	frijol; porotos
Spanish (Canary Islands)	habichuelas; vainas verdes
Spanish (Caribbean)	habichuelas
Spanish (Chile)	frijol; porotos; tabla
Spanish (Colombia)	blanquillo; cabecita negra; fríjol; frisol; grano; habichuelas; vainas verdes
Spanish (Costa Rica)	cubaces; frijol; frijoles; vainica
Spanish (Cuba)	ejote; frijol; frijoles; habichuelas; vainas verdes
Spanish (Dominican Republic)	habichuela
Spanish (Ecuador)	fréjol; poroto
Spanish (El Salvador)	ejote; frijol; frijoles
Spanish (Guatemala)	ejote; frijol; frijoles
Spanish (Honduras)	balas; balines; ejote; frijol; frijoles
Spanish (Mexico)	alubias; ejote; frijol; frijoles; frijoles bayos
Spanish (Nicaragua)	ejote; frijol; frijoles
Spanish (Panama)	frijoles; habichuelas; porotos; vainas verdes
Spanish (Paraguay)	chaucha; habilla; kumanda; porotos; saporo
Spanish (Peru)	frejol; frijol; poroto
Spanish (Uruguay)	chaucha; porotos
Spanish (Venezuela)	caraotas; frijoles
Sranan	bonki; katjang bonki
Stein	boean
Styrian German	Bohnschoten
Sundanese	buncis; kacang buncis
Swahili	maharagwe; haragi; haragwe; pojo; ukunde
Swedish	böna; buskböna; trädgårdsböna
Tagalog	sitaw

(*Continued*)

TABLE 13.4 (*Continued*)
Popular Names Denoting *Phaseolus vulgaris* L. in Some World Languages and Other Linguistic Taxa

Language/Taxon	Name
Tajik	lūbiё
Tatar	fasol'; nogıt borçagı
Tetum	fore; koto nurak
Thai	thạ̀w
Tigrinya	balidenigwa; fajolī
Tongan	ko'eni; piini
Triestine	fasol
Tswana	nawa
Turkish	fasulye
Turkmen	noýba
Tuvan	fasol
Twi	asɛ
Udmurt	fasol'
Ukrainian	kvasolia zvichaina
Umbric (Romance)	fasciolo
Urdu	biins
Uyghur	purtcaq
Uzbek	loviya
Valencian	bajoca; fesol; garrofo
Venetian	fasól; fasioi; fasiól; fasiòl; fasoler; fasolo; faxiołi destegołàr; faxiolo; faxołéti o cornéti
Venlo	boeën
Venray	boën
Viestan	fasul'
Vietnamese	đậu cô ve; đậu que; đậu ve
Waanrode	boën
Walshoutem	boën
Well	boën
Welsh	ffeuen ffrengig
West Frisian Dutch	bòòòn; bòòòòn; bòòòòòn
Xavante	uhi
Xhosa	imbotyi
Yavapai	mri'ka
Yiddish	bob
Yoruba	ẹwa
Yucatec	bu'ul
Zaza	lobiye; lovıke
Zulu	ubhontshisi
Zwartebroek	boon; kruuper

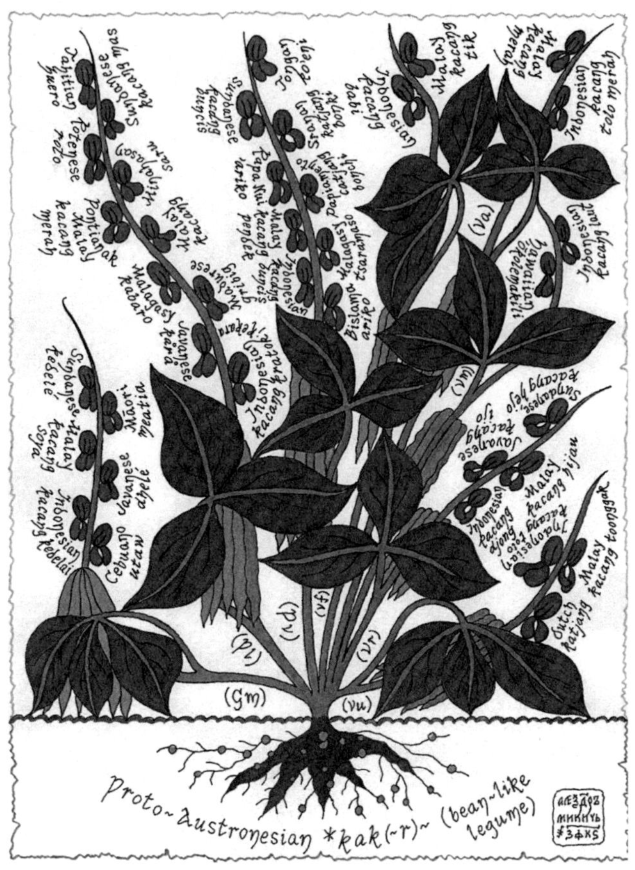

FIGURE 13.1 **(See color insert.)** One of the possible evolutions of the hypothetical Proto-Austronesian morpheme denoting a bean-like legume into its contemporary descendants and some Indo-European languages; the six basic meanings are interpreted as branches with pods, with (Gm) for *Glycine max*, (Pl) for *Phaseolus lunatus*, (Pv) for *Phaseolus vulgaris*, (Vf) for *Vicia faba*, (Vr) for *Vigna radiata*, and (Vu) for *Vigna unguiculata*, while their present words are illustrated as flowers.

Indonesian, Javanese, Madurese, Malay, Pontianak Malay, and Sundanese, the Oceanic with Rapa Nui and Tongan, and the creole languages with Bislama (Figure 13.1).

As may be seen in the case of many other pulse crops, some names did make a true voyage around the world, making full circles over an immensely vast geographic space (see, among others, Chapter 3). One such case is present in many Austronesian names denoting the *Phaseolus* and *Vigna* species, especially in those of the languages spoken in Indonesia and neighboring countries, for a few centuries effectually governed by the Dutch East India Company. The more or less common Malayo-Polinesian word *kacang*, denoting various kinds of bean-like pulses, was borrowed into Dutch as *katjang* together with cowpea, while the Dutch *boontjes*, referring to green bean, was imported along with this crop into Indonesian, Javanese,

Malay, and Sundanese, as well as to Portuguese, the Dutch archenemy in the Indonesian archipelago in the seventeenth century, and various Dutch-based creole languages, such as Kupang Malay, Papiamento, or Sranan (Tables 13.3 and 13.4).

The Linnean name *Phaseolus* is based upon the Latin noun *phasēlus* (Linnaeus 1753, 1758), which, in turn, was an adoption of the Ancient Greek *fáselis*, *phásēlos*, denoting cowpea (*Vicia unguiculata* [L.] Walp.). However, the etymology of this Ancient Greek name remains unknown. According to some, it may be of Pre-Greek or Mediterranean origin, having been borrowed by the Proto-Hellenic tribes at the time of their settlement in the Balkans (Beekes et al. 2010). We will take liberty and propose a candidate for this still empty place of the ultimate proto-word: it is the Proto-Indo-European root **bhask(')-*, denoting bundle and evolving into, among others, the Latin *fascis* and the Ancient Greek *phásko-s* (Nikolayev 2012). Perhaps this Ancient Greek word referred to a (cowpea) pod, where the grains are stuck together, just like another attested Proto-Indo-European root, *leg'-*, meaning *to gather*, brought forth the well-known *legūmen* for a pod (Mikić 2012). In any case, *phasēlus* and *Phaseolus* have become the basis for the names relating to the species of the eponymous genus, with a verily extraordinary abundance and diversity among those belonging to the Romance group, as testified by the following (Tables 13.1 through 13.4):

- The Afroasiatic, with the Semitic Arabic, Maltese, and Tigrinya;
- The Altaic, with the Turkic Bashkir, Tatar, Turkish, and Tuvan;
- The Indo-European, with Albanian, with the Germanic Austrian German and Bavarian, with the Hellenic Greek and Griko, with the Iranian Central and Northern Kurdish and Romani, with the Romance Abruzzan, Apulian, Aromanian, Bergamasque, Bolognese, Brescian, Calabrian, Catalan, Caterisani, Dalmatian, Emilian, Emilian-Romagnol, Ferraresi, Friulian, Galician, Genoese, Istriot, Ladin, Ladino, Ligurian, Lombard, Eastern Lombard, Mantovani, Marchigiano, Mirandolesi, Modenese, Neapolitan, Occitan, Parmesan, Piedmontese, Portuguese, Reggiano, Reggiano Arsàve, Romagnol, Romanesco, Romanian, Salentino, Campidanese and Logudorese Sardinian, Sicilian, Spanish, Triestine, Umbric, Valencian, Venetian and Viestan, and with the Slavic Belarusian, Bulgarian, Czech, Polish, Russian, Serbian, Moravian Serbian, Slovak, Slovenian, and Ukrainian;
- The Uralic, with Hungarian and Udmurt;
- The constructed languages, with Esperanto, Ido, and Interlingua.

If we try to simplify all this remarkably extensive and abundant set of data, given in the previous two paragraphs, it may become quite obvious that the names based upon the Latin *phasēlus* dominate in the southern half of Europe, while those akin to the words denoting faba bean prevail in the northern part. This may be easy to explain, since cowpea is a warm season annual legume and thus was unknown to, for example, Baltic, Celtic, and Germanic peoples, who, on the other hand, cultivated faba bean as a typical cool season legume. Thus, cowpea used to be widely spread from the Iberian Peninsula to Asia Minor and left its linguistic traces. The border between these two European halves, that is, faba bean *versus* cowpea, may pass through modern France, over southern Germany and Austria, and end amidst the Slavic homeland in East Europe. This is the reason why we find *phasēlus*-based words solely in Austrian German and Bavarian and why almost all Slavic languages, with borrowings to neighboring Uralic languages, have the names founded on both faba bean and *phasēlus* (Tables 13.1 through 13.4).

Since its arrival in the Old World, following the West European conquests of what today is Latin America, the *Phaseolus* beans were widely domesticated in Europe and to such extent that Europe may easily be considered the second center of diversity of this rich and economically highly important genus (Angioi et al. 2010). Similar events happened in Africa (Gepts and Bliss 1988) and Asia (Rahmianna and Radjit 2000). Several *Phaseolus* species, most notably common bean, partially or completely replaced numerous native Eurasian pulse crops similarly used, such as chickpea (*Cicer arietinum* L.), grass pea (*Lathyrus sativus* L.) and especially faba bean and cowpea. Thus, today, the word denoting edible bean in the Eurasian ethnolinguistic families is almost or fully identical for common and faba bean (see Chapter 7). The examples in various languages may be found in the supplementary tables, such as:

- The Afroasiatic Maltese;
- The Altaic, with the Turkic Azerbaijani, Crimean Tatar, and Uzbek;
- The Dené-Caucasian, with Basque and with the Caucasian Avar and Chechen;
- The Indo-European, with Armenian, with the Baltic Lithuanian and Samogitian, the Celtic Breton, Irish, Manx, and Welsh, the Germanic Afrikaans, Bavarian, Danish, Dutch with its dozens of dialects and rather local speeches, Frisian, German, Icelandic, Low German, Luxembourgish, Norwegian, Styrian German, Swedish, and Yiddish, the Indo-Aryan Hindi and Punjabi, the Iranian Dari, Northern Kurdish, Pashto, Persian, Tajik, and Zaza, the Romance Asturian, French, Galician, Joratian, Portuguese, and Spanish, the Slavic Belarusian, Bulgarian, Moravian Czech, Kashubian, Russian, and Upper Sorbian;
- The Kartvelian Georgian;
- The Uralic, with Estonian, Helsinki Finnish, South-West Finnish, Meänkieli, Inari Sami, Northern Sami, and Skolt Sami;
- The creole, with Papiamento.

In diverse languages, the *Phaseolus* species resembled their speakers of other more common legume crops, such as to:

- Peanut (*Arachis hypogaea* L.), in the Amerind Guarani;
- *Lathyrus japonicus* Willd., in the Paleosiberian Nivkh, where this wild relative of grass pea was regarded and used as pea in local cuisine;
- Pea (*Pisum sativum* L.) in the Altaic with the Turkic Bashkir, Chuvash, Kazakh, Kyrgyz, Tatar, Turkmen, and Uyghur, the Eskimo-Aleut with Greenlandic, the Indo-European with the Iranian Ossetian, the Romance French, Jèrriais, and Picard, the Slavic Bunjevac, Croatian, Macedonian, Gora Serbian, Morava Serbian, and standard Serbian, in the Uralic with Hungarian, and in the creole languages with Haitian Creole;

- A kind of bean, in the Amerind with Nahuatl, the Altaic with Ainu, Korean, and Japanese, the Austroasiatic with Khmer and Vietnamese, the Indo-European with the Romance Extremaduran and the Slavic Carinthian Slovenian, the Niger-Congo with the Lingala and Swahili, the Sino-Tibetan with Cantonese, Mandarin, and Taiwan Chinese and Myanmar, and the Tai-Kadai with Thai and Lao.

Curiously, one of the Romani words relating to common bean is a hybrid of two Serbian words denoting common bean, *grah* + *pasulj* = *grasulj* (Table 13.4).

Apart from the Americas, the *Phaseolus* beans were quite correctly denoted as coming from few initial countries of their spread into the Old World, such as Spain in Greek, Italian or Portuguese, and Turkey, in Belarusian, Kazakh, Russian, or Slovenian. Among the imagined places associated with *Phaseolus* beans are the Brazil in Breton, Brazilian city Belém in Portuguese, Burma (Myanmar) in Czech, China in Malay, India in Dutch and German, Java in Italian, Judaea in Spanish, Madagascar in Bulgarian and English. The richest seems to be French, with Cape Town in South Africa, Chad, Madagascar, and Spain (Tables 13.1 through 13.4).

It may be worth mentioning some terms that are specific to the use of the *Phaseolus* species in the form of immature (or, as usually referred to, green) pods. One of them is the Afroasiatic, more precisely, Semitic Arabic *būrāniyyä*, designating immature pods of grain legumes, which, via the Turkish *borani, burani* ended in the Slavic Croatian and Serbian languages, with the preserved initial meaning, but mostly referring to common bean (Vujaklija 1980, Mikić-Vragolić et al. 2007). Another is the Indo-European Romance compound word denoting that both immature pod and immature seeds are edible, *eat-all*, or, in French, *mangetout*. This term is basically invented for the pea varieties, which are in the same way for human consumption (see Chapter 14). However, it began to denote mainly the cultivars of *Phaseolus vulgaris*, commonly known as French beans, and selected for the reduced presence or complete absence of cellulose threads in the pod tissue (Ngelenzi et al. 2016). Such names are present, apart from French, in Catalan, Italian, and Occitan. It may be curious to point out one more term relating to exclusively immature common bean pods and grains, attested only in metropolitan Spanish and Spanish in Andalusia, the Canary Islands, Colombia, Costa Rica, Cuba, and Panama, with a basic form of *vaina* and being derived from the Latin word *vāgīna*, initially signifying a kind of envelope or sheath and subsequently becoming to refer to a pod, as an *envelope* for the grains (Table 13.4).

It has been known for long that the consumption of many pulse crops, however essential in human diets throughout space and time, has an undesirable and unpleasant side effect in the form of bloating and flatulence. This has been widely assessed in diverse local cultures across the world, where common and other *Phaseolus* species have a distinguished place (Manderson 1981, Mintz and Tan 2001, Quiroga et al. 2012). Among countless humorous anecdotes relating to this peculiar aspect of eating beans, thoroughly recorded by ethnologists, there is one from the region of Užice in modern-day western Serbia: "Milojko podiže visoko jednu nogu i odvali iz puške grašare"—"Milojko (a typical male name in those parts) lifted a leg high and fired a shot from his bean-gun (*derrière*) as strongly as he could" (Cvijetić 2014).

14 *Pisum* L.

14.1 LIST OF TAXA SCIENTIFIC AND POPULAR NAMES

The genus *Pisum* may be credited for several milestones in plant research, pointing out its uniqueness among other pulses and grain crops. First of all, it is one of the most ancient domesticated plants in the world (Kislev and Bar-Yosef 1988, Zohary et al. 2012). Secondly, the science of plant genetics was founded on the studies of the mechanism of inheritance of this species a century and a half ago (Mendel 1866, Ellis et al. 2011). Finally, it is regarded as the pioneering legume species in extracting ancient DNA (Medović et al. 2010, 2011, Jovanović et al. 2011, Smýkal et al. 2014, Mikić 2015c). Yet, despite all these extraordinary achievements, there is a lot that remains to be assessed in numerous research fields, from crop history (Mikić et al. 2014a) and taxonomy (Sinjushin and Belyakova 2010, Sinjushin and Demidenko 2010, Jing et al. 2012, Schaefer et al. 2012, Mikić et al. 2014c, Vishnyakova et al. 2016), over genomics (Kosterin and Bogdanova 2008, Kosterin et al. 2010, Smýkal et al. 2015) to agronomy and industrial uses (Mikić et al. 2015a). To put some more weight to these diverse future tasks, we may add that the genus *Pisum* L., with its most cultivated representative, is one of the pioneering crops in demonstrating how historical linguistics may assist in an integrated and multidisciplinary approach to cast more light onto domestication (Mikić 2011c, Mikić et al. 2014a). In a way, it is no wonder that despite the fact that it contains just a few species, and along with all these still addressing issues, the genus *Pisum* L. demonstrates an immense richness in popular names denoting its species and various subtaxa, offering more material for various scientific efforts in the future (Makasheva 1979, ISTA 1982, Rehm 1994, Maxted and Ambrose 2001, Gledhill 2008, Porcher 2008, The Plant List 2013, Ecocrop 2017, EPPO 2017, Ethnologue 2017, IBIS 2017, ILDIS 2017, Logos 2017, NPGS 2017, Wikipedia 2017, Wiktionary 2017).

- *Pisum abyssinicum* A. Braun

Synonyms: *Pisum abyssinicum* var. *vacilaianum* A. Braun; *Pisum sativum* subsp. *abyssinicum* (A. Braun) Govorov

English: Abyssinian pea; Ethiopian pea

Estonian: Abessiinia hernes

Italian: pisello d'Abyssinie

Kurdish (Northern): Şoqil

Occitan: pese d'Abissinia

Serbian: abisinski grašak; etiopski grašak

- *Pisum formosum* (Steven) Alef.

Synonyms: *Alophotropis aucheri* (Jaub. & Spach) Grossh.; *Alophotropis aucheri* (Jaub. & Spach) Grossh.; *Alophotropis formosa* (Stev.) Grossh.;

Lathyrus frigidus Schott & Kotschy; *Orobus formosus* Steven; *Orobus formosus* var. *microphyllus* Ser.; *Pisum aucheri* Jaub. et Spach.; *Pisum formosum* Alef. var. *microphyllum* Ser.; *Pisum formosum* Alef. var. *pubescens* C. C. Towns.; *Pisum formosum* Alef. var. *typicum* Gov.; Vavilovia *formosa* (Steven) Fedorov; *Vicia aucheri* Boiss.; *Vicia variegata* var. *aucheri* (Jaub. & Spach) Bornm.

English: Aucher's pea; Aucher's vetch; beautiful pea; beautiful vavilovia; cold vetchling; vavilovia

Kazakh: äsem burşaq

Russian: gorokh mnogoletnyi; gorokh Oshe; krasivyi gorokh; prekrasnyi gorokh; sochevichnik krasivyi; vavilovia; vavilovia prekrasnaia

Serbian: prekrasna vavilovija; prekrasni grašak; vavilovija

- *Pisum fulvum* Sm.

Synonyms: *Pisum fulvum* var. *amphicarpum* Warb. & Eig

English: Middle-Eastern wild climbing pea; red-yellow pea; tawny pea

Hebrew: afun matzuy

Serbian: mrkožuti grašak

- *Pisum sativum* L.

Synonyms: *Lathyrus oleraceus* Lam.; *Pisum arvense* L.; *Pisum biflorum* Raf.; *Pisum elatius* M. Bieb.; *Pisum humile* Boiss. & Noe; *Pisum sativum* subsp. *tibetanicum* ined.; *Pisum vulgare* Jundz.

- *Pisum sativum* L. subsp. *asiaticum* Govorov

English: Asian pea

Kurdish (Northern): nokekesk an baqilê xatûnî ya rihayê

- *Pisum sativum* L. subsp. *elatius* (M. Bieb.) Asch. & Graebn.

Armenian: volorr bardzr

English: tall pea

Hebrew: afun kippeach; afun kipper

Kazakh: biïk burşaq

Serbian: visoki grašak

- *Pisum sativum* L. subsp. *elatius* (M. Bieb.) Asch. & Graebn. var. *brevipedunculatum* P. H. Davis & Meikle

English: short-peduncle tall pea

- *Pisum sativum* L. subsp. *elatius* (M. Bieb.) Asch. & Graebn. var. *elatius* (M. Bieb.) Alef.

Synonyms: *Pisum elatius* M. Bieb.

English: common tall pea

- *Pisum sativum* L. subsp. *elatius* (M. Bieb.) Asch. & Graebn. var. *pumilio* Meikle

Synonyms: *Pisum arvense* subsp. *humile* Holmboe; *Pisum humile* Boiss. & Noë; *Pisum pumilio* (Meikle) Greuter; *Pisum sativum* subsp. *humile* (Holmboe) Greuter et al.; *Pisum sativum* subsp. *syriacum* A. Berger; *Pisum syriacum* (A. Berger) C. O. Lehm.

English: Syrian fodder pea; Syrian pea

French: pois fourrager de Syrie

Hebrew: afun namuch

Kazakh: jatağan burşaq

Kurdish (Northern): polik
- *Pisum sativum* L. subsp. *jomardii* (Schrank) Kosterin
Synonyms: *Pisum jomardii* Schrank
English: Jomard's pea
- *Pisum sativum* L. subsp. *sativum*
Synonyms: *Pisum arvense* var. *hibernicum* O. Schwarz; *Pisum arvense* var. *vernale* L.; *Pisum commune* Clavaud; *Pisum sativum* subsp. *commune* (Clavaud) Govorov
English: cultivated pea
- *Pisum sativum* L. subsp. *sativum* var. *arvense* (L.) Poir. (Table 14.1)
Synonyms: *Pisum arvense* L.
- *Pisum sativum* L. subsp. *sativum* var. *macrocarpum* Ser. (Table 14.2)
Synonyms: *Pisum macrocarpum* (Ser.) Sturtev.; *Pisum saccharatum* (Ser.) hort. ex Rchb.; *Pisum sativum* subsp. *sativum* var. *macrocarpon* Ser.; *Pisum sativum* var. *saccharatum* Ser.; *Pisum sativum var.-gr. axiphium* Alef.
- *Pisum sativum* L. subsp. *sativum* var. *sativum* (Table 14.3)
- *Pisum sativum* L. subsp. *transcaucasicum* Govorov
Synonyms: *Pisum transcaucasicum* (Govorov) Stankov
English: Caucasian pea
Kurdish (Northern): bezelî
Russian: barda

14.2 ORIGIN OF SCIENTIFIC AND POPULAR TAXA NAMES

The words denoting common pea (*Pisum sativum*), which firmly established the status of one of the primeval cultivated plants in the world millennia ago, are immensely numerous and can be found in nearly all the ethnolinguistic families of North Africa, Asia, and Europe (Tables 14.1 through 14.3).

The available resources on the names relating to pea in the native languages of the Americans, which are traditionally associated with the first wave of inhabitation of the continent from Asia (Cavalli-Sforza and Seielstad 2001) and that are, mostly by

TABLE 14.1
Popular Names Denoting *Pisum sativum* L. Subsp. *sativum* var. *arvense* (L.) Poir. in Some World Languages and Other Linguistic Taxa

Language/Taxon	Name
Armenian	volorr dashtayin
Bulgarian	polski furazhen grah
Chinese	tian wan dou
Chuvash	hir părśa
Croatian	krmni grašak
Czech	hrách setý rolní; peluška
Danish	carlin; helsæd; markært

(Continued)

TABLE 14.1 (*Continued*)
Popular Names Denoting *Pisum sativum* L. Subsp. *sativum* var. *arvense* (L.) Poir. in Some World Languages and Other Linguistic Taxa

Language/Taxon	Name
Dutch	grauwe erwt
English	Austrian winter pea; dun pea; field pea; grey pea; mutter pea; partridge pea; peluskins
Estonian	hall hernes
Finnish	peltoherneiseen
French	pois de champs; pois fourrager
German	Ackererbse; Felderbse; Futtererbse; Grünfuttererbse; Peluschken
Italian	pisello da foraggio; pisello foraggero
Japanese	ao endo; kō kyōshu endō
Kazakh	may burşaq
Korean	bulgeunwandu
Kurdish (Northern)	polika mêrgê
Macedonian	stochen grashok
Occitan	pese dels camps; pese ferratgièr
Polish	groch błękitnopurpurowy; groch polny; peluszka
Portuguese	ervilha griséus
Russian	gorokh kormovyi; peliushka
Serbian	krmni grašak; poljski grašak
Sicilian	pisedda furaggera
Slovak	hrach siaty roľný; peluška
Slovenian	krmni grah
Sorbian (Lower)	rolny groch
Sorbian (Upper)	rólny hroch
Spanish	guisante de campo; guisante de huerta; guisante forragero
Turkish	tarla bezelyesi

non-mainstream linguists, designated as Amerind (Greenberg and Ruhlen 1992), are rather scarce, especially in the northern part of this vast continent. This is somewhat expectable, since pea is not an American native crop. It may be possible that the Proto-Amerind root root **icu*, which primarily denotes seed (Greenberg and Ruhlen 2007) and, also, peanut (*Arachis hypogaea* L.), such as in several modern Southern Amerind languages (see Chapter 3), was preserved in the names referring to pea in Cheyenne and Plain, West, and Woods Cree (Table 14.3). This cautiously proposed suggestion could be supported by the analogy between the morphemes *-kemo-* in Cheyenne and *-îcîmi-* and *- iichiimi-* in Cree, as well as the fact that the Cheyenne word literally means *round bean (seed).*

There are two Proto-Afroasiatic roots relating, to a smaller or greater extent, to pea (Militarev and Stolbova 2007): the **lay/w-* ~ **ʔVll-* ~ **w/yVlal-*, designing a

TABLE 14.2
Popular Names Denoting *Pisum sativum* L. Subsp. *sativum* var. *macrocarpum* Ser. in Some World Languages and Other Linguistic Taxa

Language/Taxon	Name
Bulgarian	kulinaren grah
Catalan	estirabec; tirabec; pèsol Caputxí; pèsol fi
Chinese	tian cui wan dou; he lan dou
Czech	hrách dřeňový; hrách cukrový; hrášek
Danish	sukkerært
Dari	nxud frngii
Dutch	peul; peulen; peultjes; suikererwt; suikerpeul; vleeserwt; vleespeul
English	eat-all pea; edible-pod pea; flat-podded snow pea; sickle pea; snap pea; snow pea; sugar pea; sugar snap pea
Esperanto	sukerpizo
Estonian	udi-suhkruhernes
Finnish	sokeriherne
Flemish (West)	mangetout
French	croquetout; mangetout; pois gourmand; pois mangetout à cosse plate; pois mangetout à gousse plate
Galician	tirabeque
German	Gemeine Zuckererbse; Kaiserschoten; Kiefelerbsen; Kefe; Mange-tout; Zuckerbrecherbse; Zuckererbse; Zuckerpalerbse; Zuckerschwerterbse
Hebrew	afunat hagina
Hungarian	cukorborsó
Icelandic	snjóertur
Indonesian	kacang kapri; kapri
Italian	pisello mangiatutto; taccola
Japanese	saya endou
Korean	baegseolkong
Kurdish (Northern)	polika zadeganan
Lak	s:iru
Latvian	cukurzirņi
Nahali	malkā
Norwegian (Bokmål)	sukkerert
Occitan	pese manjatot
Russian	gorokh sakharnyi
Serbian	grašak šećerac
Slovenian	grah cukrar; sladki grah
Sorbian (Lower)	cukorowy groch
Sorbian (Upper)	cokorowy hroch
Spanish	arveja; cometodo; miracielo; tirabeque
Sundanese	kacang kapri; kapri
Swahili	mnjegere sukari
Swedish	sockerärt
Tetum	ervilha
Vietnamese	đậu hoà lan
Welsh	pysen felys

TABLE 14.3
Popular Names Denoting *Pisum sativum* L. Subsp. *sativum* var. *sativum* in Some World Languages and Other Linguistic Taxa

Language/Taxon	Name
Abaza	k'yrk'yrlaš
Abkhaz	k'yrk'yrra
Adyghe	nekhut
Afrikaans	ertjie
Aghul	xur
Ainu	mame
Akkadian	lillânu
Akusha	qara
Albanian	bizele; biselja; modhë
Albanian (Arvanitika)	pizèlle
Antwerp	aert
Apeldoorner	ätte
Arabic	bazla
Aragonese	arbella; bisalto; guixón
Aramaic	ateri
Aranese	pèsol
Archi	čaq
Arem	do:ʔ
Armenian	volorr; volorr ts'anovi
Aromanian	grãshac; madzârea
Asturian	arbeyu
Avar	xiʎálo
Azerbaijani	bozbaş; əkin noxudu
Bagvalal	hal
Balearic (Felanitx)	estiragassó
Balearic (Llevant)	xítxero; xitxo
Balearic (Santa Margalida)	pitxo
Balearic (Sóller)	xítxol
Balti	garaz
Bashkir	borsaq
Basque	ilar; ilarra
Basque (Biscayan)	irar
Basque (Souletin)	ilhar-biríbil; ilhar-xúri
Bats	mukhudo
Belarusian	haroch pasjaŭny
Bengali	maṭaraśum̐ṭi
Bergamasque	roàia
Berlaar	aat
Beurla Reagaird	peasair
Bezhta	holo

(*Continued*)

TABLE 14.3 (*Continued*)
Popular Names Denoting *Pisum sativum* L. Subsp. *sativum* var. *sativum* in Some World Languages and Other Linguistic Taxa

Language/Taxon	Name
Bihari	kērā'i
Bilzen	êrt
Bolognese	arvajja
Bonan	pəžag
Bosniak	grašak
Botlikh	hali
Bouyei	tu
Boyko	horoch
Brescian	roaiot
Breton	piz; piz-bihan; pizenn
Brithenig	pies
Bruges	erreweete
Budels	êrt
Bulgarian	gradinski grah; grah
Bunjevac	grašak
Burgundian (Romance)	poi
Burushaski (Hunza)	ɣaráṣ
Burushaski (Nagar)	ɣaráṣ
Burushaski (Yasin)	ɣaráṣ
Buryat	būrsag
Calabrian	pisillu; pisiddru; prisedda
Catalan	pèsol
Catanian	pusedda
Caterisani	poseda
Cebuano	gisantes
Central Siberian Yupik	unátaq
Chadakolob	holó
Chechen	qöş
Cherokee	duyunasdi
Chewa	nswawa
Cheyenne	ová'kemonėškeho
Chinese (Cantonese)	qīng dòu
Chinese (Mandarin)	nen wan dou; wāndòu; wāndòu shŭ
Chirag	qara
Chulym	myrçaq
Chuvash	aka pǎrśa
Colognian	Ääze
Cornish	pýsen
Corsican	piseddu
Cree (Plains)	ayicîminak; mîcîminak
Cree (West)	iichiiminach

(*Continued*)

TABLE 14.3 (*Continued*)
Popular Names Denoting *Pisum sativum* L. Subsp. *sativum* var. *sativum* in Some World Languages and Other Linguistic Taxa

Language/Taxon	Name
Cree (Woods)	mîcîmin; splēpa
Crimean Tatar	nohut
Croatian	grašak; povrtni grašak
Cumbric	pissenn
Czech	hrách setý; polní hrachy; zahradní hrachy
Danish	almindelig ært; haveært; marvært; skalært
Dargwa	qara
Daur	borečō
Drents	aart; arft(e); art
Dutch	doperwt; erwt; gele erwt; kreukerwt; ronde groene erwt
Dutch (Middle)	aerwete; arwete; erwete
Dutch Low Saxon	atepoele
Eastern Yugur	purčag
Eeklo	irrewete
Elfdalian	ert
English	garden pea; green pea; marrowfat pea; pea; peas; pease; shelling pea
English (Early Modern)	pease; peasen
English (Middle)	grene pēse; pease; pece; peis-e; peisse; peose; pes; pēse; pesse; pise; poese
English (Old)	pise; peose; poise
English (Ulster)	pea
Erzya	ksnav
Esperanto	ĝardena pizo; medola pizo; pizo; ŝelpizo; verdaj pizoj
Estonian	harilik hernes
Extremaduran	arvilla
Eys	èëts
Faroese	ertur
Ferraresi	ruviè
Filipino	gisantes; kagyus
Finnish	herne; tarhaherneisiin
Finnish (Satakunta)	rista
Finnish (Savonian)	rokka
Finnish (South-West)	papu
Flemish	erwt
Flemish (West)	errewete; siererrewete
Forth and Bargy	piz; pizzen
Franco-Provençal	pês
French	petit pois; pois des jardins; pois cassé; pois cultivé; pois potager; pois protéagineux
French (Old)	pisos mauriscos
Frisian (North)	eert; irt

(*Continued*)

TABLE 14.3 (*Continued*)
Popular Names Denoting *Pisum sativum* L. Subsp. *sativum* var. *sativum* in Some World Languages and Other Linguistic Taxa

Language/Taxon	Name
Frisian (Saterland)	aate
Frisian (West)	dopeart; eart; urt
Friulian	bîsi; cesaron
Gagauz	borchaq
Galician	chícharo; ervella; ervello; perico
Gawwada	sumburo
Geel	aart
Genoese	poéixo; poiscétto; poîscio
Georgian	barda; ch'veulebrivi barda; erevandi; kalaki; mukhudo; nokhud; ojakhi; sisar; tsertsvi
German	Erbse; Gartenerbse; Kneifelerbsen; Markerbsen; Pahlerbsen; Palerbsen; Schalerbsen; Speiseerbse
Gestel	Et
Ghent	irwete
Giesbaargs	eirte
Giethoorn	Aarftepoelen; Greune Aarfte
Glosa	pisum
Godoberi	hali
Graauw	jèrt
Greek	arakás; bizéli; píson to émeron
Greek (Ancient)	píson; písos
Greenlandic	eertaq
Griko	acho; áho
Gronings	aard(e); aarft(e); aart
Guarani	kumandá; kumanda'i; kumandachu
Guernésiais	peis
Gujarati	vaṭāṇā
Haitian Creole	pwa
Halls	ate
Harelbeke	eirwete
Hasselt	aart
Hausa	fis
Hawaiian	pe'a
Hebrew	afun tarbuti
Herentals	èrtje
High German (Middle)	ar(e)weiʒ; arwīʒ; erbeiʒ; er(e)weiʒ
High German (Old)	arawīʒ; araweiʒ
Hindi	matar
Hinuq	hilu
Hmong Daw	taum mog
Hoeselt	ert
Hungarian	borsó; kerti borsó; kultúrborsó; kifejtő borsó; termesztett borsó; velőborsó; veteményborsó

(*Continued*)

TABLE 14.3 (*Continued*)
Popular Names Denoting *Pisum sativum* L. Subsp. *sativum* var. *sativum* in Some World Languages and Other Linguistic Taxa

Language/Taxon	Name
Hunsel	ert
Hunzib	helu
Hutsul	horoch
Icelandic	erta; garðerta; gráerta; grænar baunir; gulertur; matarerta; matbaun
Idiom Neutral	pis
Ido	pizo
Igbo	azama
Ilocano	gisántes
Indonesian	ercis; kacang ercis; kacang polong; kapri
Ingrian	herne
Ingush	gerga qeŝ
Innu	atitshimin
Interlingua	piso
Interslavic	gråh
Irish	pis; phis bheag
Istriot	biʃi
Italian	piselli novelli; piselli proteaginosi; piselli proteici; piselli; pisello; pisello commune; pisello da orto
Izegem	nairwéte
Japanese	endō; nan kyōshu endō; piisu
Javanese	kacang polong
Jèrriais	pais
Joratian	pâi
Judaeo-Spanish	arveja; erviya
K'iche'	karawan
Kabardian	cesh
Kabyle	tirifin
Kalmyk	bürcëg
Kam-Sui	tau
Kannada	baṭāṇi; pappu
Kanne	àèrt
Karachay-Balkar	burchaq
Karaim	burchax
Karakalpak	buršaq
Karelian	herneh
Kashmiri	mutter
Kashubian	zwëczajny groch
Kasseng	ntuəŋ
Katu	tatu:ŋ
Kazakh	asburşaq; as burşaq; burşaq; egistik asburşaq; ekpe burşaq; kökönis burşağı
Kerkrade	eats
Khanty	an'kəš

(*Continued*)

TABLE 14.3 (*Continued*)
Popular Names Denoting *Pisum sativum* L. Subsp. *sativum* var. *sativum* in Some World Languages and Other Linguistic Taxa

Language/Taxon	Name
Khmer	sândaèk muul
Khwarshi	hel
Kinyarwanda	amashaza; ishaza
Klarenbeeks	ate
Klingon	pea
Koksijde	erwete
Komi-Permyak	an'kytš
Komi-Yazva	an'kœtš
Komi-Zyrian	an'kytš
Konda	pap
Kongo	bankasa; nkasa
Korean	phath; wandu
Kortemark	erweete
Kortrijk	erreweete
Kri	tâ:ŋ
Kryts	xarxar
Kubachi	qā
Kui	pap
Kumyk	burchaq
Kupang Malay	kacang
Kurdish (Central)	polika
Kurdish (Northern)	baqilê xatûnî ya bidlîsê; bezeliya hişk; polik; polika glover
Kuy	tɔ:ŋ
Kven	hernet
Kyrgyz	buurçak
Ladin	arbëia
Lak	quIru
Lao	rav kab kae
Latin	pisum
Latvian	apaļie zirņi; dārza zirņi; sējas zirņi
Laz	frasuli; parʒuli
Lebbeke	eit
Lemko	horokh
Leonese	arbeyu
Lezgian	nahut; zar
Lierens	raasdonder
Ligurian	poéixo; poiscétto; poîscio
Limbu	kubuN
Limburgish	ert
Lingala	wandu
Lingua Franca Nova	pi
Lingwa de planeta	pwa
Lithuanian	sėjamasis žirnis; žalieji žirneliai

(*Continued*)

TABLE 14.3 (*Continued*)
Popular Names Denoting *Pisum sativum* L. Subsp. *sativum* var. *sativum* in Some World Languages and Other Linguistic Taxa

Language/Taxon	Name
Livonian	jernõd
Livvi-Karelian	herneh
Lojban	debrpisu; rutrpisu
Loker	Eirte
Lombard (Western)	erbion
Low German	aruten
Low German (Middle)	ērt; erwete
Lunteren	aart
Luxembourgish	erbse; iebëssen
Macedonian	grašok
Makasae	uta
Malagasy	pitipoà
Malay	kacang polong
Malayalam	parippu; payarvarggaṅṅa!
Maldivian	bahun; fehi ‘oš
Maltese	piżella
Manchu	boxori; turi
Mansi	an’kas
Mantovani	ravion
Manx	pishyr
Māori	huapī; pī
Mapundungun	allfid
Marathi	maṭāra; vāṭāṇā
Mari (Hill)	nər pərsa; pərsa
Mari (Meadow)	pursa
Meänkieli	herne
Meerhout	et; ette
Min (Eastern)	gĭng-dâu
Mingrelian	mukhudo
Minionese	beep
Mirandolesi	piśel
Modenese	pisèe
Modern Indo-European	kikēr
Mohawk	onékwa
Moksha	snav; snavnja
Mol (Belgium)	ette; etteke
Moldovan	mazere
Mondial	pise
Mong Njua	taum mog
Mongolian	buurcag; vandui
Mongolian (Khalkha)	būrcag
Mongolian (Middle)	burčax
Monguor	puźag

(*Continued*)

TABLE 14.3 (*Continued*)
Popular Names Denoting *Pisum sativum* L. Subsp. *sativum* var. *sativum* in Some World Languages and Other Linguistic Taxa

Language/Taxon	Name
Mozarabic	biššáuṭ; číčar-o
Muong	dǝ̆w
Myanmar	gark; sadaw-pè
Nanai	turi
Navajo	naa'ołi nimazi
Neapolitan	pesiello; pesiéllo
Neeroeteren	ert
Negidal	turi
Nepali	kērā'u; maṭara
Nguồn	do; ʔǝ̆w
Nieuwpoort (Belgium)	erweete
Nijswiller	eëts
Ninove	eit
Nogai	burşaq
Norwegian (Bokmål)	ert; erter; hageert
Norwegian (Nynorsk)	erter; hageert
Novial	pise
Occidental	pise
Occitan	pese; pese ortalièr; pese proteaginós
Odia	matara
Oirat	mɨrčaq
Old Norse	ert
Old Prussian	kekêrs
Old Saxon	erit; eriwit
Ordos	burčaq
Ossetian	tymbylqædur
Otomi (Northwestern)	guisante
Oudenaarde	irreweete
Papiamento	huisarts
Pashto	muh'
Pennsylvania German	Aerbs; Aereb; Aerebs; Arbs; Arrebe; Arrebse
Persian	nxud frngi
Picard	pos
Po-ai	tu
Polabian	gorch
Polish	groch zwyczajny; groch typowy
Portuguese	ervilha; ervilha verde; ervilheira
Portuguese (Brazil)	ervilha
Punjabi (Eastern)	maṭara
Punjabi (Western)	mutter
Qenya	orivaine
Quechua	allwi; allwirha; alwuirja; arwija

(*Continued*)

TABLE 14.3 (*Continued*)
Popular Names Denoting *Pisum sativum* L. Subsp. *sativum* var. *sativum* in Some World Languages and Other Linguistic Taxa

Language/Taxon	Name
Quenya	erdevaine; erdevaiya
Regiano	arviot
Rillaar	aat; aikke
Ro	lugbap
Romagnol	fisarìl
Romani	boobi; giril
Romanian	mazărea; mazărea verde; mazărea uscată
Romanid	pisos; verde pisos
Romansh	arveglia
Russian	gorokh; gorokh liushchil'nyi; gorokh mozgovoi; gorokh posevnoi; zelionyi goroshek
Rusyn (Carpathian)	horokh
Rusyn (Pannonian)	hraščok
Rutul	xar
Salentino	pesiidde
Sami (Inari)	hertâ; herttâ
Sami (Lule)	ertar; ærttar
Sami (Northern)	earta; hearta
Sami (Skolt)	jönnkåårak; kåårak̆
Samoan	pi
Samogitian	žėrnis
Sanskrit	harēṇuḥ
Santa	puča; puǯa
Sardinian	pisu
Scheveningen	urret
Schoonderbuken	aikke
Scots	pease
Scottish Gaelic	peasair
Serbian	baštenski grašak; grašak; grašak krunac; grašak za suvo zrno; konzervni grašak; mladi grašak; obični grašak; povrtarski grašak; proteinski grašak; zeleni grašak
Serbian (Dalmatia)	biži; grašak
Serbian (Gora)	graška
Sercquiais	pɒi
Sesotho	lierekisi
Shina	garā́ṣ
Shor	mɨrčaq
Sicilian	pisedda; piseda ortìcula
Silesian (Cieszyn)	groch
Silesian (Upper)	groch
Simpelvelds	eëts
Sindarin	eredhwaen; eredhwe
Sindarin (Old)	eredhwae
Sindhi	mar

(*Continued*)

TABLE 14.3 (*Continued*)
Popular Names Denoting *Pisum sativum* L. Subsp. *sativum* var. *sativum* in Some World Languages and Other Linguistic Taxa

Language/Taxon	Name
Sinhalese	æṭa
Sint-Niklaas	ert
Sittard	ert
Slovak	hrach; hrach siaty; hrach siaty pravý
Slovenian	grah; rumeni grah; zeleni grah
Slovianski	groh
Slovincian	grùọχ
Slovio	goroh
Solon	boxrō
Solresol	dosidomi
Somali	digir
Sona	pea
Sorbian (Lower)	zagrodny groch; zmaršćony groch
Sorbian (Upper)	zahrodny hroch; zmoršćeny hroch
Spanish	abejaquilla; alvilla; arbeja; arbella; arveja; arvejana; arvejo; arvejote; arvejón; bisaltera; bisaltero; bisalto; bisantes; bisarto; chicharro; chícharo; cuchillejo; disante mayor; disante menor; fasolera; garbaneta; guisante; guisante ordinario; guisantera; guisón; haberos; ilar; moros; nanos; pelailla; pequeñicos; petipuás; poa; prinsol; présoles; pésole; tabilla; tirabeques; tres reyes
Spanish (South America)	arvejón; bisalte; tirabeque
St. Ouennais	pɛi
Staphorsten	arfte
Stellingwarfs	atepoelen
Sudovian	kekeris
Sundanese	ercis; kacang polong
Svan	ghedar
Swahili	mnjegere wa kizungu
Swedish	ärt; trädgårdsärt
Ta'Oi	ʔantuaŋ
Tabasaran	harar; xar
Tagalog	patani; sitsaro; tsitsaro
Tahitian	tapunu
Tajik	naxūddona
Taliang	tuaŋ
Tamil	paruppu; paṭṭāṇi
Tatar	borçak
Telugu	pappu; baṭhānī
Tetum	ervilha musan
Thai	t̄hạw lạnteā
Tibetan	greu
Tigrinya	‘ayinī ‘ateri
Tilburg	èrt
Tongan	pī

(*Continued*)

TABLE 14.3 (*Continued*)
Popular Names Denoting *Pisum sativum* L. Subsp. *sativum* var. *sativum* in Some World Languages and Other Linguistic Taxa

Language/Taxon	Name
Triestine	biso
Tsakhur	xara
Tsez	hil
Tswana	letlhodi; nawa
Turkish	bahçe bezelyesi; bezelye
Turkmen	burčaq; nohut
Tuvan	čočak-taraa; gorox
Udmurt	kӧžy
Ukrainian	horokh; horokh posivnyi
Universalglot	piso
Uraxa-Axusha	qara
Urdu	mutter
Urkers	urte
Uropi	pize
Uyghur	nuqt; počaq
Uzbek	burčɔq; no'xat
Valencian	pesol
Veluws	aart; arfe; ärfe; arft(e); ärfte; arte; ärte
Venetian	bìsi; biso; bíxara; bixo
Venlo	ert
Veps	herneh
Veurne	errewete
Viestan	p'sidd'
Vietnamese	đậu hà lan
Volapük	pisäl
Võro	herneh
Waanrode	et
Walloon	peû
Waray	chicharo
Waregem	èrreweete
Welsh	pys; pysen
Welsh Romani	boba
Wenedyk	pies
West Frisian Dutch	skokker; urt
Western Yugur	pɨrčaq
Xhosa	ipeysi
Yiddish	arbes; piz
Yucatec	guisante
Zeelandic	erten
Zelzate	irweete
Zhuang	duhlanhdouq; duhlanjdou
Zulu	uphizi
Zuojiang Zhuang	thu

kind of corn, gave the Proto-Semitic *laylay-an- ~ *yil-t-*, referring to pea and with, unfortunately, no surviving offspring, while, the **ʕV(m)bar-*, denoting both a kind of corn and chickpea (*Cicer arietinum* L.) and with its only derivative relating to pea being a Cushitic root, the Proto-Dullay **sumbur-*, is responsible for the modern word with the same meaning in Gawwada (Table 14.3).

The proposed Altaic ethnolinguistic macrofamily has three attested roots related to pea. The first is the Proto-Altaic **bŭkrV*, with a primeval triple meaning, namely, *cone*, *nut*, and *pea*, and with a remarkable historical development, resulting in a considerable wealth of the names referring to pea in its contemporary descendants. It evolved into four main direct branch roots, all of which denote pea and one or few other terms (Starostin et al. 2003):

- The first is the Proto-Korean **phə̆s* (Starostin 2006c), referring to beans in a broader sense and pea, which, through the Old Korean **phə̆č* and the Middle Korean *phə̆s*, brought forth the modern name in Korean (Table 14.3);
- The second one is the Proto-Mongolian **buɣurčag*, designating only pea (Mudrak 2006) and which, via the Middle Mongolian *burčax*, produced the names for pea in Bonan, Buryat, Daur, Eastern Yugur, Kalmyk, Khalkha, Mongolian, Monguor, Oirat, Ordos, and Santa (Mikić and Perić 2012);
- The third one is the Proto-Tungus-Manchu **boKa-ri*, **boKa-kta*, denoting equally cone, nut, pea, and tree fungus (Dybo 2006a), which gave the words referring solely to pea in Manchu and Solon;
- The fourth and the most productive one is the Proto-Turkic **burčak*, designating both bean-like grains and pea (Dybo 2006b), that, over the Old Turkic *burčaq* and the Middle Turkic and the Karakhanid *burčaq*, was extraordinarily diversified in contemporary Azerbaijani, Bashkir, Chulym, Chuvash, Gagauz, Kalmyk, Karachai-Balkar, Karaim, Karakalpak, Kazakh, Kumyk, Kyrgyz, Nogai, Shor, Tatar, Turkmen, Tuvan, Uyghur, Uzbek, and Western Yugur, with borrowings into the neighboring Kartvelian Laz and Uralic Hungarian (Gombocz 1912), and Hill and Meadow Mari (Mikić and Perić 2011).

The second attested Proto- Altaic root, **zi̯ăbsa*, initially denoted both lentil (*Lens culinaris* Medik.) and pea. It later shifted its meaning mainly to the former (see Chapter 11).

The third root of the proposed Altaic ethnolinguistic macrofamily, the Proto-Tungus-Manchu **turi-*, designated both bean-like legumes and pea (Dybo 2006a) and, so far, has no attested corresponding Proto-Altaic root. It gave another word referring to pea in Manchu that was subsequently imported into Nanai and Negidal (Table 14.3).

Although the Austroasiatic ethnolinguistic family is positioned at the utmost east and south of the *pea-cultivating realm*, it has at least three attested proto-roots relating to both bean-like legumes and pea in the Proto-Austroasiatic and some of its direct descendants (Peiros 2005, Peiros and Starostin 2005). This is a nice testimony that this region is, despite the prejudices stigmatizing this life-flourishing region as a dangerous and savage wilderness, an impressive meeting point of both Eurasian cool season legumes, such as lentil or pea, and countless East and Southeast Asian warm season legumes, most notably soybean (*Glycine max* [L.] Merr.) and *Vigna* species (Raes et al. 2013). One of the Proto-Austroasiatic roots primarily denoting bean, *tVh*, gave the modern words referring to bean in the Vietic languages, such as Arem, Muong, Nguồn, and Vietnamese (Table 14.3). Another two roots are attested in two branches of the Austrasiatic family, namely Bahnaric and Katuic. The Proto-Bahnaric **tuaŋ*, denoting beans and pea and via the Proto-North-West-Bahnaric **tuəŋ*, produced the corresponding words in Kasseng and Taliang. In the Katuic branch, there were two separate developments: the Proto-East-Katuic **tɔ:ŋ/ *tuaŋ* brought forth the names in Katu, Kri and Ta'Oi, while the Proto-West-Katuic **tɔ:ŋ* evolved into the modern word in Kuy (Table 14.3).

The proposed Dené-Caucasian ethnolinguistic macrofamily is abundant in the terms relating to pea, demonstrating how immensely vast is its area of its distribution, from the westernmost regions of Europe to the East Asian highlands and great rivers valley and plains (Jing et al. 2010). Among numerous roots relating to grain legumes, we selected the following (Starostin 2005a,c–e, 2015, Bengtson 2015):

- The Proto-Dené-Caucasian **cwə́rV*, signifying dried fruit or grass, that evolved first into the Proto-Caucasian **c_wɨrV*, denoting a kind of fruit, then into the Proto-Lak *s:iru* and, finally, into the morphologically identical modern Lak word referring to immature pods of beans and pea (Table 14.2), what is rather rarely attested meaning in comparison to the terms associated with mature grains;
- The Proto-Dené-Caucasian **hVwƚV*, with a primeval meaning of *bean* (see Chapter 7), which evolved into the Proto-Basque **iƚha-r̄*, responsible for the names associated with pea in Biscayan and Souletin (Table 14.3), and the Proto-Caucasian **hōwƚ[ā]*, which, through the Proto-Avar-Andi-Dido **ħoli* and the Proto-Tsezic **hel(u)*, gave the words designating pea in the modern Avar-Andi-Dido, such as Bagvalal, Botlikh, Chadakolob and Godoberi, and Tsezic, such as Bezhta, Hinuq, Hunzib, Khwarshi, and Tsez, respectively (Mikić 2011b);
- The Proto-Dené-Caucasian **xqŏrʔắ (~-rɦ-)*, denoting a kind of cereal, which brought forth three direct descendants referring to pea: firstly, the Proto-Burushaski **ɣaráṣ*, remaining morphologically identical in all three modern Burushaski languages and being borrowed into the neighboring Sino-Tibetan Balti and Indo-European Shina (Table 14.3); secondly, the Proto-Caucasian **qŏrʔā (~-rɦ-)*, designating pea and preserving this primeval meaning into the Proto-Dargwa **qara*, with Akusha, Chirag, Kubachi, and Uraxa-Axusha, into the Proto-Lak *quIru*, with Lak, into the Proto-Lezgian **χara*, with Aghul, Archi, Kryts, Lezgic, Rutul, and Tabasaran, and into the Proto-Nakh **qo(w)e ~ *qe(w)u*, with Chechen and Ingush (Mikić and Vishnyakova 2012)—this root was borrowed by other Caucasian languages, such as Abaza and Abkhaz, and the geographically close Indo-European Ossetian and Kartvelian Svan; and, thirdly, the Proto-Sino-Tibetan **krā (~g-)*, denoting a kind of grain, as in Burmese and Tibetan;
- The Proto-Caucasian **qäwx_wV*, referring to a nut-like fruit and without attested Proto-Dené-Caucasian ancestor, that produced the

Proto-Avar-Andi-Dido **qiʎV- (~ *χ-)*, denoting pea, as in modern Avar (Starostin 2003b);

- The unattested Proto-Na-Dené root, which brought forth, through its Athabaskan descendants, the word *naa'ołi*, denoting pod and being a morpheme in the compound word designating pea in Navajo (Table 14.3);
- The currently still unknown Proto-Kiranti root and its Sino-Tibetan ancestor, giving the word for pea in modern Limbu language (Van Driem 2005).

Pea was certainly known to the ancient Dravidian peoples (Krishna and Morrison 2009), although the words denoting it are often closely associated with other grain legumes, such as pigeon pea (*Cajanus cajan* [L.] Huth.), mostly because of the same way of use, especially as dal, either as simply split grains or their soup. The linguistic evidence of the presence of pea in the diets of the Dravidian speakers could be nicely illustrated with the Proto-Dravidian root **parup-*, equally denoting dal, a pulse crop and pea (Starostin 2006), which evolved into the Proto-South-Dravidian **par-up-*, denoting peas and split pulse in Kannada, Malayalam, and Tamil, and the Proto-Telugu **papp-*, referring to dal, shelled kernels, and split pulse in modern Telugu and with a possible borrowing into the Proto-Konda-Kui, as **pap-*, in both Konda and Kui (Table 14.3).

Perhaps the most curious of the accumulated compendium of the names denoting pea worldwide is that of the Eskimo-Aleut ethnolinguistic family, where it was assessed that the Proto-Eskimo root **unata-*, designating edible plant or root (Mudrak 2005) and through the Proto-Yupik **unata-*, gave the word in Central Siberian Yupik (Table 14.3).

The attested words referring to pea in the Hmong-Mien languages, such as Hmong Daw or Mong Njua, could have the same origin as a few other East and Southeast Asian families, such as the Austronesian, the Sino-Tibetan, or the Tai-Kadai, at some hypothetical long-range level (Starostin 2006d), or, perhaps, simply and more likely, were borrowed from one of them.

The greatest treasury of the names relating to pea is found in the Indo-European ethnolinguistic family. On one hand, its abundance is most remarkable, but it is equally realistic that the Indo-European is surely the one that is studied in the most thorough way and for centuries (Lockwood 1977). In the following paragraphs, we shall try to present the attested roots linked with pea in the Proto-Indo-European treasury (Pokorny 1959, Nikolayev 2012):

The Proto-Indo-European root **arenko-*, **arn k(')-*, signifying both a kind of cereal and a leguminous-like plant, evolved in the Ancient Greek word *árak-s*, *árako-s*, *ărakos* and, finally, into the modern Greek and Griko words for pea (Mikić 2011a), with a hypothetical correspondence to the Sanskrit (Table 14.3); this root was used by Linnaeus to name his genus *Arachis* L. (see Chapter 3)

- The Proto-Indo-European root **erəgʷ[h]-*, denoting a legume grain, had a complex evolution and served as a basis for naming the Linnean genera *Ervum* L. and *Orobus* L. (see Chapter 6); one of its direct descendants was the Proto-Germanic **arwait=*, **arwīt=*, designating solely pea, which preserved this meaning in its offspring and their modern representatives, namely, (1) in the unattested Old Frisian root, giving North, Saterland, and West Frisian words; (2) in the Old High German *arawīȝ, araweiȝ*, succeeded by the Middle High German *ar(e)weiȝ, arwīȝ, erbeiȝ, er(e)weiȝ*, with Colognian, High or Standard German, Luxembourgish, Pennsylvania German, and Yiddish, which, via Bavarian, entered the Rhaeto-Romance Ladin and Romansh; (3) in the Old Norse *ert*, with Danish, Elfdalian, Faroese, Icelandic, Norwegian, and Swedish and with borrowings into the Eskimo-Aleut Greenlandic and the Uralic Inari, Lule, Northern, and Skolt Sami languages; (4) in the Old Saxon *erit, eriwit*, succeeded by the Middle Dutch *aerwete, arwete, erwete*, with Afrikaans, Dutch, Dutch Low Saxon, Flemish, East Flemish, and tens of dialects and local speeches in Belgium and the Netherlands, as well as with the exports into the Austronesian Indonesian and Sundanese and the creole Papiamento; (5) in the Middle Low German *ērt, erwete*, with modern Low German (Tables 14.1 through 14.3); through the Lombardic language, that vanished more than twelve centuries ago, the Proto-Germanic root **arwait=, *arwīt=*, with a fully conserved meaning, have survived in numerous tongues of present northern Italy, like Bergamasque, Bolognese, Brescian, Ferraresi, Western Lombard, Mantovani, and Regiano (Table 14.3); to conclude this excessive account, the words referring to pea in some Iberian Romance languages, such as Aragonese, Extremaduran, Galician, Judaeo-Spanish, Leonese, Portuguese, and Spanish, along with the borrowings of the last one into the Amerind K'iche' and Quechua (Table 14.3), may equally have its roots in the Latin *ervum*, with its derivation *ervilia*, and another Germanic language, once spoken in Hispania, namely Gothic of the Visigoths (Dworkin 2012);

- The Proto-Indo-European root **g'er[a]n-, *grān-*, an ultimate source of the well- known words *corn* and *grain*, brought forth numerous descendants with the same meaning, such as the Proto-Baltic **ǯir̂n-ia-, *ǯirn-iā̃*, the Proto-Celtic **grāno*, the Proto-Germanic **kirn-ō, *kurn-a-, *kurn-ia-, *kurn-il-a-*, the Latin *grānum*, or the Slavic *zĕrno* (Pokorny 1959, Nikolayev 2012). Interestingly enough, in the Eastern Baltic languages a shift of meaning happened from *grain* to *pea*, as seen in Latvian, Lithuanian, and Samogitian (Mikić 2014b), as well as in the neighboring Uralic languages (Mikić and Stoddard 2013), such as Estonian, Finish, Ingrian, Karelian, Kven, Livonian, Livvi-Karelian, Meänkieli, Veps, and Võro (Table 14.3);
- The Proto-Indo-European root **ghArs-*, referring to a leguminous plant (Pokorny 1959, Nikolayev 2012) and via the Proto-Slavic **gorxŭ* (Vasmer 1953), became the word denoting pea in all the extinct and living Slavic languages (Mikić 2013b, 2014a), dialects, and speeches, with Belarusian, Boyko, Bulgarian, Bunjevac, Croatian, Czech, Hutsul, Kashubian, Lemko, Macedonian, Polabian, Polish, Russian, Carpathian and Pannonian Rusyn, Standard and Gora Serbian, Cieszyn and Upper Silesian, Slovak, Slovenian, Slovincian, Lower and Upper Sorbian, and Ukrainian (Tables 14.1 through 14.3); the majority of Slavic peoples use its derivative or diminutive forms to denote the species of the genera *Lathyrus* L. (see Chapter 10) and *Vicia* L. (see Chapter 15); it was also borrowed by the geographically close

non-Slavic speakers, in order to name pea, such as in the Altaic Tuvan and the Romance Aromanian, as well as in the constructed Interslavic, Slovianski and Slovio (Table 14.3); curiously enough, the Slavic words denoting pea could contribute to a hypothesis of contacts between the Slavs and other Indo-Europeans with the Burusho people in some undefined time in the past, mirrored in the similarity between the Common Burushaski *ɣarsáṣ* and the Proto-Slavic **gorxŭ*, although the ongoing discussion *pro* (Čašule 2009) and *contra* (Bengtson and Blažek 2011) of this theory seems to be developing in the favor of the latter;

- The Proto-Indo-European root **kek-*, **k'ik'-*, with a primeval designation of oat (*Avena sativa* L.) and pea and subsequent evolution into the terms relating mostly to chickpea (see Chapter 5), retained the second of its initial meaning in the extinct Western Baltic languages, namely Old Prussian and Sudovian (Table 14.3), pointing a remarkable importance of these languages for comparative Indo-European linguistics (Ringe et al. 2002); this attested Proto-Indo-European root was also recorded in the Old Indo-Aryan, as *śiśnā́* (Pokorny 1959, Nikolayev 2012) and used to name the pea crop in the constructed Modern Indo-European language (Mikić et al. 2010);
- The Proto-Indo-European root **k(')now-* for nut (Pokorny 1959; Nikolayev 2012), and its descendant, the Old Persian *noḫud* (see Chapter 5), are the source of the names designating pea in the languages of several Eurasian ethnolinguistic families, such as the Altaic Azerbaijani, Crimean Tatar and Turkmen, the Caucasian Adyghe and Bats, the Indo-European Dari, Pashto, Persian, and Tajik, and the Kartvelian Georgian and Mingrelian (Table 14.3);
- The Proto-Indo-European root **mAis-*, denoting skin (Nikolayev 2012) may be equally responsible for the words denominating lentil (see Chapter 11) and pea in numerous Indo-Aryan languages, such as, in the case of the latter, Bengali, Hindi, Kashmiri, Marathi, Nahali, Nepali, Odia, Eastern and Western Punjabi, Sindhi, Urdu, and, possibly, Sinhalese (Tables 14.2 and 14.3); it could also be a substratum of the words referring to pea in the modern Albanian and the Daco-Romanian languages, with Aromanian, Moldovan, and Romanian (Mikić 2009, Ungureanu 2014) and thus contribute to the viewpoint of the existence of Daco-Mysian, a hypothetical Indo-European branch of the Indo-European family (Georgiev 1981); interestingly enough, one of the direct derivatives of the Proto-Indo-European root **mAis-* is the Proto-Slavic **mēhŭ*, denoting bag or sack (Vasmer 1955) and which gave forth a variety of the terms associated with legume pods in Serbian and other South Slavic languages, such as *mahuna*, *mauna*, *meuna*, *meunica*, and that found its definite botanical taxonomic position as a name for the entire family of *Fabaceae*: *mahunarke* or *mahunjače* (Mikić-Vragolić et al. 2007);
- The originally French name *mange tout* or *mangetout*, meaning literally *eat all*, denotes an agronomic type of *Pisum sativum*, used in the form of unripe pods with just conceived grains, without fibrous layers in the pod tissue and usually rich in sugar (Carrouée 1993, Mihailović et al. 2004); the

same name refers to the market class of common bean (*Phaseolus vulgaris* L.), with the same use and properties (see Chapter 13); this type of pea was widely cultivated in West Europe at least since the end of the sixteenth century (Gerard 1597, Myers et al. 2001), with, most likely, *sickle pea* as its first English name (Ray 1686), and was also included in Mendel's pioneering hybridization trials (Ellis et al. 2011); along French, the reference to *eating all* is also present in West Flemish, German, Italian, Occitan, and Spanish (Table 14.2);

- The renown Linnean genus name *Pisum* has its basis in the Latin verb *pīnsere*, meaning *to bray, peel, to pound*, and the Ancient Greek noun *píson, písos*, referring to pea (Mikić et al. 2015c), and the ultimate origin in the Proto-Indo-European root **peys-, *pis-*, meaning *to crush* and *to thresh*, respectively (Pokorny 1959, Nikolayev 2012), which represents another and frequently encountered example of a name with a descriptive nature, in this case, the act of shelling out the pea grains from their pods; its modern representatives are numerous and denote the pea crop in different languages, belonging mainly to the Indo-European family, such as Albanian, the Hellenic, with Greek, and the fascinatingly abundant Romance, with Aragonese, Aranese, Burgundian, Calabrian, Catalan, Catanian, Caterisani, Corsican, Franco-Provençal, French, Friulian, Istriot, Genoese, Guernésiais, Italian, Jèrriais, Joratian, Ligurian, Mirandolesi, Modenese, Neapolitan, Occitan, Picard, Romagnol, Salentino, Sardinian, Sercquiais, Sicilian, Spanish, St. Ouennais, Triestine, Valencian, Venetian, Viestan, and Walloon (Tables 14.1 through 14.3); either via Latin or through various Romance languages, *Pisum* was exported and locally modified into the Afroasiatic Arabic and Maltese, the Altaic Turkish, the Austronesian Malagasy, the Indo-European, with the Celtic Beurla Reagaird, Cornish, Cumbric, Irish, Manx, Scottish Gaelic, and Welsh, with the Germanic Old, Middle, and modern English, Forth and Bargy, Scots, Ulster English, and Yiddish, with the Iranian Northern Kurdish and the Slavic Dalmatian Serbian (Table 14.3); it also entered the creole Haitian Creole and was extensively used as a pattern for the names referring to pea in rather numerous constructed languages, such as Brithenig, Esperanto, Glosa, Idiom Neutral, Ido, Interlingua, Lingua Franca Nova, Lingwa de planeta, Lojban, Klingon, Mondial, Novial, Occidental, Romanid, Sona, Universalglot, Uropi, Volapük, and Wenedyk (Table 14.3); finally, the Latin term *pisum sapidum* produced *biššáuṭ* in the extinct Iberian Romance Mozarabic, which, impacted by its verb *guisar*, meaning *to stew*, resulted in one of the most specific words denoting pea in Aragonese and probably the most widely used name for pea in Castilian Spanish, with subsequent exports into the Amerind Otomi and Yucatec, and the Austronesian Cebuano, Filipino, and Ilocano (Table 14.3).

Among extraordinarily scarce lexical evidence on the use of pea among the speakers of the suggested Indo-Pacific family is the Trans-New Guinea Makasae (Table 14.3), where it also denotes peanut. We may only presume that this Makasae word initially denoted pea, which may be somewhat ambiguous, since, on the one hand, New Guinea is geographically beyond all conventionally defined boundaries of the pea distribution, but, one the other hand, pea is an ancient Old World crop, which dispersed over an immensely vast space, and this may contribute to its history in the utmost southeastern regions of Asia. In addition, peanut, newly introduced in these regions by the West European naval powers from the Americas, could bring a certain resemblance to pea by some features of its aboveground growth habit, especially the pairs of large leaflets. Whatever, the origin of the Makasae word needs essentially more support in the form of the corresponding names in other Trans-New Guinean languages and an appropriate comparative linguistic analysis for assessing a hypothetical Proto-Trans-New Guinea root relating to pea (Pawley 2012).

Although the linguistic evidence demonstrates that the members of the Kartvelian ethnolinguistic family mostly borrowed their terms relating to pea from the neighboring Altaic, Caucasian, or Indo-European languages, there is a possibility that there are some attested Proto-Kartvelian roots, which could produce at least a few genuine words denoting this crop, cultivated across the Caucasus for millennia (Bussmann et al. 2016); one of the potential candidates is the Proto-Kartvelian root **ber-*, meaning *blow up*, *to blow* (Starostin 2005b) and resembling the inflated pea pods, which, through the morphologically identical Proto-Georgian, could give the modern Georgian *barda*, while another is the Proto-Kartvelian **car-/*cr-*, meaning *to sift*, *to sow* and via the Proto-Georgian *cer-/cr-* might result in the Georgian *tsertstvi* (Table 14.3).

The available resources on the terminology relating to pea in the great Niger-Congo family are considerably poorer in comparison to the other languages. This may be ascribed to several facts, such as that it is the third largest language family regarding the number of its speakers (see Chapter 2), its classification is still far from being firmly established (Stewart 2002, Olson 2004, de Filippo et al. 2010), its etymological research is faced with numerous insufficiently solved issues for decades (Greenberg 1972, Williamson and Blench 2000, Hyman 2011), and, in the end, because pea is not a native crop in the Sub-Saharan Africa (Rippke et al. 2016), suggesting that its rare recorded local names might result from mutually independent descriptive associations. One of the proposed links is between the words denoting pea in Kongo and Lingala, along with the common adjective *nguba*, *kidney-like*, in Kimbundu and Kongo, and with a transoceanic export into the Amerind Guarani (Table 14.3), carried out by the West European slave traders (see Chapter 3). More highly hypothetical suggestions, which certainly require considerably more thorough comparative linguistic analyses, may include the morphological and semantic similarities associated with pea between Chewa and Tswana and between Xhosa and Zulu (Table 14.3).

One of the still undetermined Proto-Tai-Kadai roots gave the Proto-Zhuang-Tai **d(h)ōs*, relating to a bean-like grain legume and pea (Peiros 2009), which produced the words referring to pea in some of its contemporary offspring, such as Bouyei, Kam-Sui, Po-ai, Thai, Zhuang, and Zuojiang Zhuang (Table 14.3).

From the viewpoint of mainstream linguistics, it may not be expected to assess the existence of the genuine words denoting pea or other pulses in the Uralic ethnolinguistic family. This is usually explained by stressing that its languages and its supposed homeland are simply too north to make arable farming competitive enough in comparison to fishery, animal husbandry, foraging, and other traditional economic activities that are not associated with tilling earth. Even today, despite the numerous advances in agronomy, such a harsh cold environment brings many challenges to the agronomists and breeders specialized for boreal climates (Honkola et al. 2013, Lizarazo et al. 2015). According to the mainstream linguistics, such constrained crop farming would limit the cultivation of many crops and development of the genuine words denoting them. All this results in the fact that the vast majority of the names referring to pea in the Uralic languages are of either Germanic or Balto-Slavic origin, as mentioned in the previous paragraphs of this chapter, with numerous examples in the Finnic and Sami branches and a complete absence of the terms relating to any kind of crop in the Samoyedic subfamily. Nevertheless, the Uralic languages that had developed on both slopes of the Ural Mountains and which speakers did not migrate far from the *urheimat* seem to have in common the attested Proto-Permic root **kɜǯs* (Lytkin and Gulyaev 1970) that gave the names referring to pea in all three Komi languages and Udmurt. It fully corresponds to the Proto-Mordvinic root **kɜsnav*, in Erzya and Moksha, and to the names for pea in Khanty and Mansi (Table 14.3). The evidence, that despite their natural habitats for the past few millennia, the Uralic peoples have their own word associated with pea may be a small contribution to the assumption that their family used to be one of several members of the supposed Eurasiatic ethnolinguistic supergroup (Greenberg 2000, 2002, Pagel 2013). According to the suggested position of the Eurasiatic homeland, the ancestors of the Proto-Uralic people had migrated from the *deep* south, perhaps more than hundred centuries ago, to their new home on the eastern slopes of the Urals. There, they gradually shifted from arable agriculture to other less intensive forms of providing food, but with a preserved memory of pea, once a crop they had cultivated and now merely a plant they meet in local flora or which grain they gain by trading with neighboring peoples living in less cold climates (Mikić and Stoddard 2013). At this point, considering the possibility that there is at least one potential Proto-Uralic term relating to pea, which brought forth the above-mentioned mediating proto-words and modern names, we dare to suggest three possible roots (Starostin 2005f): the first one is **kača*, designating a hole, a cavity, and a wooden vessel; the second is **keśV*, meaning *to rip*, *to tear*; and the third is **kopa*, associated with bark and skin (Figure 14.1). All three could easily have a descriptive nature and contain a resemblance to either the vessel-like form of a pea pod or the act of hollowing out the pea seeds or the skin-like nature of pods, which, at any rate, still remains to be processed in details by comparative and historical linguistic analysis.

The Savonian Finnish word denoting pea originated from the Proto-Uralic **rokka*, referring to fat meal, porridge, or soup (Starostin 2005f), with no attested counterparts associated with pea in other Finnish dialects or Uralic languages, where it designates mainly soup.

As many other pulse crops, pea resembles some other grain legumes in various languages. As a result, the morphemes relating to other legume species or quality properties are incorporated in certain vernacular names for pea. Such are the cases referring to chickpea in the Indo-European Balearic Friulian, Galician, and

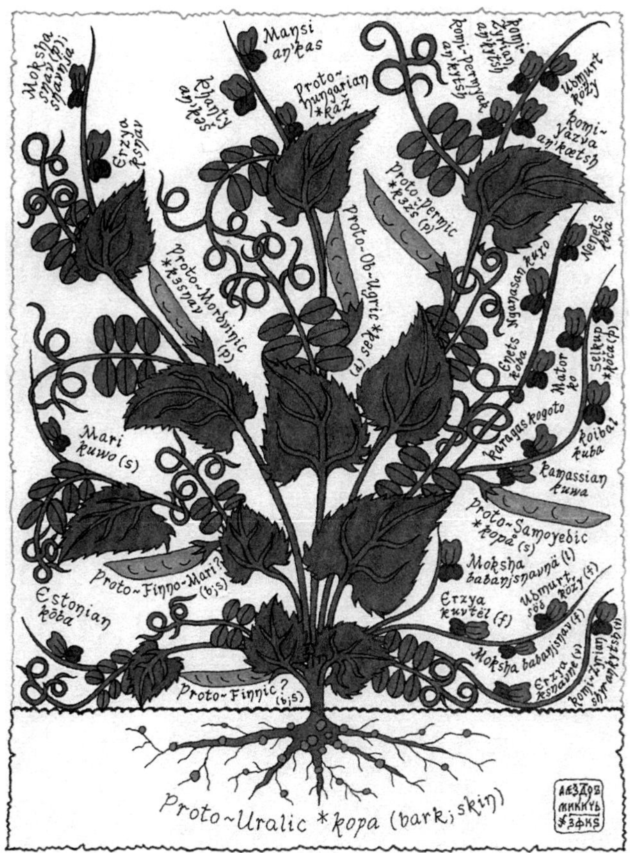

FIGURE 14.1 **(See color insert.)** One of the possible evolutions of the Proto-Uralic root **kopa*, denoting bark and skin (Starostin 2006f), into its direct derivatives, drawn as pods, and contemporary descendants, rendered as flowers, in the Finnic, Finno-Mari, Mordivinic, Ob-Ugric, Permic, and Samoyedic languages; the meanings of the proto-words are given within brackets and the meaning of each modern word, if given without brackets, is the same in its proto-word, while with the bracketed abbreviations, such as (b) for *bark*, (f) for *faba bean*, (l) for lentil, (p) for *pea*, (s) for *skin*, and (v) for *vetches*, are given to mark the distinction where needed.

Spanish, with exports from the latter into the Austronesian Tagalog and Waray, and in the Kartvelian Georgian; to faba bean (*Vicia faba* L.) in the Iranian Romani and Welsh Romani and South-West Finnish, all actually borrowed from the corresponding Slavic word; and to an unspecified kind of legume bean in the Austroasiatic Khmer, the Indo-European Icelandic and Ossetian, and the Sino-Tibetan Cantonese and Mandarin Chinese and Eastern Min (Table 14.3); and to a high content of sugar, such as in *mangetout* (Table 14.2).

Although several previous paragraphs demonstrated how numerous are the constructed languages, which carefully developed all their words together with those relating to pea, we shall provide two of them with specific attention. Those in question are two Elvish languages, namely Quenya and Sindarin, invented by one of the most famous authors in English literature, John Ronald Reuel Tolkien (1892–1973). Following his *leitmotif* of building up a world of its own simply because his languages needed a place where they could be spoken, Tolkien thoroughly elaborated not only the grammar or lexicon of his languages, but also their *evolution.* What the readers of his works, such as *The Lord of the Rings*, know under the names of Quenya and Sindarin had their proto-forms, termed as Elphin or Qenya and Gnomish or (Old) Sindarin (Hostetter 2007a, b). In the posthumously published material, it was discovered that Tolkien, as early as 1915, invented the Qenya names for various plants (Salo 1999) and, among them, to pea and few other legume species (Mikić and Stoddard 2009, Mikić 2013a). A detailed etymological and comparative linguistic analysis of the attested Qenya *orivaine*, a compound of *ori* and *vaine* literally meaning *seed-sheath*, proposes the words for pea in Quenya and Sindarin (Rausch 2017, pers. comm.). The later development of Qenya into Quenya changed *ori* into *erde*, while *vaine* remained more or less stable, producing the Quenya **erdevaine* and **erdevaiya*. All this may correspond to the Sindarin *eredh* for seed and **gwae*, **waen* for envelope or sheath, giving the names for pea **eredhwaen* and **eredhwe*, with a possibly more archaic **eredhwae* (Table 14.3).

15 *Vicia* L.

Synonyms: *Abacosa* Alef.; *Arachus* Medik.; *Atossa* Alef.; *Bona* Medik; *Coppoleria* Todaro; *Cracca* Medik.; *Cujunia* Alef.; *Endiusa* Alef.; *Endusia* Benth. and Hook f.; *Ervilia* Link; *Ervum* L.; *Faba* Adans; *Faba* Mill.; *Hypechusa* Alef.; *Orobella* C. Presel; *Parallosa* Alef.; *Rhynchium* Dulac; *Sellunia* Alef.; *Swantia* Alef.; *Tuamina* Alef.; *Vicilla* Schur; *Viciodes* Moench; *Wiggersia* Gaertn.

15.1 LIST OF TAXA SCIENTIFIC AND POPULAR NAMES

The genus *Vicia* L. is one of the most abundant and the most widespread in the world with a vast number of species belonging to both the Old World and New World and ranging remarkably wide across contrasting environments. Its taxonomy is currently far from being definitely established and has been in a prominent dynamic state since the beginnings; while other research topics are also in the process of contant accumulation of gathered knowledge on its species' biodiversity and genetics (Davies 1970, Potokina et al. 1999, Frediani et al. 2004, Endo et al. 2008, Gianfranco et al. 2008, Shiran et al. 2014). The abundance of the names denoting both cultivated crops and wild species and their scientific and vernacular names is presented in the following paragraphs, with what we hope is a detailed and clear overview (Hermann 1960, ISTA 1982, Allkin et al. 1986, Rehm 1994, Davies and Jones 1995, Maxted 1995, Van de Wouw et al. 2001, Leht 2005, Gledhill 2008, Porcher 2008, Bryant and Hughes 2011, The Plant List 2013, Ecocrop 2017, EPPO 2017, Ethnologue 2017, IBIS 2017, ILDIS 2017, Logos 2017, NPGS 2017, Wikipedia 2017, Wiktionary 2017).

- *Vicia abbreviata* Fisch. ex Spreng.
English: shortened vetch
Russian: goroshek ukorochennyi
- *Vicia acutifolia* Elliott
English: sand vetch
- *Vicia aintabensis* Boiss. & Hausskn.
English: Aintab vetch; Gaziantep vetch
- *Vicia alpestris* Steven
English: alpine vetch
Kazakh: alpi sïırjoñışqası
Russian: goroshek gornyi
- *Vicia altissima* Desf.
Czech: vikev nejvyšší
English: tallest vetch
Spanish: veza la más alta
- *Vicia americana* Muhl. ex Willd.

Synonyms: *Lathyrus linearis* Nutt.; *Vicia americana* var. *linearis* (Nutt.) S. Watson; *Vicia americana* var. *oregana* (Nutt.) A. Nelson; *Vicia linearis* (Nutt.) Greene; *Vicia oregana* Nutt.
Arabic: biqi mriki
English: American vetch; club-leaf vetch; mat vetch; narrow-leaf American vetch; narrow-leaf vetch; peavine; purple vetch; stiff-leaf vetch; trellis-leaf vetch; wild vetch
Greek: amerikanós víkos; mov víkos; víkos psátha
Lakota: tasúsu
Navajo: ch'iidą́ą́ts' óóz
Persian: mashk ahmrikaii
- *Vicia amoena* Fisch.
Synonyms: *Vicia amoena* var. *oblongifolia* Regel
English: lovely vetch
Japanese: tsurufujibakama
Kazakh: süykim sïırjoñışqa
Korean: malgullepul
Polish: wyka piękna
Russian: goroshek priatnyi
- *Vicia amurensis* Oett.
Synonyms: *Vicia pallida* Turcz.
English: Amur vetch
Japanese: noharakusafuji
Kazakh: amur sïırjoñışqası
Polish: wyka amurska
Russian: goroshek amurskii
- *Vicia anatolica* Turrill
Synonyms: *Vicia hajastana* Grossh.
English: Anatolian vetchling
- *Vicia andicola* Kunth
English: Andean vetch
- *Vicia angustipinnata* Nakai
English: narrow-pinate vetch
Japanese: hosobanoendou
- *Vicia argentea* Lapeyr.
English: silvery vetch
Spanish: arveja plateada; veza
- *Vicia articulata* Hornem. (Table 15.1)
Synonyms: *Cracca monanthos* Gren. & Godron; *Ervum monanthos* L.; *Vicia monanthos* (L.) Desf.
- *Vicia assyriaca* Boiss.
English: Assyrian vetch
- *Vicia bakeri* Ali
English: Baker's vetch
- *Vicia balansae* Boiss.
English: Balansa's vetch

Russian: goroshek Balanzy
- *Vicia barbazitae* Ten. & Guss.
English: Barbazita's vetch
- *Vicia basaltica* Plitmann
Arabic: biqi bazlti
English: basalt vetch
Hebrew: vkt hvzlt
- *Vicia benghalensis* L. (Table 15.2)
Synonyms: *Cracca atropurpurea* (Desf.) Gren. & Godr.; *Vicia albicans* Lowe; *Vicia atropurpurea* Desf.; *Vicia loweana* Steud.; *Vicia micrantha* Lowe
- *Vicia benthamiana* Ali
English: Bentham's vetch
- *Vicia biennis* L.
Synonyms: *Vicia picta* Fisch. & C. A. Mey.
English: two-year vetch
Russian: goroshek dvuletnyi
Slovak: vika dvojročná
Swedish: högvicker
- *Vicia bifolia* Nakai
English: two-leaved vetch
Japanese: miyamataniwatashi
- *Vicia bijuga* Hook. & Am.
Synonyms: *Lathyrus anomalus* Phil.; *Vicia solisii* Phil.; *Vicia sericella* Speg.; *Vicia saffordii* Britton
English: two-leaflet vetch
- *Vicia bithynica* (L.) L.
Synonyms: *Lathyrus bithynicus* L.
Arabic: biqi Bithini
Czech: vikev maloasijská
English: Bithynian vetch
Esperanto: vicio fortika
Greek: víkos Bithynikós
Hebrew: vk'h ntvlt
Polish: wyka bityńska
Russian: goroshek vifinskii
Serbian: maloazijska grahorica; vitinijska grahorica
Swedish: turkvicker
Welsh: ffugbysen ruddlas arw-godog
- *Vicia bungei* Ohwi
English: Bunge's vetch
Japanese: touendou
- *Vicia caesarea* Boiss. & Balansa
English: Caesarean vetch
- *Vicia californica* Greene
English: Californian vetch
- *Vicia canescens* Labill.

Arabic: biqi rmadi
Chinese: hui ye guang bu ye wan dou
English: grey vetch
Japanese: ke kusa fuji; kusa fuji
Slovak: vika sivá
- *Vicia canescens* Labill. subsp. *gregaria* (Boiss. & Heldr.) P. H. Davis
Synonyms: *Vicia gregaria* Boiss. & Heldr.
English: population vetch
- *Vicia canescens* Labill. subsp. *variegata* (Willd.) P. H. Davis
Synonyms: *Vicia nissoliana* L.; *Vicia persica* Boiss.; *Vicia variegata* Willd.
English: variegated vetch
Polish: wyka upstrzona
- *Vicia cappadocica* Boiss. & Balansa
Synonyms: *Ervum paucijugum* Trautv.; *Vicia paucijuga* (Trautv.) B. Fedtsch.
English: Cappadocia vetch
- *Vicia caroliniana* Walter
Cherokee: altsa'sti
English: Carolina vetch; Carolina wood vetch; pale vetch; wood vetch
Greek: víkos ksílou Karolínas
Russian: goroshek maloparnyi
Serbian: karolinska grahorica
- *Vicia cassia* Boiss.
Arabic: albiqi alkasiusi
English: Cassia vetch
- *Vicia cassubica* L. (Table 15.3)
- *Vicia chinensis* Franch.
English: chinese vetch
- *Vicia chosenensis* Ohwi
English: Korean vetch
Japanese: chousen'ebirafuji
- *Vicia ciliatula* Lipsky
English: hair-fringe vetch
Russian: goroshek resnitchatyi
- *Vicia cirrhosa* C. Sm. ex Webb & Berthel.
English: many-tendril vetch
- *Vicia cornigera* Chaub.
English: horned vetch
- *Vicia costata* Ledeb.
Synonyms: *Vicia sinkiangensis* H. W. Kung
English: ribbed vetch
Russian: goroshek rebristyi
- *Vicia cracca* L.(Table 15.4)
- *Vicia cracca* L. subsp. *cracca*
Synonyms: *Vicia cracca* subsp. *grossheimii* (Ekutim.) Hashimov; nom. inval.; *Vicia grossheimii* Ekutim.
English: true cow vetch

- *Vicia cracca* L. subsp. *incana* (Gouan) Rouy
Synonyms: *Vicia cracca* subsp. *gerardi* Bonnier & Layens; *Vicia cracca* [unranked] *gerardi* Gaudin; nom. inval.; *Vicia incana* Gouan
English: hoary vetch
- *Vicia cracca* L. subsp. *japonica* Miq.
English: Far East tufted vetch
- *Vicia cretica* Boiss. & Heldr.
English: Crete vetch
- *Vicia crocea* (Desf.) B. Fedtsch.
Synonyms: *Orobus aurantius* Steven ex M. Bieb.; *Orobus croceus* Desf.; *Vicia aurantia* (Steven ex M. Bieb.) Boiss.
English: saffron vetch
Russian: goroshek oranzhevyi
- *Vicia cusnae* Foggi et Ricceri
English: Mount Cusna vetch
- *Vicia cuspidata* Boiss.
Arabic: biqi mdbb
English: stiff-point vetch
Hebrew: vk'h chdvdh
- *Vicia cypria* Kotschy
Arabic: biqi qbrsi
English: Cyprus vetch
Hebrew: vkt kfrsn
- *Vicia dionysiensis* Mouterde
English: Dionysius vetch
- *Vicia disperma* DC.
Synonyms: *Vicia parviflora* Loisel.
English: small French tare; two-seed vetch
Finnish: hentovirvilä
Italian: veccia a due semi
Kazakh: dual burşaq
- *Vicia dissitifolia* (Nutt.) Rydb.
English: separated-leaf vetch
- *Vicia dumetorum* L.(Table 15.5)
- *Vicia epetiolaris* Burkart
English: petioleless vetch
- *Vicia eristalioides* Maxted
English: Eristalis-like vetch
- *Vicia ervilia* (L.) Willd.
See Chapter 6.
- *Vicia esdraelonensis* Warb. & Eig
English: Jezreel Valley vetch
Hebrew: vkt zr'l
- *Vicia faba* L.
See Chapter 7.
- *Vicia fauriei* Franch.

English: Faurie's vetch
Japanese: tsugarufuji
- *Vicia ferreirensis* Goyder
English: Ferreira vetch
- *Vicia ferruginea* Boiss.
English: Ferruginous vetch
- *Vicia filicaulis* Webb & Berthel.
Synonyms: *Vicia bifoliolata* J. J. Rodr.
English: thread-stem vetch
Spanish: veza con tallo delgado
- *Vicia floridana* S. Watson
English: flowering vetch
- *Vicia fulgens* Batt.
Arabic: albiqi alsat
English: scarlet vetch
- *Vicia galeata* Boiss.
English: helmeted vetch
Hebrew: vkt hvtzvt
- *Vicia galilaea* Plitmann & Zohary
Synonyms: *Vicia galilaea* subsp. *faboidea* Plitmann & Zohary; *Vicia galilaea* var. *faboidea* (Plitmann & Zohary) H. I. Schäf.; *Vicia galilaea* var. *galilaea* Plitmann & Zohary
Arabic: albiqi aljlili
English: Galilee vetch
Hebrew: vkt hgll
- *Vicia glauca* C. Presl
English: bright vetch
- *Vicia graminea* Sm.
Synonyms: *Vicia selloi* Vogel
English: grass-like vetch
- *Vicia grandiflora* Scop. (Table 15.6)
Synonyms: *Vicia grandiflora* var. *kitaibeliana* W. D. J. Koch
- *Vicia hassei* S. Watson
English: Hasse's vetch; slender vetch
- *Vicia hirsuta* (L.) Gray (Table 15.7)
Synonyms: *Cracca minor* Godr.; *Cracca minor* var. *eriocarpa* Godr.; *Endiusa hirsuta* (L.) Alef.; *Ervilia hirsuta* (L.) Opiz; *Ervilia vulgaris* Godr.; *Ervum hirsutum* L.; *Ervum terronii* Ten.; *Vicia hirsuta* var. *terronii* (Ten.) Burnat; *Vicia leiocarpa* Moris; *Vicia mitchellii* Raf. *Vicia parviflora* Lapeyr. (non Cav.: preoccupied)
- *Vicia hololasia* Woronow
English: Entirely woolly vetch
- *Vicia hulensis* Plitmann
English: Hula Valley vetch
Hebrew: Vkt hchvlh
- *Vicia humilis* Kunth

English: low-growing vetch
- *Vicia hyaeniscyamus* Mouterde
Arabic: biqi dbi
English: hyaena-bean vetch
- *Vicia hybrida* L.
Arabic: albiqi alhjin
Czech: vikev zvrhlá
English: hairy yellow vetch
Finnish: kalvasvirna
Hebrew: vkt chlm
Russian: goroshek pomesnyi
Swedish: solvicker
Welsh: ffugbysen cymysgryw; ffugbysen felen flewog
- *Vicia hyrcanica* Fisch. & C. A. Mey.
English: Hyrcanian vetch
Kazakh: gïrkan burşaq
Russian: goroshek girkanskii
- *Vicia incisa* M. Bieb.
Synonyms: *Vicia incisiformis* Stef.; nom. inval.; *Vicia sativa* subsp. *incisa* (M. Bieb.) Arcang.
English: deep-cut vetch
Russian: goroshek nadreznyi
- *Vicia japonica* A. Gray
Synonyms: *Vicia japonica* var. *oblongifolia* A. Gray
English: Far East vetch
Japanese: hirohakusafuji
Russian: goroshek iaponskii
- *Vicia johannis* Tamamsch.
Arabic: biqi iuhani
English: Johann's vetch
- *Vicia kalakhensis* Khattab et al.
Arabic: biqi klkhi
English: Kalakh vetch
- *Vicia kokanica* Regel & Schmalh.
English: Kokand vetch
Russian: goroshek kokandskii
- *Vicia lathyroides* L. (Table 15.8)
Synonyms: *Ervum lathyroides* (L.) Stank.; *Ervum soloniense* L.; *Vicia lathyroides* subsp. *olbiensis* (Reut.) Smejkal; *Vicia olbiensis* Timb.-Lagr.
- *Vicia leucantha* Biv.
English: white-flowered vetchling
- *Vicia lilacina* Ledeb.
Synonyms: *Vicia neglecta* Hanelt & Mett.
English: lilac vetch
Russian: goroshek lilovatyi
- *Vicia linearifolia* Hook. & Arn.

English: linear-leafed vetch
- *Vicia loiseleurii* (M. Bieb.) Litv.
Synonyms: *Vicia litvinovii* Boriss.; *Vicia meyeri* Boiss.
English: Loiseleur's vetch
- *Vicia ludoviciana* Nutt. ex Torr. & A. Gray
English: deer-pea vetch; Louisiana vetch
- *Vicia ludoviciana* Nutt. ex Torr. & A. Gray subsp. *ludoviciana*
Synonyms: *Vicia caroliniana* var. *texana* Torr. & A. Gray; *Vicia exigua* Nutt.; *Vicia ludoviciana* var. *texana* (Torr. & A. Gray) Shinners; *Vicia texana* (Torr. & A. Gray) Small
English: slim vetch
- *Vicia ludoviciana* Nutt. ex Torr. & A. Gray subsp. *leavenworthii* (Torr. & A. Gray) Lassetter & C. R. Gunn
Synonyms: *Vicia leavenworthii* Torr. & A. Gray
English: Leavenworth's vetch
- *Vicia lutea* L. (Table 15.9)
Synonyms: *Vicia laevigata* Sm.; *Vicia lutea* var. *laevigata* (Sm.) Boiss.
- *Vicia lutea* L. subsp. *lutea*
English: common yellow vetch; true yellow vetch; yellow vetch
- *Vicia lutea* L. subsp. *vestita* (Boiss.) Rouy
Synonyms: *Vicia hirta* Balb. ex DC.; *Vicia lutea* var. *hirta* (Balb. ex DC.) Loisel.; *Vicia lutea* var. *violascens* Rouy; *Vicia vestita* Boiss.
English: clothed yellow vetch
- *Vicia macrograminea* Burkart
English: big grass-like vetch
- *Vicia magellanica* Hook. f.
Synonyms: *Vicia kingii* Hook. f.
English: Magellan's vetch
- *Vicia megalotropis* Ledeb.
English: Big-keel vetch
- *Vicia melanops* Sm.
Czech: vikev černavá
English: black-eyed vetch
Finnish: mustatäplävirna
Polish: wyka czarna
Serbian: crna grahorica
Slovak: vika černastá
Sorbian (Lower): zelenokwětkata wójka
Sorbian (Upper): zelenokwětkata woka
- *Vicia menziesii* Spreng.
English: Hawaiian vetch
Greek: havanézikos víkos
- *Vicia michauxii* Spreng.
Arabic: biqi mishu
English: Michaux's vetch
- *Vicia minutiflora* D. Dietr.

Synonyms: *Vicia micrantha* Nutt. ex Torr. & A. Gray
Arabic: biqi dqiq alzhar
English: pygmyflower vetch; smallflower vetch
Greek: víkos louloúdi-pygmaíos
- *Vicia mollis* Boiss. & Hausskn.
Arabic: biqi nam
English: soft vetch
- *Vicia monantha* Retz.
Synonyms: *Vicia biflora* Desf.; *Vicia calcarata* Desf.
Arabic: biqi uhid alsda; duhhrayg; kharig; 'udays
English: bard vetch; single-flowered vetch; spurred vetch; square-stem vetch
French: vesce uniflore
German: einblütige Wicke
Hebrew: vk'h mdvrvnt
Japanese: arechinoendou
Spanish: algarroba
Swedish: vimpelvicker
- *Vicia monantha* Retz. subsp. *monantha*
Synonyms: *Vicia cinerea* M. Bieb.
English: single-flower vetch
- *Vicia monantha* Retz. subsp. *triflora* (Ten.) B. L. Burtt & P. Lewis
Synonyms: *Vicia triflora* Ten.
English: three-flower vetch
- *Vicia monardii* Boiss.
Arabic: biqi munar
English: Monard's vetch
- *Vicia montbretii* Fisch. & C. A. Mey.
Synonyms: *Lens montbretii* (Fisch. & C. A. Mey.) P. H. Davis & Plitmann
English: Montbret's vetch
- *Vicia montevidensis* Vogel
Synonyms: *Vicia obscura* Vogel
English: Montevideo vetch
- *Vicia multicaulis* Ledeb.
English: many-stem vetch
Japanese: tachinoendou
- *Vicia nana* Vogel
English: dwarf vetch
- *Vicia narbonensis* L. (Table 15.10)
Synonyms: *Vicia serratifolia* f. *integrifolia* Beck
- *Vicia narbonensis* L. var. *aegyptiaca* Asch. & Schweinf.
English: Egyptian bean; Egyptian vetch
- *Vicia nataliae* U. Reifenb. & A. Reifenb.
English: Christmas vetch
- *Vicia nervata* Sipliv.
English: veined vetch
- *Vicia nigricans* Hook. & Arn.

Arabic: biqi suda
Azerbaijani: qara lərgə
English: black vetch; blackish vetch
Greek: maúros víkos
- *Vicia nigricans* Hook. & Arn. subsp. *gigantea* (Hook.) Lassetter & C. R. Gunn
Synonyms: *Vicia gigantea* Hook.
English: giant vetch; large vetch; Sitka vetch
Greek: gigantiaíos víkos
Vietnamese: vetch khổng lồ
- *Vicia nigricans* Hook. & Arn. subsp. *nigricans*
Synonyms: *Vicia apiculata* Phil.; *Vicia macraei* Hook. & Arn.
English: common blackish vetch
- *Vicia nipponica* Matsum.
English: Japanese vetch
Japanese: yotsubahagi
- *Vicia noeana* Reut. ex Boiss.
Arabic: albiqi alsri
English: Noë's vetch
- *Vicia ocalensis* R. K. Godfrey & Kral
English: Ocala vetch
- *Vicia ochroleuca* Ten.
English: pale yellow-white vetch
Polish: wyka bladożółta
- *Vicia ochroleuca* Ten. subsp. *baborensis* (Batt. & Trab.) Quézel & Santa
Synonyms: *Vicia baborensis* Batt. & Trab.
English: Babor vetch
- *Vicia onobrychioides* L.
Arabic: albiqi alnbrisi
Czech: vikev vičencovitá
English: sainfoin vetch
Spanish: alverjana; arveja cana; berzas; esparceta falsa; garrandas; veza; veza de montaña
Swedish: esparsettvicker
- *Vicia oreophila* Zertova
Czech: vikev horská
English: mountain vetch
Slovak: vika horská
- *Vicia oroboides* Wulfen
English: orobus-like vetch
Polish: wyka groszkowata
- *Vicia orobus* DC. (Table 15.11)
- *Vicia palaestina* Boiss.
Arabic: biqi flstini; Karsanah barr-
English: palestine vetch
French: vesce de Palestine

Hebrew: vk'h rtz-shrlt
Vietnamese: liênđậu đẳm
- *Vicia pallida* Hook. & Arn.
Synonyms: *Vicia vicina* var. *pallida* (Hook. & Arn.) Reiche
English: pale vetch
- *Vicia pannonica* Crantz (Table 15.12)
- *Vicia pannonica* Crantz subsp. *pannonica*
English: common Hungarian vetch
- *Vicia pannonica* Crantz subsp. *striata* (M. Bieb.) Nyman
Synonyms: *Vicia striata* M. Bieb.
Czech: vikev červená
Danish: stribet vikke
English: striped Hungarian vetch
- *Vicia parviflora* Cav.
Synonyms: *Ervum gracile* DC.; *Vicia gracilis* Loisel.; *Vicia laxiflora* Brot.; *Vicia tenuissima* auct.; *Vicia tetrasperma* subsp. *gracilis* (DC.) Hook. f.
Arabic: biqi sghir alzhar
Dutch: slanke vikke
English: slender tare
Esperanto: vicio kvarsema; vicio svelta
Greek: leptós víkos
Hebrew: vk'h dkk'h
Serbian: sitnocvetna grahorica
Sorbian (Lower): šwižna wójka
Sorbian (Upper): šwižna woka
Swedish: finvicker
Welsh: corbysen fain
- *Vicia pectinata* Lowe
English: comb-like vetch
- *Vicia peregrina* L. (Table 15.13)
- *Vicia pisiformis* L. (Table 15.14)
- *Vicia pseudo-orobus* Fisch. & C. A. Mey.
English: orobus-like vetch
Japanese: oobakusafuji
Polish: wyka nibygroszkowa
Russian: goroshek lzhesochevnikovyi
- *Vicia pubescens* (DC.) Link
Synonyms: *Ervum pubescens* DC.
Czech: vikev pýřitá
English: pubescent vetch
Hebrew: vk'h ktnh
Russian: goroshek pushistyi
- *Vicia pulchella* Kunth
English: pretty vetch
- *Vicia pyrenaica* Pourr.
Arabic: biqi bransi

English: Pyrenean vetch
French: vesce des Pyrénées
Polish: wyka pirenejska
Serbian: pirenejska grahorica
Spanish: arveja de los Pirineos; veza del Pirineo
- *Vicia qatmensis* Gomb.
English: Qatma vetch
- *Vicia ramuliflora* (Maxim.) Ohwi
Synonyms: *Orobus ramuliflorus* Maxim.
English: branching-flowered vetch
Russian: goroshek razvetvlionnyi
- *Vicia rigidula* Royle
English: inflexible vetch
- *Vicia sativa* L.
Synonyms: *Vicia abyssinica* Alef.; *Vicia alba* Moench; *Vicia amphicarpa* Dorthes; *Vicia amphicarpa* L.; *Vicia angustifolia* L.; *Vicia angustifolia* Reichard; *Vicia bacla* Moench; *Vicia bobartii* E. Forster; *Vicia bobartii* E. Forst.; *Vicia bobartii* Koch; *Vicia canadensis* Zuccagni; *Vicia communis* Rouy; *Vicia consobrina* Pomel; *Vicia cordata* Hoppe; *Vicia cornigera* Chaub.; *Vicia cornigera* St.-Amans; *Vicia cosentini* Guss.; *Vicia cuneata* Gren. & Godr.; *Vicia cuneata* Guss.; *Vicia debilis* Perez Lara; *Vicia erythosperma* Rchb.; *Vicia glabra* Schleich.; *Vicia globosa* Retz.; *Vicia heterophylla* C. Presl; *Vicia incisa* M. Bieb.; *Vicia incisaeformis* Stef.; *Vicia intermedia* ViV; *Vicia lanciformis* Lange; *Vicia lentisperma* auctor ign.; *Vicia leucosperma* Moench; *Vicia macrocarpa* Bertol.; *Vicia maculata* C. Presl; *Vicia maculata* Rouy; *Vicia melanosperma* Rchb.; *Vicia morisiana* Boreau; *Vicia nemoralis* Boreau; *Vicia nemoralis* Ten.; *Vicia notota* Gilib.; *Vicia pallida* Baker; *Vicia pilosa* M. Bieb.; *Vicia pimpinelloides* Mauri; *Vicia sativa* var. *maculata* (C. Presl) Burnat; *Vicia segetalis* Thuill.; *Vicia subterranea* Dorthes; *Vicia terana* Losa; *Vicia vulgaris* Uspensky
English: common vetch
- *Vicia sativa* L. subsp. *amphicarpa* (L.) Batt.
Synonyms: *Vicia amphicarpa* L.; *Vicia angustifolia* var. *amphicarpa* (L.) Alef.; *Vicia sativa* var. *amphicarpa* (L.) Boiss.
English: subterranean vetch
Spanish: veza con semillas de todas partes
- *Vicia sativa* L. subsp. *cordata* (Wulfen ex Hoppe) Batt.
Synonyms: *Vicia cordata* Wulfen ex Hoppe
Czech: vikev srdčitá
English: heart-leaf common vetch
Serbian: srcolisna grahorica
Slovak: vika srdcovitá
Sorbian (Lower): wutšobojta wójka
Sorbian (Upper): wutrobojta woka
- *Vicia sativa* L. subsp. *macrocarpa* (Moris) Arcang.

Synonyms: *Vicia macrocarpa* (Moris) Bertol.; *Vicia sativa* var. *macrocarpa* Moris
Chinese: da zi ye wan dou
English: large-fruit common vetch
- *Vicia sativa* L. subsp. *nigra* (L.) Ehrh. (Table 15.15)
Synonyms: *Vicia angustifolia* L.; *Vicia angustifolia* var. *segetalis* (Thuill.) W. D. J. Koch; *Vicia cuneata* Guss.; *Vicia heterophylla* C. Presl; *Vicia nigra* (L.) Dubois; *Vicia pilosa* M. Bieb.; *Vicia sativa* subsp. *cuneata* (Guss.) Maire; *Vicia sativa* subsp. *segetalis* (Thuill.) Čelak.; *Vicia sativa* subsp. *terana* (Losa) Benedí & Molero; *Vicia sativa* var. *angustifolia* L.; *Vicia sativa* var. *nigra* L.; *Vicia segetalis* Thuill.; *Vicia terana* Losa
- *Vicia sativa* L. subsp. *sativa* (Table 15.16)
Synonyms: *Vicia alba* Moench; *Vicia communis* Rouy; *Vicia leucosperma* Moench; *Vicia sativa* var. *leucosperma* (Moench) Ser.; *Vicia sativa* L. subsp. *notata* Asch. & Graebner; *Vicia sativa* var. *linearis* Lange; *Vicia sativa* var. *obovata* Ser.
- *Vicia scandens* R. P. Murray
English: climbing vetch
- *Vicia semiglabra* Rupr. ex Boiss.
English: half-smooth vetch
Russian: goroshek polugolyi
- *Vicia sepium* L. (Table 15.17)
- *Vicia sericocarpa* Fenzl
English: silky-pod vetch
Hebrew: vkt hmsh
- *Vicia sericocarpa* Fenzl var. *microphylla* Boiss.
English: small-leafed silky-pod vetch
- *Vicia sericocarpa* Fenzl var. *sericocarpa*
English: true silky-pod vetch
- *Vicia serratifolia* Jacq.
Synonyms: *Vicia narbonensis* f. *serratifolia* (Jacq.) F. J. Herm.; *Vicia narbonensis* var. *serratifolia* (Jacq.) Ser.
Arabic: biqi mnshari aluraq
English: French vetch
Finnish: sahavirna
Greek: gallikós víkos
Russian: goroshek zubchatolistnyi; pilolistnaia vika
Serbian: zubolisna grahorica
Sorbian (Lower): zubičkata wójka
Sorbian (Upper): zubičkata woka
Swedish: tandvicker
- *Vicia sessiliflora* Clos
English: stalkless-flower vetch
- *Vicia setifolia* Kunth
English: bristly-leafed vetch

- *Vicia setifolia* Kunth var. *setifolia*
English: common bristly-leafed vetch
- *Vicia setifolia* Kunth var. *bonariensis* Burkart
English: Buenos Aires vetch
- *Vicia sicula* (Raf.) Guss.
Synonyms: *Orobus siculus* Raf.
English: Sicilian vetch
- *Vicia sparsiflora* Ten.
English: sparsely-flowered vetch
Slovak: vika riedkokvetá
- *Vicia sylvatica* L. (Table 15.18)
Synonyms: *Cracca sylvatica* (L.) Opiz; *Ervilia sylvatica* (L.) Schur; *Ervum sylvaticum* (L.) Peterm.; *Vicilla sylvatica* (L.) Schur; *Vicioides sylvatica* (L.) Moench; *Wiggersia sylvatica* (L.) Gaertn.; Mey. & Scherb.
- *Vicia tenuifolia* Roth (Table 15.19)
Synonyms: *Vicia elegans* Guss.; *Vicia variabilis* Freyn
- *Vicia tenuifolia* Roth subsp. *dalmatica* (A. Kern.) Greuter
Synonyms: *Vicia cracca* subsp. *stenophylla* auct.; *Vicia dalmatica* A. Kern.; *Vicia tenuifolia* subsp. *stenophylla* auct.; *Vicia tenuifolia* var. *dalmatica* (A. Kern.) Asch. & Graebn.; *Vicia tenuifolia* var. *laxiflora* Griseb.
Czech: vikev dalmatská
German: Dalmatinische Vogel-Wicke; Dalmatiner Wicke
English: Dalmatian vetch
Serbian: dalmatinska grahorica
Slovak: vika dalmatská
Turkish: dilice otu
- *Vicia tenuifolia* Roth subsp. *tenuifolia*
Synonyms: *Vicia brachytropis* Kar. & Kir.; *Vicia cracca* subsp. *tenuifolia* (Roth) Bonnier & Layens; *Vicia cracca* [unranked] *tenuifolia* (Roth) Gaudin; nom. inval.
English: fine-leaved vetch; slender-leaved vetch
German: Dünnblättrige Wicke; Feinblättrige Wicke; Schmalblättrige Wicke
- *Vicia tenuifolia* Roth subsp. *villosa* (Batt.) Greuter
Synonyms: *Vicia tenuifolia* f. *villosa* Batt.
English: hairy bramble vetch
- *Vicia tetrasperma* (L.) Schreb. (Table 15.20)
Synonyms: *Ervum tenuissimum* Pers.; *Ervum tetraspermum* L.; *Vicia agrestis* Scheele; *Vicia gemella* Crantz
- *Vicia tigridis* Mouterde
English: Tigris vetch
- *Vicia tsydenii* Malyschev
English: Tsyden's vetch
- *Vicia unijuga* A. Braun
English: two-leaf vetch; two-leaved vetch
Finnish: siberianvirna
Japanese: nanten-hagi

Limburgish: prachwèk
Polish: wyka jednogrzbietowa
Russian: goroshek odnoparnyi
Swedish: praktvicker
- *Vicia venosa* (Willd. ex Link) Maxim.
Synonyms: *Orobus venosus* var. *baicalensis* Turcz.; *Vicia baicalensis* (Turcz.) B. Fedtsch.
English: veined vetch
Russian: goroshek zhilkovyi
Slovak: vika pruhovaná
- *Vicia venosa* (Willd. ex Link) Maxim. var. *cuspidata* Maxim.
Synonyms: *Vicia sexajuga* Nakai
English: stiff-point veined vetch
Japanese: ebirafuji
- *Vicia venosa* (Willd. ex Link) Maxim. var. *venosa*
Synonyms: *Orobus venosus* Willd. ex Link
English: common veined vetch
- *Vicia venosissima* Nakai
English: very veined vetch
- *Vicia vicioides* (Desf.) Cout.
Synonyms: *Ervum vicioides* Desf.
English: vetch-like tare
- *Vicia villosa* Roth
Synonyms: *Vicia bivonea* Raf.
English: hairy vetch
- *Vicia villosa* Roth subsp. *eriocarpa* (Hausskn.) P. W. Ball
Synonyms: *Vicia eriocarpa* (Hausskn.) Halácsy; *Vicia varia* var. *eriocarpa* Hausskn.
English: woolly-fruited vetch
Russian: goroshek pushistoplodnyi
- *Vicia villosa* Roth subsp. *pseudocracca* (Bertol.) Rouy
Synonyms: *Cracca bertolonii* Gren. & Godr.; *Cracca elegantissima* Shuttlew nom. inval.; *Vicia ambigua* Guss.; *Vicia elegantissima* Shuttlew ex Rouy; *Vicia pseudocracca* Bertol.; *Vicia villosa* subsp. *ambigua* (Guss.) Kerguélen; *Vicia villosa* subsp. *elegantissima* (Shuttley ex Rouy) G. Bosc & Kerguélen
English: false tufted vetch
Russian: goroshek lozhnomyshinnyi
- *Vicia villosa* Roth subsp. *varia* (Host) Corb. (Table 15.21)
Synonyms: *Cracca varia* (Host) Gren. & Godr.; *Vicia dasycarpa* Ten.; *Vicia glabrescens* (W. D. J. Heimerl); *Vicia plenigera* Formánek; *Vicia pseudovillosa* Schur nom. illeg.; *Vicia varia* Host; *Vicia villosa* subsp. *dasycarpa* (Ten.) Cavill.; *Vicia villosa* var. *Glabrescens* W. D. J. Koch
- *Vicia villosa* Roth subsp. *villosa* (Table 15.22)
Synonyms: *Cracca villosa* (Roth) Godr. & Gren.; *Ervum villosum* (Roth) Trautv. nom. illeg.; *Vicia boissieri* Heldr. & Sartori; *Vicia glabrescens* A. Kern. nom. Illeg.; *Vicia godronii* (Rouy) A. W. Hill; *Vicia plumose* Martin-Donos;

Vicia polyphylla Waldst. & Kit.; *Vicia reuteriana* Boiss. & Buhse; *Vicia unguiculata* subsp. *villosa* (Roth) Bonnier & Layens; *Vicia varia* subsp. *villosa* (Roth) H. J. Coste; *Vicia varia* var. *villosa* (Roth) Arcang.; *Vicia villosa* subsp. *euvillosa* Cavill.; *Vicia villosa* var. *godronii* Rouy

- *Vicia villosa* Roth var. *gore*

English: Gore vetch

15.2 ORIGIN OF SCIENTIFIC AND POPULAR TAXA NAMES

There are many similarities between the genera vetchlings (*Lathyrus* L.) and vetches (*Vicia* L.) from etymological and lexicological viewpoints. Both have few hundred species with accepted botanical status (Duke 1981), both are spread across contrasting environments in the Old World and New World (Chalup et al. 2014, Hechenleitner 2015), and both are considerably rich in neglected and underutilized crop wild relatives and locally cultivated and maintained landraces (Mikić and Mihailović 2014a, b). Nevertheless, the number of genuine roots denoting them in various ethnolinguistic families is rather scarce and, as we shall see in the next paragraphs, the names denoting the *Vicia* species are mostly associated with other annual and perennial forage or grain legume species from agricultural and wild flora.

The Indo-European language family has three attested roots relating to the *Vicia* species:

- One of the descendants of the Proto-Indo-European root **darəw-/*derəw-*, equally referring to field and a kind of wild cereals (Nikolayev 2012), gave the Proto-Germanic **tarwō*, **tarwōn*, meaning simply *a plant*; in its turn, it gave several derivatives with various connotations, with the Old English **taru*, followed by the Middle English *tāre* and the word peculiar solely to several vetch species in modern English (Tables 15.4, 15.7, 15.16, and 15.20), and the Old Norse *tari*, also referring to *Vicia*, but without survivors among its contemporary offspring;
- The Proto-Indo-European root **mAis-*, denoting skin (Nikolayev 2012) and being an ultimate origin of the words denoting lentil (*Lens culinaris* Medik.) and, especially, pea (*Pisum sativum* L.) in numerous Indo-Aryan languages (see Chapters 11 and 14), could be the source of the words generally designating the *Vicia* species in the Iranian Persian (Tables 15.2, 15.4, 15.16, 15.18, 15.20, and 15.22);
- With a clear association with the vetch growth habit and with resemblances to bending or something pliable, the attested Proto-Indo-European **weig-*, **weik* (Pokorny 1959, Nikolayev 2012) had a diverse evolution and was a forefather of the Ancient Greek *bíkion*, *bíkos*, referring to vetch, which, through the Latinized form of *vicia*, gave the names for various species of the eponymous Linnean name in various Indo-European languages (Linnaeus 1753, 1758), such as (Tables 15.1 through 15.22) (1) Armenian; (2) Baltic, with Latvian, Lithuanian, Old Prussian, Samogitian, and Sudovian;

TABLE 15.1
Popular Names Denoting *Vicia articulata* Hornem. in Some World Languages and Other Linguistic Taxa

Language/Taxon	Name
Arabic	biqi mfsli
Czech	vikev článkovaná
Danish	monanthavikke; linsevikke; vikkelinse
English	bard vetch; monantha vetch; one-flower vetch; one-leaved vetch; single-flowered vetch
Finnish	yksikukkavirvilä
French	jarosse d'Auvergne; lentille d'Auvergne; lentille à une fleur; vesce uniflore
German	algaroba-Linse; Einbluetige Erve; Einblütige Wicke; Wicklinse
Greek	víkos várdos
Italian	veccia articolata
Limburgish	wèklinse
Lithuanian	vienažiedis vikis
Polish	wyka członowana; wyka jednokwiatowa
Portuguese	lentilha-parda
Portuguese (Brazil)	ervilhaca-algaroba; ervilhaca parda
Serbian	člankovita grahorica; jednocvetna grahorica
Slovak	vika článkovaná
Sorbian (Lower)	jadnokwětkata wójka
Sorbian (Upper)	jednokwětkata woka
Spanish	alberja; algalrroba; algarroba; algarroba de Castilla; algarrobas; alverja; arveja; arveja articulada; arveja cuadrada; arvejana; arvejera; arvejona; cuadrado; garroba; garroba de Castilla; garrobas; garrubia; lenteja; lenteja de Aragón; paja herbaliza; vicia
Swedish	monanthavicker; spanskvicker

TABLE 15.2
Popular Names Denoting *Vicia benghalensis* L. in Some World Languages and Other Linguistic Taxa

Language/Taxon	Name
Arabic	biqi bnghali
Catalan	matagat; veça de ramelles; veça vermella
Czech	vikev bengálská
Danish	bengalsk vikke
English	algerian vetch; Bengali vetch; hairy vetch; purple vetch; winter vetch
French	vesce de Bengale; vesce pourpre foncé
German	Purpurwicke
Hebrew	vk'h vnglt

(Continued)

TABLE 15.2 (*Continued*)
Popular Names Denoting *Vicia benghalensis* L. in Some World Languages and Other Linguistic Taxa

Language/Taxon	Name
Italian	veccia del Bengala
Persian	mashk arghuani
Polish	wyka bengalska
Russian	goroshek bengal'skii
Serbian	bengalska grahorica
Spanish	alberjacón; alverjanas; alverjilla de Bengala; arbejancas; arbenjacón; arveja roja; veza púrpura; veza purpúrea
Swedish	purpurvicker
Welsh	ffacbysen borffor

TABLE 15.3
Popular Names Denoting *Vicia cassubica* L. in Some World Languages and Other Linguistic Taxa

Language/Taxon	Name
Arabic	biqi kashubi
Czech	vikev kašubská
Danish	kassubisk vikke
English	Danzig vetch; Kashubian vetch
Erzya	kashubon' ksnavne
Esperanto	kaŝuba vicio
Estonian	püstine hiirehernes
Finnish	pommerinvirna
French	vesce de Kashoubie
German	Kassuben-Wicke
Greek	víkos kasouvías
Italian	veccia dei cassubi
Kashubian	kaszëbskô wika
Lithuanian	kašubinis vikis
Norwegian (Nynorsk)	sørlandsvikke
Polish	wyka kaszubska
Russian	goroshek kashubskii
Serbian	kašupska grahorica
Slovak	vika kašubská
Sorbian (Lower)	kašubska wójka; sucha wójka
Sorbian (Upper)	sucha woka
Swedish	backvicker
Welsh	ffugbysen fer-godog

TABLE 15.4
Popular Names Denoting *Vicia cracca* L. in Some World Languages and Other Linguistic Taxa

Language/Taxon	Name
Arabic	arab afâqi; biqi mnqd; dandarân
Catalan	garlanda comuna
Chinese	duō huāyě wāndòu; guang bu ye wan dou
Czech	vikev ptačí
Danish	musevikke
Dutch	vogelwikke
English	bird tare; bird vetch; blue vetch; boreal vetch; cow vetch; tufted vetch
Erzya	cheeren' ksnavne
Esperanto	birda vicio
Estonian	harilik hiirehernes
Finnish	hiirenherne; hiirenvirna
French	pois à crapauds; vesque craque; vesce à épis; vesce de Cracovie
French (Quebec)	vesce jargeau; vesce multiflore; vesce sauvage
Frisian (West)	wikje
German	vogelwicke
Greek	víkos fountotós
Guernésiais	hazette; véchaon
Icelandic	umfeðmingur
Istriot	fava salvàdega
Italian	cracca; veccia delle siepi; veccia moltiflora; veccia montanina; veccia spicata
Japanese	kusafuji
Jèrriais	véchon
Kashubian	ptôsznik
Kazakh	tışqan burşaq; tışqan sïırjoñışqa
Komi-Zyrian	shyrankytsh; shyrgyorysh
Latvian	vanagu vīķi
Limburgish	krejjewèk; vogelwèk
Lithuanian	mėlynžiedis vikis
Mari (Hill)	kalia pərsa
Moldovan	mezerike shoarechului
Norwegian (Bokmål)	fuglevikke
Norwegian (Nynorsk)	fuglevikke
Ossetian	qæddag qædur
Persian	mashk klaghi
Polish	wyka ptasia
Portuguese	ervilhaca
Portuguese (Brazil)	cisirão; ervilhaca-dos-passarinhos
Romanian	măzăriche pasareasca

(*Continued*)

TABLE 15.4 (*Continued*)
Popular Names Denoting *Vicia cracca* L. in Some World Languages and Other Linguistic Taxa

Language/Taxon	Name
Russian	goroshek myshinyi
Sami (Northern)	fierbmerássi; sáhpal
Serbian	ptičja grahorica
Slovak	vika vtáčia
Sorbian (Lower)	ptaškowa wójka
Sorbian (Upper)	dźiwja woka; foglica; ptača woka; sočk; wočka
Spanish	arveja; arveja silvestre; veza de pájaro; veza francesa
Swedish	fågelvicker; kråkvicker; tranärt
Turkish	kara burçak; küshéne
Ukrainian	goroshok mishachii; vyka bagatokvitkova
Vietnamese	điêu tử; liên đậu
Welsh	ffacbysen y berth; ffacbysen y berth gwyg; gwygbysen; pys y gath; pys y llygod bach; tagwyg; tagwyg bysen

TABLE 15.5
Popular Names Denoting *Vicia dumetorum* L. in Some World Languages and Other Linguistic Taxa

Language/Taxon	Name
Arabic	biqi altkhum
Czech	vikev křovištní
Danish	kratvikke
English	thorn-bush vetch
Esperanto	vicio bluviola
Finnish	pensaikkovirna
German	Hecken-Wicke
Kazakh	buta burşaq
Limburgish	bóswèk
Lithuanian	krūmelinis vikis
Polish	wyka zaroślowa
Russian	goroshek kustarnikovyi; goroshek Zaroslevyi
Serbian	trnova grahorica
Slovak	vika krovisková
Sorbian (Lower)	kerčna wójka
Sorbian (Upper)	kerčna woka
Swedish	buskvicker

TABLE 15.6
Popular Names Denoting *Vicia grandiflora* Scop. in Some World Languages and Other Linguistic Taxa

Language/Taxon	Name
Arabic	biqi kbir alzhar
Czech	vikev velkokvětá
English	large-flowered vetch
Erzya	poksh cecia ksnavne
German	Großblütige Wicke
Japanese	kibanakarasunoendou
Polish	wyka brudnożółta
Russian	goroshek krupnotsvetkovyi
Serbian	krupnocvetna grahorica
Slovak	Vika Kitaibelova; vika veľkokvetá
Sorbian (Lower)	wjelikokwětna wójka
Sorbian (Upper)	wulkokwětna woka
Swedish	storvicker

TABLE 15.7
Popular Names Denoting *Vicia hirsuta* (L.) Gray in Some World Languages and Other Linguistic Taxa

Language/Taxon	Name
Arabic	biqi shri
Chinese	que ye wan dou; xiăo cháo cài; ye can dou; ying mao guo ye wan dou
Czech	vikev chlupatá
Danish	laadden vikke; tofrøet vikke
Dutch	ringelwikke
English	hairy tare; tare vetch; tiny vetch
English (Australia)	hairy vetch
Erzya	ponav ksnavne
Esperanto	vicio dusema; vicio vila
Estonian	karvane hiirehernes
Finnish	peltovirvilä
French	ers lentille; ers velu; vesce hérissée; vesceron
German	Acker-Wicke; Beeharte Wicke; Ervenlinse; Rauhhaarwicke; Rauhaarige Wicke; Zitterlinse; Zitterwicke
Hindi	munmana
Italian	tentennino; veccia irsuta
Japanese	suzumeno endō
Kashubian	drobnokwiatowô wika

(Continued)

TABLE 15.7 (*Continued*)
Popular Names Denoting *Vicia hirsuta* (L.) Gray in Some World Languages and Other Linguistic Taxa

Language/Taxon	Name
Limburgish	doevewèk; ringelwèk
Lithuanian	gauruotasis vikis
Norwegian (Bokmål)	tofrøvikke
Norwegian (Nynorsk)	tofrøvikke
Polish	wyka drobnokwiatowa
Portuguese	cigerão; unhas-de-gato
Portuguese (Brazil)	ervilhaca-pilosa
Russian	goroshek volosistyi; vika volosistaia
Serbian	kosmata grahorica; sitna grahorica
Slovak	vika chlpatá
Sorbian (Lower)	mólička wójka
Sorbian (Upper)	mólička woka
Spanish	achicoria; alberja; alberjon; alberjón; alberjón silvestre; alverja; alverja erizada; alverjón; veza; veza hirsuta
Swedish	duvvicker
Welsh	corbys blewog; corbysen flewog; ffacbysen flewog

TABLE 15.8
Popular Names Denoting *Vicia lathyroides* L. in Some World Languages and Other Linguistic Taxa

Language/Taxon	Name
Arabic	biqi rbii
Czech	vikev hrachorovitá
Danish	vår-vikke; vårvikke
Dutch	lathyruswikke
English	spring vetch; vetchling-like vetch
Esperanto	vicio kuŝanta
Estonian	väikeseõieline hiirehernes
Finnish	nätkelmävirna
French	vesce fausse gesse; vesce printanière
German	Frühlings-Zwerg-Wicke, Kicher-Wicke; Platterbsen-Wicke; Sand-Wicke
Greek	anoiksiátikos víkos
Hebrew	vk'h tvfchn
Japanese	hinakarasunoendou
Limburgish	hawèk

(*Continued*)

TABLE 15.8 (*Continued*)
Popular Names Denoting *Vicia lathyroides* L. in Some World Languages and Other Linguistic Taxa

Language/Taxon	Name
Lithuanian	pavasarinis vikis
Norwegian (Nynorsk)	vårvikke
Polish	wyka lędźwianowata
Russian	goroshek chinovidnyi
Serbian	grahorasta grahorica; grahorolika grahorica
Slovak	vika hrachorovitá
Sorbian (Lower)	tšawna wójka
Sorbian (Upper)	trawna woka
Swedish	vårvicker
Welsh	ffacbys y gwanwyn; ffacbysen y gwanwyn; ffugbysen y gwanwyn

TABLE 15.9
Popular Names Denoting *Vicia lutea* L. in Some World Languages and Other Linguistic Taxa

Language/Taxon	Name
Arabic	biqi sfra
Catalan	galabarç; galavars o veça groga
Czech	vikev žlutá
Danish	gul vikke
Dutch	gele wikke
English	smooth yellow vetch; yellow vetch
Esperanto	vicio flava
Finnish	keltavirna
French	vesce jaune
German	Gelbe Wicke
Greek	kítrinos víkos
Hebrew	vkt tzhvvh
Italian	cicerchia pelosa; veccia gialla
Japanese	onikarasunoendou
Limburgish	gaele wèk
Norwegian (Bokmål)	gulvikke
Norwegian (Nynorsk)	gulvikke
Polish	wyka żółta
Portuguese	ervilhaca amarela
Russian	goroshek zholtyi
Serbian	žuta grahorica

(*Continued*)

TABLE 15.9 (*Continued*)
Popular Names Denoting *Vicia lutea* L. in Some World Languages and Other Linguistic Taxa

Language/Taxon	Name
Slovak	vika žltá
Sorbian (Lower)	žołta wójka
Sorbian (Upper)	žołta woka
Spanish	Abrejaca; abrejacón; alberjacas; alberjana; alberjmeana; algarrobón; almejana; alvehón; alvejana; alvejón; alverja; alverja silvestre; alverjaca; alverjana; alverjana de gente; alverjaque; alverjotes; alverjón; arbejaca; arbejana; arveana; arveja; arveja amarilla; arvejaca; arvejana; arvejancas; arvejona; arvejón; arverja; arverjas; beza loca; cuchillejo; erbiaca; guisantera; guisantes silvestres; habón; pelailla; veza
Swedish	gulvicker
Ukrainian	horoshek zhovtyi
Welsh	eurbys; eurlys; ffacbysen felen; ffugbysen felen arw-godog

TABLE 15.10
Popular Names Denoting *Vicia narbonensis* L. in Some World Languages and Other Linguistic Taxa

Language/Taxon	Name
Arabic	biqi nrbuni
Czech	vikev narbonská
Danish	narbonevikke
English	French vetch; moor's pea; narbon bean; Narbonne vetch; purple broad bean
Esperanto	vicio franca; vicio larĝfolia
Finnish	purppuravirna
French	vesce de Narbonne
Georgian	tzkhenis tzertzvela
German	Französische Wicke; Maus-Wicke; Schwarze Ackerbohne
Greek	fasóli narbon; mpizéli tou hersótottou
Hebrew	vk'h tzrftt
Kazakh	rïm burşağı
Polish	wyka narbońska
Portuguese	ervilha-de-Narbona
Russian	goroshek narbonskyi; narbonskaia vika; rimskaia vika
Serbian	narbonska grahorica
Slovak	vika narbonská
Sorbian (Lower)	myšaca wójka
Sorbian (Upper)	myšaca woka
Spanish	alverijón; alverjón; haba loca
Swedish	klotvicker

TABLE 15.11
Popular Names Denoting *Vicia orobus* DC. in Some World Languages and Other Linguistic Taxa

Language/Taxon	Name
Arabic	biqi almruj
Danish	lyngvikke; lyng-vikke
Dutch	heidewikke
English	upright vetch; wood bitter vetch
Esperanto	vicio staranta
Finnish	nummivirna
German	Heide-Wicke
Greek	ksýlo pikrós-víkos; órthios víkos
Limburgish	heiwèk
Norwegian (Bokmål)	vestlandsvikke
Norwegian (Nynorsk)	vestlandsvikke
Portuguese	ervilhaca-do-prado
Serbian	Šumski urov
Sorbian (Lower)	holanska wójka
Sorbian (Upper)	holanska woka
Swedish	ljungvicker
Welsh	chwerbys y coed; ffacbys chwerw; ffacbysen chwerw; pys y garanod; pys y maes; pysen y coed; pysen yr aran

TABLE 15.12
Popular Names Denoting *Vicia pannonica* Crantz in Some World Languages and Other Linguistic Taxa

Language/Taxon	Name
Arabic	biqi banuni
Catalan	veça pannonica
Chinese	he mao ye wan dou; xiong ya li ye wan dou
Czech	vikev panonská
Danish	ungarsk vikke
Dutch	hongaarse wikke
English	Hungarian vetch
Esperanto	hungara vicio
Finnish	unkarinvirna
French	vesce de Hongrie; vesce de Pannonie
Georgian	ungruli tzvertzvela
German	Pannonisch-Wicke; pannonische Wicke; Ungarische Wicke

(*Continued*)

TABLE 15.12 (*Continued*)
Popular Names Denoting *Vicia pannonica* Crantz in Some World Languages and Other Linguistic Taxa

Language/Taxon	Name
Greek	ouggrikós víkos
Italian	veccia della Pannonia; veccia d'Ungheria
Kazakh	pannon burşaq
Norwegian (Bokmål)	ungarsk vikke
Polish	wyka pannońska
Romanian	măzăriche panonica
Russian	goroshek pannonskii; goroshek vengerskii
Serbian	panonska grahorica
Slovak	vika panónska
Sorbian (Lower)	hungorska wójka
Sorbian (Upper)	wuherska woka
Spanish	veza de Hungria; veza panonica; veza húngara
Swedish	ungersk vicker
Welsh	ffacbysen Hwngaria

TABLE 15.13
Popular Names Denoting *Vicia peregrina* L. in Some World Languages and Other Linguistic Taxa

Language/Taxon	Name
Arabic	biqi jnbi; gulubb'n
Czech	vikev cizí
Danish	vandrevikke
English	broad-pod vetch; rambling vetch
Finnish	muukalaisvirna
French	vesce à gousses larges
Hebrew	vk'h mtzvh
Italian	veccia viaggiatrice
Kazakh	bögde burşaq
Russian	goroshek inozemnyi
Serbian	širokoplodovita grahorica
Turkish	culban
Vietnamese	liênđậu ngao

TABLE 15.14
Popular Names Denoting *Vicia pisiformis* L. in Some World Languages and Other Linguistic Taxa

Language/Taxon	Name
Arabic	biqi bazlai
Czech	vikev hrachovitá
English	pale-flower vetch; pea vetch; pea-flowered vetch
Esperanto	vicio pizoforma
Finnish	hernevirna
French	vesce à feuilles de pois
German	Erbsen-Wicke
Greek	mpizelo-loúloudos víkos
Kazakh	burşaqtärizdi sïır joñışqa
Limburgish	ertwèk
Lithuanian	Žirnialapis vikis
Norwegian (Bokmål)	ertevikke
Norwegian (Nynorsk)	ertevikke
Polish	wyka grochowata
Russian	goroshek gorokhovidnyi
Serbian	graškolika grahorica
Slovak	vika hrachovitá
Sorbian (Lower)	grochowa wójka
Sorbian (Upper)	hrochowa woka
Swedish	ärtvicker

TABLE 15.15
Popular Names Denoting *Vicia sativa* L. subsp. *nigra* (L.) Ehrh. in Some World Languages and Other Linguistic Taxa

Language/Taxon	Name
Arabic	albiqi rfi aluraq
Chinese	niao ye wan dou; que ye wan dou; wu ye wan dou; xia ye ye wan dou; zhai ye ye wan dou
Czech	vikev úzkolistá
Danish	smalbladet vikke
Dutch	akkerwikke; smalle wikke; vergeten wikke
English	black-podded vetch; narrow-leaved vetch
English (U.S.)	black-pod vetch; narrow-leaf vetch
Erzya	teine lopa ksnavne
Estonian	ahtalehine hiirehernes
Finnish	kaitavirna

(*Continued*)

TABLE 15.15 (*Continued*)
Popular Names Denoting *Vicia sativa* L. subsp. *nigra* (L.) Ehrh. in Some World Languages and Other Linguistic Taxa

Language/Taxon	Name
French	vesce à feuilles étroites; vesce à gousses noires; vesce noire
German	Schmalblättrige Wicke
Greek	stenófyllos víkos
Icelandic	sumarflækja
Italian	veccia a foglie strette; veccia angustifoglia; veccia near
Japanese	karasu no endo; yahazuendou
Kazakh	ayıljapıraq sïirjoñışqa; tarjapıraqtı sïirburşaq
Lithuanian	siauralapis vikis
Norwegian (Bokmål)	sommarvikke
Norwegian (Nynorsk)	sommarvikke
Portuguese	larica
Russian	goroshek uzkolistniy
Scots	fitchy pease
Serbian	uskolisna grahorica
Slovak	vika úzkolistá
Sorbian (Lower)	wuska wójka
Sorbian (Upper)	wuska woka
Spanish	alberja; alverjana borriquera; alverjilla; arvejana; averijilla; veza; veza almerejana; veza de hoja estrecha
Swedish	liten sommarvicker; stor sommarvicker
Welsh	ffugbysen faethol gyfreddin; ffugbysen gulddail ruddog; ffugbysen wyllt

TABLE 15.16
Popular Names Denoting *Vicia sativa* L. subsp. *sativa* in Some World Languages and Other Linguistic Taxa

Language/Taxon	Name
Afrikaans	wieke
Albanian (Arvanitika)	vik
Arabic	biqi mzru; bizillet iblâs; duhhrayg; ‘udaysah
Armenian	vik tzanovi
Asturian	alverja; veza
Azerbaijani (North)	lərgə; yem noxudu; əkin lərgəsi
Azerbaijani (South)	lərgə; ij’m nuxudu
Bashkir	kəreşkə
Basque	txirta; zalke
Basque (Biscayan)	idar
Basque (Gipuzkoan)	illar

(*Continued*)

TABLE 15.16 (*Continued*)
Popular Names Denoting *Vicia sativa* L. subsp. *sativa* in Some World Languages and Other Linguistic Taxa

Language/Taxon	Name
Basque (Lapurdian)	illar
Basque (Souletin)	illar
Belarusian	harošak; vika jaravaja; vika pasjaŭnaja
Bengali	ankari
Breton	beñs; gweg
Bulgarian	fiy; glushina
Catalan	erb; veça; veça comuna
Chechen	bürtig; qöš
Chinese	chao cai; da chao cai; jiù huāngyě wāndòu
Croatian	grahorica
Czech	vikev; vikev seta
Danish	fodervikke
Dutch	voeder wikke; wikke
English	common vetch; Eurasian vetch; garden vetch; spring tare; summer vetch; tare; vetch
English (Middle)	fecche; fiche; vecche
Erzya	videma ksbavne; videma vika
Esperanto	kultiva vicio
Estonian	kurehernes; suvivikk; vikk
Finnish	elovirna; rehuvirna; vikkeri
Forth and Bargy	vetch
French	vesce commune; vesce cultivée; vesce fourragère
French (Old)	vece
Friulian	vece
Galician	arvellaca; brenza; herba da fame; herba do crego; herbellaca; ervellaca; leriquia; liriquia; livica; nichela; pan do cuco; veza; verza
Georgian	chveulebrivi tzertzvela; sagazap'khulo ts'erts'vela
German	ackerwicke; Futterwicke; Saatwicke; Saat-Wicke
Greek	agrióvikos; koinós víkos; víkos
Greek (Ancient)	bíkion; bíkos
Hebrew	vk'h trvvtt
High German (Old)	wiccha
Hindi	akra; akta; ankra
Hungarian	bükköny
Icelandic	akurflækja
Ido	vecho
Irish	peasair
Istriot	dénti de vècia; seʃemanarìn
Italian	veccia angustifoglia; veccia commune; vecca coltivata; veccia dolce
Japanese	karasu no endō; ooyahazu-endō
Jèrriais	vèche

(*Continued*)

TABLE 15.16 (*Continued*)
Popular Names Denoting *Vicia sativa* L. subsp. *sativa* in Some World Languages and Other Linguistic Taxa

Language/Taxon	Name
Kabardian	qhašč
Kashubian	zwëczajnô wika
Kazakh	egistik sïırjoñıșqa; ekpe sïırjoñıșqa; ekpeli sïırburșaq
Korean	salgalkwi
Kyrgyz	jazdık jer buurçak; jer buurçak
Latin	vicia
Latvian	sējas vīķi
Limburgish	voorwèk
Lithuanian	sėjamasis vikis
Macedonian	glushina
Mari (Meadow)	koliavursa
Moldovan	mazerike; mezerike
Mongolian	gonzgoi navchit gish; tarimal gish
Nepali	kutulee kosa
Norman (Old)	veche
Norwegian (Bokmål)	fôrvikke
Old Prussian	wikis
Ossetian	qædurhos
Persian	mashk mymuli
Polish	wyka siewna
Portuguese	ervilhaca-comum; ervilhaca vulgar
Romagnol	vezza
Romanian	măzăriche
Romansh	vetscha
Russian	goroshek kormovyi; goroshek posevnoi; vika obyknovennaia; vika posevnaia
Rusyn (Pannonian)	liednik
Samogitian	vėkis; žėrnėks
Sardinian (Campidanese)	pappasolu
Scots	horse pease
Serbian	grahorica; grahorika; graorica; graorika; graševina; gravorika; obična grahorica
Serbian (Bosnian Frontier)	graor'ca; lanjik
Serbian (Gallipoli)	grȁorak
Serbian (Great Morava Valley)	meúnice
Serbian (Užice)	mèūnka
Sicilian	vizza
Slovak	vika obyčajná; vika siata
Slovenian	grašica; navadna grašica
Sorbian (Lower)	rolna wójka
Sorbian (Upper)	rólna woka

(*Continued*)

TABLE 15.16 (*Continued*)
Popular Names Denoting *Vicia sativa* L. subsp. *sativa* in Some World Languages and Other Linguistic Taxa

Language/Taxon	Name
Spanish	abrejacón; albajaca; albajana; albejana; alberja; alberjana; alberjanca; alberjón; albilla; algarroba; algarroba común; algarroba de burricu; algarroba-veza; algarrobes; algarrobilla; alveja; alvejas; alvejones; alverja; alverja montesa; alverjacón; alverjana; alverjana andaluza; alverjana de Castiella; alverjaque; alverjas; alverjas de burru; alverjilla; alverjones; alverjón; arabeja; arbeicha; arbeja; arberjones; arbexaca; arbeya; arbeyes de pollín; arrica; arveaca; arvechaca; arbeyera; arbeyera colorada; arbeyera común; arbeyera negra; arvejaca; arvejana; arbeyeres; arvejilla; arvejo; arvejón; arverja; arverjana; arveyo; arvichaca; avejaca; avejacón; avejones; avejácara; averjaca; avesa; aveza; bejancones; berzo; berzón; borricón; burricón; caballuna; cagayón de ratu; carballón; clarín; corvina; francu; garrobilla; guijeta; guixeta; madrasta triguera; pesolilo; pesolillo cantu; pimpinela; piseo; presule cantu; riques; titarra; titarrina; pedrete redondu; titón; verza; veza; veza blanca; veza común; veza forrajera; veza negra; veza montesa; vezas; vicia; yapina; yerbo; yero; yeros
Sudovian	vikis
Swedish	fodervicker; liten sommarvicker
Tatar	käreškä
Turkish	burçak; fiğ
Ukrainian	horoshok posivnyi; vika posiina; vika siina; vika sivna; vika zvichaina
Uzbek	ekma vika
Vietnamese	đậu răng ngựa nhỏ
Võro	hiireherneh
Welsh	ffugbysen cyffredin; ffugbysen faethol; ffugbysen faethol gyffredin; ffugbysen gulddail ruddog; ffugbysen wyllt; gwŷg; pupys

TABLE 15.17
Popular Names Denoting *Vicia sepium* L. in Some World Languages and Other Linguistic Taxa

Language/Taxon	Name
Arabic	biqi alsiaj
Armenian	gyulul
Belarusian	ljada
Chinese	dian ye wan dou; ye wan dou
Czech	vikev plotní
Danish	gærdevikke
Dutch	heggevikke; heggenvikke; vitsen
English	bush vetch; hedge vetch

(*Continued*)

TABLE 15.17 (*Continued*)
Popular Names Denoting *Vicia sepium* L. in Some World Languages and Other Linguistic Taxa

Language/Taxon	Name
Erzya	piriavks langa kasycia ksnavne
Esperanto	vicio heĝa
Estonian	aed-hiirehernes
Finnish	aitovirna
French	vesce des haies; vesce sauvage
German	Zaun-Wicke
Greek	thamnódis víkos
Icelandic	giljaflækja
Italian	veccia delle siepi
Japanese	ibukinoendou
Kazakh	qora sïırjoñışqa
Limburgish	hègkewèk
Lithuanian	patvorinis vikis
Norwegian (Bokmål)	gjerdevikke
Norwegian (Nynorsk)	gjerdevikke
Polish	wyka płotowa
Portuguese (Brazil)	ervilhaca-de-cerca
Russian	goroshek zabornyi
Serbian	krajputaška grahorica; živična grahorica
Slovak	vika plotná
Sorbian (Lower)	lucna wójka
Sorbian (Upper)	łučna woka
Spanish	alverja; alverjana; arveja de los setos; arveja silvestre; mata trepadera; vera; veza; vicia de los vallados
Swedish	häckvicker
Welsh	ffugbys y clawdd; ffucbysen y cloddiau; ffugbysen y cloddiau; pys y berth

TABLE 15.18
Popular Names Denoting *Vicia sylvatica* L. in Some World Languages and Other Linguistic Taxa

Language/Taxon	Name
Arabic	biqi hraji
Czech	vikev lesní
Danish	skovvikke; skov-vikke
English	wood vetch
Erzya	viren' ksnavne
Esperanto	vicio arbara

(*Continued*)

TABLE 15.18 (*Continued*)
Popular Names Denoting *Vicia sylvatica* L. in Some World Languages and Other Linguistic Taxa

Language/Taxon	Name
Estonian	mets-hiirehernes
Finnish	metsävirna
French	vesce des bois
German	Wald-Wicke
Greek	ksylóvikos
Kashubian	lasowô wika; lesnô wika
Kazakh	orman burşaq; orman sïırjoñışqa
Lithuanian	miškinis vikis
Norwegian (Bokmål)	skogvikke
Persian	mashk dcngli
Polish	wyka leśna
Russian	goroshek lesnoi
Serbian	šumska grahorica
Slovak	vika lesná
Sorbian (Lower)	lěsna wójka
Sorbian (Upper)	lěsna woka
Swedish	skogsvicker
Welsh	ffacbys; ffacbys y wig; ffacbysen y coed; ffacbysen y wig; ffagbysen y wig; ffugbysen y wig

TABLE 15.19
Popular Names Denoting *Vicia tenuifolia* Roth in Some World Languages and Other Linguistic Taxa

Language/Taxon	Name
Arabic	biqi rqiq aluraq
Chinese	hei zi wan dou; san chi cao teng; xi ye ye wan dou
Czech	vikev tenkolistá
Danish	langklaset vikke
Dutch	stijve wikke
English	bramble vetch; fine-leaf vetch
Esperanto	vicio mallarĝfolia
Estonian	peenelehine hiirehernes
Finnish	tuoksuvirna
Greek	oraiófyllos víkos
Hebrew	vk'h dkt-'lm
Kazakh	jiñişke japıraq sïırjoñınqa; jiñişke japıraqtı sïırjoñışqa
Limburgish	stieve wèk; stinkendjen

(*Continued*)

TABLE 15.19 (*Continued*)
Popular Names Denoting *Vicia tenuifolia* Roth in Some World Languages and Other Linguistic Taxa

Language/Taxon	Name
Lithuanian	smulkialapis vikis
Norwegian (Nynorsk)	luktvikke
Polish	wyka długożagielkowa
Russian	goroshek tonkolisntyi
Serbian	nežnolisna grahorica; tankolisna grahorica
Slovak	vika tenkolistá
Sorbian (Lower)	drobna wójka
Sorbian (Upper)	drobna woka
Swedish	luktvicker
Vietnamese	liên đậu nhỏ
Welsh	ffugbysen feinddail

TABLE 15.20
Popular Names Denoting *Vicia tetrasperma* (L.) Schreb. in Some World Languages and Other Linguistic Taxa

Language/Taxon	Name
Arabic	biqi rbai alhbub
Chinese	duo zi ye wan dou; si zi cao teng; sì zǐ yě wāndòu
Czech	vikev čtyřsemenná
Danish	taddervikke; tadder-vikke
Dutch	gladde wikke; vierzadige wikke
English	four-seeded vetch; lentil tare; lentil vetch; many-seeded slender tare; multi-seeded slender vetch; slender tare; slender vetch; smooth tare; smooth vetch; sparrow vetch
Erzya	nile vid'men' ksnavne
Estonian	neljaseemnene hiirehernes
Finnish	mäkivirvilä
French	vesce à quatre graines
German	Viersamige Wicke; Zierliche Wicke
Greek	leíos víkos
Hebrew	vk'h dnh
Italian	veccia tetrasperma
Japanese	kasuma-gusa
Kazakh	taspa burşaq; taspan japıraq; taşpan japıraq
Korean	eolchigiwandu
Limburgish	mösjewèk
Lithuanian	ketursėklis vikis

(*Continued*)

TABLE 15.20 (*Continued*)
Popular Names Denoting *Vicia tetrasperma* (L.) Schreb. in Some World Languages and Other Linguistic Taxa

Language/Taxon	Name
Norwegian (Bokmål)	firfrøvikke
Norwegian (Nynorsk)	firfrøvikke
Persian	mashk tcx'ardanx'ai
Polish	wyka czteronasienna
Russian	chetyriohsemiannyi goroshek
Serbian	četvorosemena grahorica
Slovak	vika štvorsemenná
Sorbian (Lower)	šańka wójka
Sorbian (Upper)	ćeńka woka
Spanish	vícia tetrasperma
Swedish	sparvvicker
Welsh	corbysen lefn bedair-ronynnog; corbysen lefn bedwar-ronynnog; corbysen lefn ronynnog; ffacbysen lefn

TABLE 15.21
Popular Names Denoting *Vicia villosa* Roth subsp. *varia* (Host) Corb. in Some World Languages and Other Linguistic Taxa

Language/Taxon	Name
Arabic	biqi mubr nui almzrksh
Chinese	rong jia ye wan dou
Czech	vikev olysalá
Danish	glatvikke
English	hairy vetch; purple vetch; sand vetch; smooth vetch; winter vetch; woollypod vetch
Finnish	myllyruisvirna
French	vesce à gousse velue
German	bunte Wicke
Japanese	birodo kusa fuji; nayo kusa fuji
Polish	wyka pstra
Russian	goroshek izmenchivyi
Serbian	raznobojna grahorica; šarena grahorica
Slovak	vika olysalá
Sorbian (Lower)	pisana wójka
Sorbian (Upper)	pisana woka

TABLE 15.22
Popular Names Denoting *Vicia villosa* Roth subsp. *villosa* in Some World Languages and Other Linguistic Taxa

Language/Taxon	Name
Arabic	biqi mubr
Armenian	vik t'avot
Asturian	veza vellosa; vezo piloso
Belarusian	vika azimaja; vika kasmataja
Bulgarian	valnetasta glushina
Catalan	veça villosa
Chinese	chang rou mao ye wan dou; mao er tiao zi; mao tiao cai; mao ye tiao zi; ou zhou tiao zi; rong mao chao cai; zhǎng róu máo yě wāndòu
Czech	vikev huňatá
Danish	sandvikke; sand-vikke
Dutch	bonte wikke; zachte wikke; zandwikke
English	downy vetch; fodder vetch; hairy vetch; Russian vetch; sand vetch; winter vetch; woolly-pod vetch
English (Ulster)	mouse-pea
Erzya	ozimen' vika; ponav ksnavne
Esperanto	vicio mola
Estonian	põld-hiirehernes
Finnish	peltoruisvirna; ruisvirna
French	vesce de Cerdagne; vesce de Russie; vesce des sables; vesce velue
Georgian	banjgvliani ts'erts'vela; sashemodgomo ts'erts'vela
German	Sandwicke; Winterwicke; Zottelwicke; Zottige Wicke
Greek	heimerinós víkos; trihotós víkos; zootrofikós víkos
Hebrew	vk'h sh'rh
Italian	veccia dasicarpa; veccia pelosa; veccia vellutata; veccia villosa
Japanese	birōdo kusa fuji; ke yahazu endō; nayo kusa fuji
Kashubian	szadô wika
Kazakh	jazdıq sïırjoñışqa; tükti sïırjoñışqa; tükti burşaq
Korean	teolgalkwideonggul
Limburgish	bóntje wèk; zaachte wèk; zandjwèk
Lithuanian	ruginis vikis
Mongolian	üsleg gish
Norwegian (Bokmål)	lodnevikke
Norwegian (Nynorsk)	lodnevikke
Persian	mashk zmstani
Polish	wyka kosmata; wyka ozima
Portuguese	ervilhaca-dos-cachos-roxos; ervilhaca-peluda; ervilhaca-vilosa
Romanian	măzăriche paroasa

(*Continued*)

TABLE 15.22 (*Continued*)
Popular Names Denoting *Vicia villosa* Roth subsp. *villosa* in Some World Languages and Other Linguistic Taxa

Language/Taxon	Name
Russian	goroshek mokhnatyi; goroshek Sherstistoplodnyi; vika mokhnataia; vika mokhnataia ozymaia
Scots	mice pease; moose pease
Serbian	kosmata grahorica; maljava grahorica
Serbian (Upper)	kosmata woka
Slovak	vika huňatá
Sorbian (Lower)	kósmata wójka
Spanish	arvejilla velluda; vezo piloso; veza de arena; veza vellosa; veza velluda
Swedish	kvarnvicker; luddvicker
Ukrainian	horoshok volokhatyi; vika kosmata; vika volokhata
Uzbek	tukli vika
Welsh	ffacbysen y tir âr; ffugbysen yr âr

(3) Celtic, with Breton and Welsh; (4) Germanic, with Afrikaans, Danish, Dutch, English, Forth and Bargy, West Frisian, German, Norwegian, and Swedish, and with exports into several Slavic languages and the Uralic Hungarian via the Old High German *wiccha*; (5) Hellenic, with Greek; (6) Romance, with Asturian, Catalan, French, Friulian, Istriot, Jèrriais, Guernésiais, Italian, Portuguese, Romagnol, and Spanish, and with borrowing of the Old French vece and its form in the Old Norman, *veche*, into the Middle English, as *vecche*, *fecche*, and *ficche*; and (7) Slavic, with Belarusian, Czech, Kashubian, Polish, Russian, Slovak, Lower and Upper Sorbian, and Ukrainian (Vasmer 1958); this translingual Indo-European word began to designate the vetches in general in other language families, such as Afroasiatic with Arabic and Hebrew, Altaic with Turkish, which, quite curiously made a kind of geographical "comeback" as one of two names in Bulgarian (Table 15.16), and Uzbek, Uralic, with Erzya, Estonian and Finnish, and constructed ones, with Esperanto and Ido (Figure 15.1).

Among all the collected terms relating to vetches in the Uralic ethnolinguistic family, there seems to be only one genuine, while the other are borrowings from its language neighbors. The mopheme in question is found in Finnish (Table 15.1), which represents an adessive plural form of *virpi*, referring to sapling, seedling, and sprig (Itkonen and Joki 1978).

FIGURE 15.1 **(See color insert.)** One of the possible evolutions of the Proto-Indo-European root **weig-*, **weik-*, denoting bending and something pliable (Militarev and Stolbova 2007), into its direct and indirect derivatives, shown as pods, and contemporary descendants, depicted as flowers, in the Indo-European, some non-Indo-European and constructed languages; the meanings of each proto-word and modern word is *vetch*.

As already said, in a vast majority of the world's languages, vetches are usually considered similar to grain legumes, often as the diminutive of the names relating to the latter or a kind of their derivatives, such as (Tables 15.1 through 15.22):

- Chickpea (*Cicer arietinum* L.), based upon Proto-Indo-European root **kek-*, **k'ik'-* (see Chapter 5), in the Germanic German and the Romance Italian, Portuguese, and Brazilian Portuguese;
- Chickpea, based upon the Proto-Indo-European root **k(')now-* (see Chapter 14), in the Altaic North and South Azerbaijani;
- Lentil, based upon the Proto-Indo-European root **lent-*, **lent-s-* (see Chapter 11), in the Germanic Danish, German, and Limburgish, and in the Romance French and Spanish;

- Lentil, based upon the Proto-Slavic root **sočevica* (see Chapter 11), in the Slavic Upper Sorbian;
- Lentil, as a derivation of the Latin *pisum* (see Chapter 14), in the Celtic Welsh;
- Birdsfoot trefoil (*Lotus corniculatus* L.), based upon the Proto-Slavic root **lędo* (see Chapter 10), in the Slavic Belarusian and Pannonian Rusyn;
- Common bean (*Phaseolus vulgaris* L.), based upon Proto-Indo-European root **bhask(')-* (see Chapter 13), in the Hellenic Greek and the Romance Campidanese Sardinian;
- Pea, based upon the Proto-Altaic **bŭkrV* (see Chapter 14), in the Turkic Kazakh, Kyrgyz, and Turkish, and with exports to the Caucasian Chechen and the Uralic Hill and Meadow Mari;
- Pea, based upon the Proto-Dené-Caucasian **hVwɫV* (see Chapter 14), in the Biscayan, Gipuzkoan, Lapurdian, and Souletin Basque;
- Pea, based upon the Proto-Dené-Caucasian **xqŏrʔā́ (~-rħ-)* (see Chapter 14), in the Caucasian Chechen and Kabardian, and with borrowings into the Altaic Tatar and the Indo-European Ossetian;
- Pea, based upon the Proto-Indo-European root **arenko-*, **arn k(')-* (see Chapter 3), in the Latin/Linnean cracca and the Romance French and Italian;
- Pea, based upon the Proto-Indo-European root **erəgʷ[h]-* (see Chapter 14), in the Germanic Swedish and the exports, via Vizigothic Gothic, into the Iberian Romance Asturian, Catalan, Portuguese, Brazilian Portuguese, and Spanish;
- Pea, based upon the Proto-Indo-European root **g'er[a]n-*, **grān-* (see Chapter 14), in the Baltic Samogitian and with imports into the Uralic Estonian and Finnish;
- Pea, based upon the Proto-Indo-European root **ghArs-* (see Chapters 10 and 14), in the Slavic Belarusian, Croatian, Russian, Serbian, Slovenian, and Ukrainian;
- Pea, based upon the already listed Proto-Indo-European root **mAis-* (see Chapter 14), in the Romance Moldovan and Romanian;
- Pea, based upon the Proto-Indo-European root **peys-*, **pis-* (see Chapter 14), in the Germanic English, Scots, and Ulster English, in the Hellenic Greek, and in the Romance French and Spanish;
- Pea, based upon the Proto-Kartvelian **car-/*cr-* (see Chapter 14), in the Georgian;
- Pea, based upon the Proto-Uralic roots **kača*, **keś V* or **kopa* (see Chapter 14), in Erzya and Komi-Zyrian;
- Bitter vetch (*Vicia ervilia* [L.] Willd., see Chapter 6) in the Germanic German, in the Romance Catalan and French, and in the Slavic Serbian (Table 15.11);
- Faba bean, based upon the Proto-Indo-European root **bhabh-*, *bhabhā* (see Chapter 7), in the Romance French and Istriot (Tables 15.4 and 15.10), and Spanish (Tables 15.9 and 15.10) and its Latin derivation *faba Aegyptia* (see Chapter 10) in the Romance French, Catalan, Brazilian Portuguese, and Spanish;

- Faba bean, based upon the Ancient Greek *kókkos* (see Chapter 7), in the Hellenic Greek and the Romance Spanish;
- Japanese wisteria (*Wisteria floribunda* [Willd.] DC.), in the Altaic Japanese (Tables 15.4, 15.21, and 15.22);
- Both pea and faba bean, in the Celtic Welsh;
- A kind of legume bean, in the Altaic Japanese and Korean, in the Austroasiatic Vietnamese, and in the Sino-Tibetan Chinese;
- A kind of legume pod, in the Na-Dené Navajo for *Vicia americana*.

One of the outcomes of the author's unpublished interviews with the Serbian farmers from the region of Bosnian Frontier is, perhaps, the only term that links the lexicology of vetches and their cultivation practices, *lanjik* (Table 15.16). According to the popular account, it is derived from the noun *lane*, denoting *the prevous year*, meaning that it is quite appropriate for this very low-input crop to use the self-produced seed reserve from the previous season. In addition, this intriguing name may also refer to a mixture of vetches with cereals, either for forage or grain production.

In some languages, there are the names of a highly descriptive nature, such as something round, in the Altaic Kazakh, which was exported into Central and Northern Kurdish along with a shift of meaning to lentil (see Chapter 11), and entanglement and tangle, in the Germanic Icelandic, and overgrow, in Bulgarian and Macedonian, resembling vetches' recognizable growth habit and considerably abundant aboveground biomass (Mikić et al. 2014b).

Among the speakers of the Amerind Lakota, *Vicia americana* resembles buffalo hide, and in certain Iberian Romance languages, such as in Catalan and Portuguese, there are the names like *cat's nails* (Table 15.19), and Galician, where *Vicia sativa* subsp. *sativa* is, among other designations and for a reason probably fallen into oblivion long ago, regarded as *cuckoo's bread* (Table 15.16).

The vetches, with their flowers, seem to be the synonyms of beauty for more than one human sense. In the Amerind Cherokee, it is easy to imagine *Vicia caroliniana* as a decoration for maidens or future brides, since it is literally called *wreath for the head*, while, in the Uralic Finnish, one may feel a peculiar sweet scent, which invisibly vapors from nectaries, filling the air of some Eurasian meadow.

And, for the very end of this chapter and to confirm the often challenged status of common and other vetches as pulses, we cite an anonymous housewife from Istria, today a border peninsula with Italy, in her native Romance Istriot language: "Cui dénti de vècia se fa la menèstra"—"You made the soup with common vetch" (Cergna 2015).

16 *Vigna* Savi

Synonyms: *Azukia* Takah. *ex* Ohwi; *Condylostylis* Piper; *Dolichovigna* Hayata; *Haydonia* R. Wilczek; *Liebrechtsia* De Wild.; *Plectrotropis* Schumach.; *Scytalis* E. Mey.; *Voandzeia* Thouars.

16.1 LIST OF TAXA SCIENTIFIC AND POPULAR NAMES

From a botanical point of view, the genus *Vigna* Savi may be considered one with the most complex systematics, among the taxa encompassing the economically most important pulse crops in the world (Maxted et al. 2004, Tomooka et al. 2011). This chapter segment presents the species of this rich genus with their currently most widely accepted scientific names and abundant synonyms. We hope it may contribute to the long-term issue of clarifiying a subtle and fluctuating distinction between the genera *Phaseolus* L. (see Chapter 13) and *Vigna* Savi (Maréchal et al. 1978, Budanova 1982, Beyra and Reyes Artiles 2004) as well as to a remarkable and insufficiently resolved intrageneric diversity of the latter (Pasquet 1999, Doi et al. 2002, Simon et al. 2007, Vijaykumar et al. 2009, Javadi et al. 2011). In addition, the following paragraphs present an extensive set of vernacular names referring to the *Vigna* species, with an emphasis on those of the Sub-Saharan Africa (AuSIL 2017, ISTA 1982, Zander et al. 1993, Rehm 1994, Gledhill 2008, Porcher 2008, Delgado-Salinas et al. 2011, Botanic Garden Meise 2013, The Plant List 2013, Ecocrop 2017, EPPO 2017, Ethnologue 2017, IBIS 2017, ILDIS 2017, Logos 2017, NPGS 2017, Wikipedia 2017, Wiktionary 2017).

- *Vigna aconitifolia* (Jacq.) Maréchal (Table 16.1)
Synonyms: *Phaseolus aconitifolius* Jacq.
- *Vigna ambacensis* Welw. ex Baker
Synonyms: *Vigna ambacensis* var. *pubigera* (Baker) Maréchal et al.; *Vigna pubigera* Baker; *Vigna stuhlmannii* Harms
English: Ambaca bean
Lokutu: djembe lokumu
Luba-Kasai: kasala
Zande: bambalemba
- *Vigna angivensis* Baker
English: avoko bean; Madagascar bean
Malagasy: avoko
- *Vigna angularis* (Willd.) Ohwi & H. Ohashi
English: adzuki bean
- *Vigna angularis* (Willd.) Ohwi & H. Ohashi var. *angularis* (Table 16.2)
Synonyms: *Azukia angularis* (Willd.) Ohwi; *Dolichos angularis* Willd.; *Phaseolus angularis* (Willd.) W. Wight

- *Vigna angularis* (Willd.) Ohwi & H. Ohashi var. *nipponensis* (Ohwi) Ohwi & H. Ohashi
Synonyms: *Phaseolus angularis* var. *nipponensis* (Ohwi) Ohwi; *Phaseolus nipponensis* Ohwi
Chinese: ri ben chi dou
English: Japanese wild adzuki bean
Japanese: yabu tsuru azuki
- *Vigna antunesii* Harms
English: Antunes' bean
- *Vigna aridicola* N. Tomooka & Maxted
English: dry-land bean
- *Vigna bequaertii* R. Wilczek
Bukhara (DR Congo): Umuharakuku
English: Bequaert's bean
Kinyarwanda: baharakubuye
- *Vigna bosseri* Du Puy & Labat
English: Bosser's bean
- *Vigna comosa* Baker
English: hairy bean
- *Vigna comosa* Baker var. *comosa*
Synonyms: *Vigna micrantha* Harms
English: common hairy bean; small-flowered hairy bean
- *Vigna comosa* Baker var. *lebrunii* (Baker f.) Verdc.
Synonyms: *Vigna lebrunii* Baker f.
English: Lebrun's bean
- *Vigna dalzelliana* (Kuntze) Verdc.
Synonyms: *Phaseolus dalzellii* Cooke; *Phaseolus dalzellianus* O. Kuntze; *Phaseolus pauciflorus* Dalzell, non G. Don
English: Dalzell's bean
French: haricot Dalzell
- *Vigna dalzelliana* (O. Kuntze) Verdc. var. *dalzelliana*
English: wild Dalzell's bean
- *Vigna dalzelliana* (O. Kuntze) Verdc. var. *elongata* Thuan
English: sandy Dalzell's bean
French: haricot Dalzell des sables
- *Vigna debilis* Fourc.
English: infirm bean
- *Vigna decipiens* Harv.
English: deceiving bean; trap bean
- *Vigna desmodioides* R. Wilczek
English: Desmodium-like bean
- *Vigna dinteri* Harms
English: Dinter's bean
- *Vigna dolichoides* Roxb. (Baker).
English: dolichos-like vigna bean

- *Vigna exilis* Tateishi & Maxted
English: slender bean
- *Vigna filicaulis* Hepper
English: thread-stem bean
- *Vigna filicaulis* Hepper var. *filicaulis*
English: common thread-stem bean
- *Vigna filicaulis* Hepper var. *pseudovenulosa* Maréchal et al.
English: false-veined bean
- *Vigna fischeri* Harms
English: Fischer's bean
- *Vigna friesiorum* Harms
English: Fries' bean
- *Vigna frutescens* A. Rich.
English: shrubby bean
- *Vigna frutescens* A. Rich. subsp. *frutescens* var. *buchneri* (Harms) Verdc.
Synonyms: *Librechtsia katangensis* De Wild.; *Vigna buchneri* Harms; *Vigna esculenta* Auct. non (De Wild.) De Wild.; *Vigna katangensis* (De Wild.) T. & H. Durand
English: Buchner's bean
- *Vigna frutescens* A. Rich. subsp. *frutescens* var. *frutescens*
English: true shrubby bean
- *Vigna frutescens* A. Rich. subsp. *incana* (Taub.) Verdc.
English: hoary bean
- *Vigna gazensis* Baker f.
English: Gaza bean
- *Vigna gracilis* (Guill. & Perr.) Hook. f.
Synonyms: *Dolichos gracilis* Guill. & Perr.
English: thin bean
- *Vigna grandiflora* (Prain) Tateishi & Maxted
Synonyms: *Phaseolus sublobatus* var. *grandiflora* Prain; *Vigna radiata* var. *grandiflora* (Prain) Niyomdham
English: large-flowered bean
- *Vigna heterophylla* A. Rich.
English: different-leaved bean
- *Vigna hirtella* Ridl.
English: slight-haired bean
- *Vigna hosei* (Craib) Backer
Synonyms: *Dolichos hosei* Craib; *Vigna oligosperma* Backer, nom. nud.
English: Sarawak-bean
- *Vigna juncea* Milne-Redh.
English: rushy bean
- *Vigna juncea* Milne-Redh. var. *major* Milne-Redh.
English: large rushy bean
- *Vigna juruana* (Harms) Verdc.
English: Juru bean

- *Vigna keraudrenii* Du Puy & Labat
English: Keraudren bean
- *Vigna khandalensis* (Santapau) Sundararagh. & Wadhwa
English: Khandala bean
- *Vigna kirkii* (Baker) J. B. Gillett
Synonyms: *Phaseolus kirkii* Baker; *Vigna schliebenii* Harms
English: Kirk's bean
Lombo: inaolo a kwakwa
- *Vigna lanceolata* Benth. (Table 16.3)
- *Vigna lanceolata* Benth. var. *filiformis*
Alyawarre: atnwelarr
English: Maloga bean, thread-leafed pencil yam
Warlpiri: kupurturru
- *Vigna lanceolata* Benth. var. *lanceolata*
English: common pencil yam
- *Vigna lanceolata* Benth. var. *latifolia*
Alyawarre: arlatyey
English: Maloga bean, native bean, parsnip bean, wide-leafed pencil yam
- *Vigna lasiocarpa* (Mart. ex Benth.) Verdc.
Synonyms: *Dolichos jacquinii* DC.; *Phaseolus diversifolius* Pittier; *Phaseolus hirsutus* Mart. ex Benth.; *Phaseolus lasiocarpus* Mart. ex Benth.; *Phaseolus pilosus* Kunth
English: wooly-fruited bean
- *Vigna laurentii* De Wild.
English: laurent's bean
- *Vigna lobata* (Willd.) Endl.
English: lobed bean
- *Vigna lobatifolia* Baker
English: lobe-leaved bean
- *Vigna longifolia* (Benth.) Verdc.
Synonyms: *Phaseolus longifolius* Benth.
English: long-leaved bean
- *Vigna luteola* (Jacq.) Benth. (Table 16.4)
Synonyms: *Calopogonium pendunculatum* Standl.; *Dolichos gangeticus* Roxb.; *Dolichos luteolus* Jacq.; *Dolichos niloticus* Delile; *Dolichos repens* L.; *Orobus trifoliatus* Sessé & Moc.; *Phaseolus luteolus* (Jacq.) Gagnep.; *Phaseolus maritimus* Hassk.; *Scytalis helicopus* E. Mey.; *Vigna brachystachys* Benth.; *Vigna bukobensis* Harms; *Vigna fischeri* Harms; *Vigna glabra* Savi; *Vigna helicopus* (E. Mey.) Walp.; *Vigna jaegeri* Harms; *Vigna longepedunculata* Taub.; *Vigna marina* subsp. *oblonga* sensu Padulosi; *Vigna nigerica* A. Chev.; *Vigna nilotica* (Delile) Hook. f.; *Vigna oblonga* sensu Hook. f.; *Vigna repens* (L.) Kuntze; *Vigna villosa* Savi
- *Vigna macrodon* Robyns & Boutique
English: large-toothed bean
Lendu: dedjalo

- *Vigna marina* (Burm.) Merr. (Table 16.5)
Synonyms: *Dolichos luteus* Sw.; *Phaseolus marinus* Burm.; *Scytalis retusa* E. Mey.; *Vigna lutea* (Sw.) A. Gray; *Vigna retusa* (E. Mey.) Walp.
- *Vigna membranacea* A. Rich.
English: membranous bean
- *Vigna membranacea* subsp. *caesia* (Chiov.) Verdc.
English: blue-grey bean
- *Vigna membranacea* subsp. *membranacea* A. Rich.
English: true membranous bean
- *Vigna membranaceoides* Robyns & Boutique
English: membrane-like bean
Hunde: kioraorwa; mushibanyuma
Kinyarwanda: eha musambi; igishimbo; ruharamba
- *Vigna mildbraedii* Harms
English: Mildbraed's bean
- *Vigna minima* (Roxb.) Ohwi & H. Ohashi
Synonyms: *Phaseolus minimus* Roxb.
Chinese: shan dou; ye xiao dou
English: smallest bean
- *Vigna minima* (Roxb.) Ohwi & Ohashi subsp. *gracilis*
English: slender smallest bean
- *Vigna minima* (Roxb.) Ohwi & Ohashi subsp. *minima*
English: true smallest bean
- *Vigna monantha* Thulin
English: one-flowered bean
- *Vigna monophylla* Taub.
Synonyms: *Haydonia monophylla* (Taub.) R. Wilczek
English: one-leaf bean
- *Vigna multiflora* Hook. f.
English: many-flowered bean
Lingala: imbizi; nzilo
Lombo: kwakwa lo lowe
Luba-Katanga: kakukoro; kantumbatumba
Moanda (DR Congo): mukhasa khasa
Oroko: impompo motani
Yambata: aluta; tandanda
- *Vigna multinervis* Hutch. & Dalziel
English: many-veined bean
Kirundi: agajaganzi
- *Vigna mungo* (L.) Hepper
English: black gram
- *Vigna mungo* (L.) Hepper var. *mungo* (Table 16.6)
Synonyms: *Azukia mungo* (L.) Masam.; *Phaseolus hernandezii* Savi; *Phaseolus max* sensu Auct.; *Phaseolus mungo* L.; *Phaseolus mungo var. radiatus* sensu Baker; *Phaseolus radiatus* Roxb. non L.; *Phaseolus roxburghii* Wight & Arn.

- *Vigna mungo* (L.) Hepper var. *silvestris* Lukoki et al.
Synonyms: *Phaseolus mungo* L.; *Phaseolus viridissimus* Ten. ex Miq., nom. inval.
- *Vigna nakashimae* (Ohwi) Ohwi & H. Ohashi
Synonyms: *Azukia nakashimae* (Ohwi) Ohwi & Ohashi; *Phaseolus minima* sensu Auct. Japon. non Roxb.; *Phaseolus nakashimae* Ohwi; *Vigna minima* (Roxb.) Ohwi & Ohashi *subsp. nakashimae* (Ohwi) Tateishi
English: Nakashima's bean
Japanese: hime tsuru azuki
- *Vigna nepalensis* Tateishi & Maxted
English: nepali bean
- *Vigna nigritia* Hook. f.
Synonyms: *Vigna luteola* Auct. non (Jacq.) Benth., De Wild.; *Vigna luteola* Auct. non (Jacq.) Benth. var. *villosa* Savi, Bak. in Oliv.; *Vigna pubigera* Baker var. *gossweileri* Baker f.; *Vigna racemosa* Auct. non Hutch. & Dalz.; *Vigna tisserantii* A.Chev.
English: Nigritia bean
- *Vigna nuda* N. E. Br.
Synonyms: *Librechtsia ringoetii* De Wild.; *Vigna ringoetii* (De Wild.) De Wild.
English: bare bean
Karavian: munkoio
Lubumbashi: lusashi
- *Vigna oblongifolia* A. Rich.
English: oblong-leaved bean
Lombo: lotombo
- *Vigna oblongifolia* A. Rich. var. *oblongifolia*
Synonyms: *Vigna lancifolia* A. Rich.; *Vigna wilmsii* Burtt Davy
English: common oblong-leaved bean
- *Vigna oblongifolia* A. Rich. var. *parviflora* (Welw. ex Baker) Verdc.
Synonyms: *Vigna parviflora* Welw. ex Baker
English: small-flowered bean
- *Vigna o-wahuensis* Vogel
English: Hawaii wild bean; Oahu cowpea
- *Vigna parkeri* Baker
Synonyms: *Dolichos maranguënsis* Taub.; *Vigna gracilis* Auct., non (Guill. and Perr.) Hook.; *Vigna maranguënsis* (Taub.) Harms.
English: creeping vigna; vigna menjalar
- *Vigna parkeri* Baker subsp. *acutifolia* Verdc.
English: sharp-leaved creeping vigna
- *Vigna parkeri* Baker subsp. *maranguensis* (Taub.) Verdc.
Synonyms: *Dolichos maranguensis* Taub.
English: Marangu creeping vigna
- *Vigna parkeri* Baker subsp. *parkeri*
English: Madagascan creeping vigna
- *Vigna pilosa* (J. G. Klein ex Willd.) Baker
Synonyms: *Dolichos pilosus* J. G. Klein ex Willd.; *Dolichovigna pilosa* (Willd.) Niyomdham; *Dysolobium pilosum* (J. G. Klein ex Willd.) Maréchal

English: shaggy bean
- *Vigna pygmaea* R. E. Fr.
English: pygmy bean; pygmy cowpea
Ganza (DR Congo): Katukwe
- *Vigna racemosa* (G. Don) Hutch. & Dalziel
Synonyms: *Clitoria racemosa* G. Don; *Vigna luteola* Auct. Non (Jacq.) Benth., De Wild.; *Vigna luteola* (Jacq.) Benth. var. *villosa* Baker, De Wild. et Th. Dur.; *Vigna luteola* (Jacq.) Benth. var. *villosa* De Wild.; *Vigna pubigera* Auct. non Baker, Micheli in Th. Dur. et De Wild.; *Phaseolus* sp. De Wild. & T. Durand
English: raceme bean
Kinyarwanda: urubebia
Luba-Kasai: kafufule; mutata
Luba-Katanga: kahunde bakishi
Luki: madezo ya seke; wandu nsinga
Lusanga: singa nzambi
Ngbandi (Northern): ngase
Tetela: kolululu
Tshagbo: dudu
- *Vigna radiata* (L.) R. Wilczek
English: mung bean
- *Vigna radiata* (L.) R. Wilczek var. *radiata* (Table 16.7)
Synonyms: *Azukia radiata* (L.) Ohwi; *Phaseolus abyssinicus* Savi; *Phaseolus aureus* Roxb.; *Phaseolus aureus* Auct. non Roxb.; *Phaseolus aureus* Auct. non Roxb. f. aureus; *Phaseolus aureus* Wall.; *Phaseolus aureus* Zuccagni; *Phaseolus chanetii* (H. Lev.) H. Lev.; *Phaseolus hirtus* Retz.; *Phaseolus mungo* Auct. non L.; *Phaseolus mungo* Gagn. non L.; *Phaseolus novo-guineense* Baker f.; *Phaseolus radiatus* L.; *Phaseolus radiatus* L. var. *aurea* Roxb.; *Phaseolus radiatus* L. var. *typica* Matsum.; *Phaseolus radiatus* L. var. *typicus* Prain; *Phaseolus setulosus* Dalzell; *Phaseolus sublobatus* Roxb.; *Phaseolus trinervius* Wight & Arn.; *Pueraria chanetii* H. Lev.; *Rudua aurea* (Roxb.) F. Maekawa; *Vigna aureus* (Roxb.) Hepper; *Vigna brachycarpa* Kurz; *Vigna opistricha* A. Rich.; *Vigna perrieriana* R. Vig.; *Vigna sublobata* (Roxb.) Babu & S. K. Sharma; *Vigna sublobata* (Roxb.) Bairig. & al.
- *Vigna radiata* (L.) R. Wilczek var. *sublobata* (Roxb.) Verdc.
Synonyms: *Phaseolus setulosus* Dalzell; *Phaseolus sublobatus* Roxb.; *Phaseolus trinervius* Wight & Arn.; *Vigna radiata* var. *setulosa* (Dalzell) Ohwi & H. Ohashi; *Vigna sublobata* (Roxb.) Bairig. et al.
English: Jerusalem pea
- *Vigna reflexopilosa* Hayata
Synonyms: *Azukia reflexo-pilosa* (Hayata) Ohwi; *Phaseolus mungo* L. sensu Forbes & Hemsley; *Phaseolus neocaledonicus* Baker f.; *Phaseolus reflexo-pilosus* (Hayata) Ohwi; *Vigna catjang* Endl. var. *sinensis* King sensu Matsum.; *Vigna glabrescens* Mar. Masch. & Stain.
Chinese: juan mao jiang dou
English: creole-bean
Japanese: ooyabu tsuru azuki

- *Vigna reflexopilosa* Hayata subsp. *glabra* (Roxb.) N. Tomooka & Maxted
Synonyms: *Phaseolus glaber* Roxb.; *Vigna glabrescens* Maréchal et al.; *Vigna radiata* var. *glabra* (Roxb.) Verdc.
English: clay pea; creole pea
French (Mauritius): lentille de créole
Vietnamese: đậu mời
- *Vigna reflexopilosa* Hayata subsp. *reflexopilosa*
English: common Creole pea
Japanese: oo-yabu-tsuru-azuki
- *Vigna reticulata* Hook. f.
Synonyms: *Vigna andongensis* Auct. non Baker, De Wild. et Staner
Bemba: mutambalesi; nkainia
English: netted bean
Katuba: lusashi
Luba-Kasai: montuba
Zande: akwamwa
- *Vigna riukiuensis* (Ohwi) Ohwi & H. Ohashi (Table 16.9)
Synonyms: *Phaseolus riukiuensis* Ohwi
English: assam bean
Japanese: hina azuki
- *Vigna sandwicensis* A. Gray
English: Sandwich bean
- *Vigna schimperi* Baker
Synonyms: *Vigna longepedunculata* Taub.
English: Schimper's bean
Kilur: Adjolo; adjura; okworokworo
- *Vigna schlechteri* Harms
Synonyms: *Dolichos reticulata* Schltr.; *Vigna galpinii* Burtt Davy; *Vigna nervosa* Markötter
English: Schlechter's bean
- *Vigna schottii* (Benth.) A. Delgado & Verdc.
Synonyms: *Phaseolus schottii* Benth.
English: Schott's bean
- *Vigna somaliensis* Baker f.
English: Somali bean
- *Vigna stenoloba* (Harv.) Burtt Davy
English: narrow-lobed bean
- *Vigna stipulacea* Kuntze
English: stipule bean
- *Vigna subramaniana* (Babu ex Raizada) Raizada
Synonyms: *Phaseolus subramanianus* Babu ex Raizada
English: Subramania's bean
- *Vigna subterranea* (L.) Verdc. (Table 16.8)
Synonyms: *Arachis africana* Burm. f.; *Glycine subterranea* L.; *Voandzeia subterranea* (L.) Thouars ex DC.

- *Vigna subterranea* (L.) Verdc. var. *spontanea* (Harms) Pasquet
Synonyms: *Voandzeia subterranea* f. *spontanea* Harms
English: spontaneous bambara groundnut
- *Vigna subterranea* (L.) Verdc. var. *subterranea*
English: common bambara groundnut
- *Vigna tenuicaulis* N. Tomooka & Maxted
English: narrow-stemmed bean
- *Vigna trichocarpa* (C. Wright) A. Delgado
Synonyms: *Phaseolus trichocarpus* C. Wright
English: hairy-fruited bean
- *Vigna trilobata* (L.) Verdc.
Synonyms: *Dolichos trilobatus* L.; *Dolichos trilobus* L.; *Phaseolus trilobatus* (L.) Schreb.; *Phaseolus trilobus* Auct.; *Vigna triloba* (L.) Verdc.
Chinese: sān liè yè jiāngdòu
English: African gram; jangli; jungle mat bean; jungle-bean; mugun; mungan; mukni; phillipesara; pillipesara; three-lobe-leaf cowpea
Hindi: mugun; mungan
Telugu: pillipesara
- *Vigna trinervia* (B. Heyne ex Wight & Arn.) Tateishi & Maxted
English: three-veined bean
- *Vigna triphylla* (R. Wilczek) Verdc.
Synonyms: *Haydonia triphylla* R. Wilczek
English: three-leaved bean
- *Vigna truxillensis* (Kunth) N. Zamora
Synonyms: *Phaseolus truxillensis* Kunth
English: Trujillo bean
- *Vigna umbellata* (Thunb.) Ohwi & H. Ohashi (Table 16.9)
Synonyms: *Azukia umbellata* (Thunb.) Ohwi; *Dolichos umbellatus* Thunb.; *Phaseolus calcaratus* Roxb.; *Phaseolus chrysanthus* Savi; *Phaseolus pubescens* Blume; *Phaseolus radiatus* var. *flexuosus*; *Phaseolus torosus* Roxb.; *Vigna calcarata* (Roxb.) Kurz
- *Vigna unguiculata* (L.) Walp.
Synonyms: *Dolichos biflorus* L.; *Dolichos catiang* L.; *Dolichos catjang* Burm. f. nom. illeg.; *Dolichos catjang* L.; *Dolichos hastifolius* Schnizl.; *Dolichos lubia* Forssk.; *Dolichos melanophthalamus* DC.; *Dolichos monachalis* Brot.; *Dolichos mungo* Auct. non L.; *Dolichos obliquifolius* Schnizl.; *Dolichos sesquipedalis* L.; *Dolichos sinensis* Forssk. nom. illeg.; *Dolichos sinensis* L.; *Dolichos sphaerospermus* (L.) DC.; *Dolichos tranquebaricus* Jacq.; *Dolichos unguiculata* L.; *Dolichos unguiculatus* L.; *Dolichos unguiculatus* Thunb.; *Liebrechtsia scabra* De Wild.; *Phaseolus cylindricus* L.; *Phaseolus sphaerospermus* L.; *Phaseolus unguiculatus* (L.) Piper; *Scytalis hispida* E. Mey.; *Scytalis protracta* E. Mey.; *Scytalis tenuis* E. Mey.; *Vigna alba* (G. Don) Baker f.; *Vigna angustifolia* Auct. non Hook. f.; *Vigna angustifoliolata* Verdc.; *Vigna baoulensis* A. Chev.; *Vigna catjang* (Burm. f.) Walp.; *Vigna coerulea* Baker; *Vigna dekindtiana* Harms; *Vigna*

hispida (E. Mey.) Walp.; *Vigna huillensis* Baker; *Vigna malosana* Baker; *Vigna protracta* (E. Mey.) Walp.; *Vigna pubescens* R. Wilczek; *Vigna rhomboidea* Burtt Davy; *Vigna scabra* (De Wild.) De Wild.; *Vigna scabra* (De Wild.) T. Durand & H. Durand; *Vigna scabrida* Burtt Davy; *Vigna sesquipedalis* (L.) F. Agcaoili nom. illeg.; *Vigna sesquipedalis* (L.) Fruwirth; *Vigna sinensis* (L.) Endl. ex Hassk. nom. illeg.; *Vigna sinensis* (L.) Savi ex Hausskn.; *Vigna* sp., De Wild. et Th. Dur.; *Vigna tenuis* (E. Mey.) F. Dietr.; *Vigna triloba* Auct. non Baker; *Vigna triloba* var. *stenophylla* Harv.
English: black-eyed pea; cowpea
- *Vigna unguiculata* (L.) Walp. subsp. *aduensis* Pasquet
English: Adu cowpea
- *Vigna unguiculata* (L.) Walp. subsp. *alba* (G. Don) Pasquet
Synonyms: *Clitoria alba* G. Don
English: white cowpea
- *Vigna unguiculata* (L.) Walp. subsp. *baoulensis* (A. Chev.) Pasquet
English: Baul cowpea
- *Vigna unguiculata* (L.) Walp. subsp. *burundiensis* Pasquet
English: Burundi cowpea
- *Vigna unguiculata* (L.) Walp. subsp. *dekindtiana* (Harms) Verdc.
Synonyms: *Vigna baoulensis* A. Chev.; *Vigna coerulea* Baker; *Vigna dekindtiana* Harms; *Vigna mensensis* Schweinf.; *Vigna unguiculata* subsp. *mensensis* (Schweinf.) Verdc.; *Vigna unguiculata* var. *dekindtiana* (Harms) Maréchal; *Vigna unguiculata* var. *mensensis* (Schweinf.) Maréchal et al.
Chinese (Cantonese): dauh gok; tau kok
Chinese (Mandarin): ye dou jiao; ye jiang dou
English: African cowpea; Ethiopian cowpea; wild cowpea
- *Vigna unguiculata* (L.) Walp. subsp. *letouzeyi* Pasquet
English: Letouzey's cowpea
- *Vigna unguiculata* (L.) Walp. subsp. *pawekiae* Pasquet
English: Pawekia cowpea
- *Vigna unguiculata* (L.) Walp. subsp. *protracta* (E. Mey.) B. J. Pienaar
Synonyms: *Vigna unguiculata* var. *protracta* (E. Mey.) Verdc.
English: prolonging cowpea
- *Vigna unguiculata* (L.) Walp. subsp. *pubescens* (R. Wilczek) Pasquet
Synonyms: *Vigna pubescens* R. Wilczek; *Vigna unguiculata* var. *pubescens* (R. Wilczek) Maréchal et al.
English: hoary cowpea
- *Vigna unguiculata* (L.) Walp. subsp. *stenophylla* (Harv.) Maréchal et al.
Synonyms: *Scytalis hispida* E. Mey.; *Scytalis protracta* E. Mey.; *Vigna angustifoliolata* Verdc.; *Vigna hispida* (E. Mey.) Walp.; *Vigna stenophylla* (Harv.) Burtt Davy; *Vigna triloba* Walp.; *Vigna triloba* var. *stenophylla* Harv.
English: narrow-leafed cowpea
- *Vigna unguiculata* (L.) Walp. subsp. *tenuis* (E. Mey.) Maréchal et al.
Synonyms: *Scytalis tenuis* E. Mey.; *Vigna coerulea* Baker
English: slender cowpea; thin cowpea
- *Vigna unguiculata* (L.) Walp. subsp. *unguiculata*

English: common cowpea
- *Vigna unguiculata* (L.) Walpers subsp. *unguiculata* (L.) Walp. var. *sanguinea*
Synonyms: *Vigna cylindrica* Skeels f. *sanguinea*; *Vigna sinensis* Endl. var. *sanguinea* Kitam.
English: red cowpea
Japanese: kintoki sasage
- *Vigna unguiculata* (L.) Walp. subsp. *unguiculata* var. *spontanea* (Schweinf.) Pasquet
English: spontaneous cowpea
- *Vigna unguiculata* (L.) Walp. subsp. *unguiculata* f. *biflora* (Table 16.10)
Synonyms: *Dolichos biflorus* L.; *Dolichos catiang* L.; *Dolichos catjang* L.; *Dolichos catjang* Burm. f.; *Phaseolus cylindricus* L.; *Vigna catjang* Endl.; *Vigna catjang* (Burm. f.) Walp.; *Vigna cylindrica* (L.) Skeels; *Vigna sinensis* var. *cylindricus* (L.) H. Ohashi; *Vigna unguiculata* subsp. *cylindrica* (L.) Eselt. ex Verdc.; *Vigna unguiculata* var. *cylindrica* (L.) H. Ohashi; *Vigna unguiculata* subsp. *cylindrica* (L.) Skeels; *Vigna unguiculata* (L.) Walpers subsp. *cylindrica* (L.) van Eseltine; *Vigna unguiculata* subsp. *cylindrica* (L.) Verdc.; *Vigna unguiculata* (biflora group); *Vigna unguiculata* subsp. *unguiculata* (cultigroup Biflora) Marechal
- *Vigna unguiculata* (L.) Walp. subsp. *unguiculata* f. *melanophthalmus*
Synonyms: *Dolichos melanophthalmus* DC.
English: blackish-eyed cowpea
- *Vigna unguiculata* (L.) Walp. subsp. *unguiculata* (L.) Walp. f. *nana* (Table 16.11)
Chinese: ai mei dou
English: dwarf cowpea
English (Australia): dwarf cowpea
English (USA): bush cowpea
- *Vigna unguiculata* (L.) Walp. subsp. *unguiculata* f. *sesquipedalis* (Table 16.11)
Synonyms: *Dolichos sesquipedalis* L.; *Vigna sesquipedalis* (L.) Fruwirth; *Vigna sinensis* subsp. *sesquipedalis* (L.) Van Eselt.; *Vigna unguiculata* subsp. *sesquipedalis* (L.) Verdc.
- *Vigna unguiculata* (L.) Walp. subsp. *unguiculata* f. *unguiculata* (Table 16.12)
Synonyms: *Dolichos sinensis* L.; *Dolichos unguiculatus* L.; *Phaseolus unguiculatus* (L.) Piper; *Vigna sinensis* (L.) Savi ex Hassk.
- *Vigna venulosa* Baker
English: small-veined bean
- *Vigna verticillata* Taub.
English: spindle-whirled bean
- *Vigna vexillata* (L.) A. Rich. (Table 16.13)
Synonyms: *Vigna capensis* Walp.; *Vigna hirta* Hook.; *Vigna reticulata* Auct. non Hook.
- *Vigna vexillata* (L.) A. Rich. var. *angustifolia* (Schumach.) Baker
Synonyms: *Plectrotropis angustifolia* Schumach.
English: narrow-leaved wild cowpea; wild cowpea
Lombo: kwakwa
- *Vigna vexillata* (L.) A. Rich. var. *davyi* (Bolus) B. J. Pienaar

Synonyms: *Vigna davyi* Bolus
English: Davy's bean
- *Vigna vexillata* (L.) A. Rich. var. *macrosperma* Maréchal et al.
Chinese: da zi ye jiang dou
English: large-seeded wild cowpea
Vietnamese: đậu cờ hột to
- *Vigna vexillata* (L.) A. Rich. var. *ovata* (E. Mey.) B. J. Pienaar, nom. inval.
Synonyms: *Phaseolus capensis* Thunb.; *Strophostyles capensis* var. *ovatus* E. Mey.; *Vigna capensis* (Thunb.) Burtt Davy
English: egg-shaped bean
- *Vigna vexillata* (L.) A. Rich. var. *vexillata*
Synonyms: *Phaseolus vexillatus* L.; *Vigna hirta* Hook.
English: small-vexillum bean
Vietnamese: đậu cở
- *Vigna vexillata* (L.) A. Rich. var. *youngiana* F. M. Bailey
English: Young's bean
- *Vigna vexillata* (L.) Benth. var. *yunnanensis* Franch
Chinese (Mandarin): Yun nan ye jiang dou
Chinese (Yunnan): Shan wu dou
English: Yunnan bean
- *Vigna wittei* Baker f.
Synonyms: *Vigna stenodactyla* Harms
English: Witte's bean
Yaka: nkasa nseke

16.2 ORIGIN OF SCIENTIFIC AND POPULAR TAXA NAMES

One of the most distinctive features of the genus *Vigna* Savi in this book is the fact that it, unlike the others presented in the previous chapters, is named after a man who verily existed and thus does not hide some ancient root-word among its letters. Domenico Vigna was an Italian botanist and director of the Botanical Garden in Pisa in the seventeenth century and was immortalized in the plant taxonomy by Gaetano Savi (1769–1844), his fellow countryman and colleague (Singh 2014). The establishment of the genus *Vigna* may also be regarded as one of the many attempts at the time with a goal of coming out of a kind of Minoan labyrinth or disentangling a botanical Gordian Knot, both referring to the aforementioned need of redefining status of the genus *Phaseolus* L. In the end, *a thin line* between the two genera was assessed by encompassing certain anatomical and morphological peculiarities, such as those of stipules, styles, and pollen grains, as well as different biochemical characteristics (Bailey and Bailey 1976). The surname Vigna and its taxonomic eponym have become a word denoting various species of the genus in diverse languages, although the prevalent ones are those from the Germanic, with Danish, English and Australian English, and the Slavic, with Czech, Russian, Serbian, or Ukrainian, branches of the Indo-European, to which we may add the constructed ones, such as Esperanto (Tables 16.1, 16.2 and 16.4–16.12).

TABLE 16.1
Popular Names Denoting *Vigna aconitifolia* (Jacq.) Maréchal in Some World Languages and Other Linguistic Taxa

Language/Taxon	Name
Arabic	lubia quniti aluraq
Chinese	e dou
English	dew bean; mat bean; matki; moth bean; Turkish gram
Esperanto	motho; mothovigno
French	haricot mat; haricot papillon
German	Mattenbohne
Gujarati	maṭha
Hindi	moth daal
Japanese	mosu biin
Malay	mitti kelu
Marathi	matakee
Serbian	akonitolisni pasulj; akonitolisna vigna; mat; mot; pasulj mot
Slovak	fazuľa prilbicolistá
Thai	matpe
Vietnamese	đậu bướm

TABLE 16.2
Popular Names Denoting *Vigna angularis* (Willd.) Ohwi & H. Ohashi var. *angularis* in Some World Languages and Other Linguistic Taxa

Language/Taxon	Name
Arabic	fasulia alazuki; lubia mqrn
Asturian	frijol; poroto; adzuki; azuki; soya colorada; xudía
Catalan	aduki; adzuki; azuki; mongeta azuki
Chinese (Cantonese)	hóng dòu
Chinese (Hakka)	fùng-theu
Chinese (Mandarin)	ao ye chi dou; chi dou; chi xiao dou; du chi dou; hong chi dou; hóng dòu; hong xiao dou; hung tou; jin hong dou; jin hong xiao dou; mi chi dou; mi dou; shi mu dou; xiao dou; xiao hong dou; xiao hong lu dou; zhu dou
Czech	azuki; fazole azuki
Danish	adzuki-bønne; atsuki-bønne
Dutch	adukiboon
English	adzuki; adzuki bean; azuki; azuki bean; cultivated azuki; red bean; red mung bean
Esperanto	aduki; adzuki; azuki-fabo; haricot rouge du Japon; soja rouge

(*Continued*)

TABLE 16.2 (*Continued*)
Popular Names Denoting *Vigna angularis* (Willd.) Ohwi & H. Ohashi var. *angularis* in Some World Languages and Other Linguistic Taxa

Language/Taxon	Name
Finnish	adsukipapu
French	haricot à feuilles angulaires; haricot anguleux; haricot azuki; haricot du Japon; haricots petits rouges; soja rouge
German	Adsukibohne; Adzukibohne
Hebrew	zvk
Hindi	guruns; rains
Hungarian	azukibab
Indonesian	kacang merah; kacang tolo merah
Italian	fagiolo adzuki; fagiolo azuki
Japanese	akamame; ankomame; anmame; antoki; azuki; gururimame; irakuri; kannome; konaremame; narazu; omame; shoumame
Korean	pat
Lithuanian	japoninė pupuolė
Macedonian	crven grav
Malay	kacang merah
Persian	lubiai ahzuki
Polish	czerwona soja; fasola azuki
Portuguese	feijão-azuqui
Russian	adzuki; fasol' uglovataia; vigna uglovataia
Serbian	azuki; čidu; crveni pasulj; hongdu
Spanish	frijol; poroto; adzuki; azuki; soya roja; xudía
Spanish (Argentina)	poroto arroz
Spanish (Chile)	frijol diablito
Spanish (Cuba)	frijol diablito
Spanish (Mexico)	judía adzuki
Swahili	muazuki
Swedish	adukiböna; adzukiböna; azukiböna
Tetum	koto mean
Ukrainian	adzuki; chervonyi bib; nezgrabna kvasolia
Vietnamese	đậu đỏ
Zhuang	cehvenz; lwglimz; maklimz

TABLE 16.3
Popular Names Denoting *Vigna lanceolata* Benth. in Some World Languages and Other Linguistic Taxa

Language/Taxon	Name
Alyawarre	akam
Anmatjirra	arlatyey; kame
Australian Aboriginal English	'am; yam
Australian Kriol	pinsal yarlma; yarlma; yem
Awabakal	kokabai
Bardi	narrga
Burarra	jaypurlga; mun-garra
Damin	kurrikurrijpi m!ii; wiiti m!ii
Darkinjung	wyong
Dhuwal	ganguri
Djapu	yukuwa'
Djinang	barlngunda
Djinba	buyurmarr
Dyirbal	ḍuguṛ
English	bush carrot; Maloga bean; native bean; parsnip bean; pencil yam; small yam
Enindhilyagwa	mwarntakirriyarra (arntaka - elbow; rriyari -forked)
Gaagudju	djugu-djoogu; moornarn
Gamberre	wanggalu
Gamilaraay	guwēai; kubbiai
Garawa	kabala
Garawa (Western)	mili
Giimbiyu	liindyi
Gooniyandi	birla
Gumbaynggir	da:m
Gundjeihmi	garrbarda
Guragone	djunja; wartbirritji
Gurindji	manaari; wayit; wayita
Guugu Yalandji	bambayal
Guugu Yimithirr	baabuunh
Iwaidja	wangkartuk
Jaminjung	gagawooli
Jawi	koolngarie
Jingulu	karrangayimi
Kalaw Lagaw Ya	gabau
Kayardild	jiwi; thadawa
Kija	ngawoonyji
Kune	kayawal
Kuninjku	karrbarda
Kunwinjku	karrbarda

(Continued)

TABLE 16.3 (*Continued*)
Popular Names Denoting *Vigna lanceolata* Benth. in Some World Languages and Other Linguistic Taxa

Language/Taxon	Name
Kwini	wegu
Laragiya	wila
Lardin	burrku
Lower Arrernte	alatyeye
Malak-Malak	yeyeynin
Malngin	manaari
Manyallaluk Mayali	karremudyi
Manyjilyjarra	mata
Marranj	warkuya
Martu Wangka	kanyjamarra
Martuthunira	mada
Matngele	belerr; derrngey; jambur; mal
Maung	arlamun; rlurrjij
Mayi-Kulan	kowar
Meriam	lewer
Miriwoong	wanmalang
Murrinh-patha	thalam; wawa
Nambu	taitu
Ngaringman	kagawuli
Ngarinyin	alamer
Ngarluma	mardirra
Nhanda	ajuga
Nunggubuyu	warda
Nyangumarta	kanyjamarra
Pilinara	wanymirra; wayida
Pintupi	taţupirrpa; wakati; wayali; yawalyurru; yunala
Pitta Pitta	ṱintama
Ritharngu	ganguŗi; mabaļpi; mawunu
Tiwi	marntinga; murani; murrunkawini
Torres Strait Creole	yam
Upper Arrernte	merne arlatyeye
Upper Arrernte (Central)	árlatyeye
Upper Arrernte (Eastern)	árlatyeye
Upper Arrernte (Western)	*latyeye / latjia / latjia*
Wagaya	menaji
Wagiman	yohyin; yoyin
Walmajarri	jirrirlpaja; kujuntu
Walmajarri (Eastern)	juwa
Wambaya	jigama
Wardaman	mordon; 'wayida

(*Continued*)

TABLE 16.3 (*Continued*)
Popular Names Denoting *Vigna lanceolata* Benth. in Some World Languages and Other Linguistic Taxa

Language/Taxon	Name
Warlmanpa	wabidi
Warlpiri	japirda; ngalajiyi; ngarlajiyi; wajaraki; wapirti; wapurtali; wijaraku; yumurnunju
Warndarang	ḍujaḍuja
Warrnambool	tjirang
Warrongo	ganyo
Warumungu	manaji; ngajarrma
Worrorran	*inkalba; karnmangku; wangkarlum; wungunimbim*
Wunambal	*ngulwana*
Yanyuwa	*yika*
Yiidji	*nalangga*
Yir-Yorront	may-wortol; thorrchonh; wanychw

TABLE 16.4
Popular Names Denoting *Vigna luteola* (Jacq.) Benth. in Some World Languages and Other Linguistic Taxa

Language/Taxon	Name
Arabic (The Sudan)	akwari: lubiya taiyib
Chinese	chǎng yè jiāng dòu
English (U.S.)	hairypod cowpea; deer pea
English (Australia)	dalrymple vigna; hairy cowpea; hairypod cowpea
Hausa	mare
Japanese	nagaba-hama-sasage
Katana	goko; masheke; mugulula
Kinyarwanda	toshimbo-shimbo umuharakuku; umurakuku
Lombo	A kwakwa; inaolo a kwakwa; indola a kwakwa; indolo
Malagasy	antaka; famehifary; telouravy; vahipoko; vahisanjy
Rusiga	umusigampfisi
Shi	kavuhivuhi
Spanish	frijol cimarrón; frijol de la playa; porotillo
Swahili	kashilika; kisukuna
Teli	gilibande

TABLE 16.5
Popular Names Denoting *Vigna marina* (Burm.) Merr. in Some World Languages and Other Linguistic Taxa

Language/Taxon	Name
Arabic	lubia bhri
Chinese	bīn jiāng dòu
English	dune-bean; nanea; notched bean; sea-bean
French	haricot de mer
Hawaiian	lemuomakili; mohihihi; nanea; nenea; ‘ōkolemakili; pūhili; pūhilihili; pūlihilihi; wahine ‘ōma‘o
Indonesian	kacang laut
Italian	vigna marina
Japanese	hama-azuki; hama-sasage
Malay	kachang laut
Portuguese	batatarana
Serbian	morska vigna
Thai	t̄hạw p̄hī thale

TABLE 16.6
Popular Names Denoting *Vigna mungo* (L.) Hepper var. *mungo* in Some World Languages and Other Linguistic Taxa

Language/Taxon	Name
Arabic	ds sud
Asturian	fréxol negru; frijol negru; llenteya negra; mungo; poroto mung; urd
Bengali	māṣakalā'i; mashkalai ḍal
Catalan	llentilla negra; mongeta mungo
Chinese (Cantonese)	sau dou
Chinese (Mandarin)	hei dou; xiao dou
Danish	urdbønne
Dutch	mungoboon
English	black gram; black lentil; lack matpe bean; minapa pappu; mungo bean; urad bean; urd-bean; white lentil
Esperanto	lentvigno; urdvigno
French	haricot mung; haricot mungo; haricot urd; soja noir; urd
French (Canada)	ambérique
German	Linsenbohne; Mungbohne; Mungobohne; Urdbohne
Gujarati	aḍad; aḍada; aḷad
Hindi	udad daal; udad dal; urad dāl; uṛad dāl urad; urd; urid; maash
Italian	fagiolo indiano nero; fagiolo mungo nero; mungo nero; urad
Japanese	ke tsuru azuki

(*Continued*)

TABLE 16.6 (*Continued*)
Popular Names Denoting *Vigna mungo* (L.) Hepper var. *mungo* in Some World Languages and Other Linguistic Taxa

Language/Taxon	Name
Kannada	uddina bēḷe; uddu
Malayalam	uḻunn; uẓunu
Maldivian	kaḷu mugu
Marathi	uḍid; uḍīda
Myanmar	mat pe
Nepali	maas; mās
Odia	biri ḍāli
Polish	fasola mungo
Portuguese	feijão-da-China; feijão-da-Índia; feijão-preto; feijão-urido
Portuguese (Brazil)	feijão-colubrino; feijão-da-Pérsia; feijão-do-Congo; feijão-mungo; feijão-peludo; grão-de-pulha; lentilha branca; lentilha preta; mugo; mungo; oloco
Punjabi (Eastern)	māsh
Punjabi (Western)	mash
Russian	chiornyi mash; fasol' mungo; fasol' vidov; mai; urd
Sanskrit	māṣa; māṣaḥ; uḍida
Serbian	crni grah; crni mungo; crni mungo pasulj; crni pasulj; crna vigna; urd; urd pasulj
Sinhalese	undu
Spanish	fréjol negro; frijol negro; judía mung; lenteja negra; mungo; poroto mung; urd
Swahili	mchooko mweusi
Swedish	mungböna
Tamil	ulundu; uḷuntu
Telugu	minumulu
Thai	t̄hạw dả; thuaa dahm
Tulu	urdu bele
Ukrainian	chorna sochevitsia; chornyi horoshok; urad; urad-dkhal
Vietnamese	đậu mười; đậu muồng ăn; đậu xanh bốn mùa

TABLE 16.7
Popular Names Denoting *Vigna radiata* (L.) R. Wilczek var. *radiata* in Some World Languages and Other Linguistic Taxa

Language/Taxon	Name
Arabic	allubia alshai; bql almash
Asturian	loctao; poroto chinu; soya verde; xudía mungo
Burmese	e-di; pe-di-sein; p'di sien; pe-nauk; to-pi-si
Catalan	mongeta mung; mongeta xinesa
Chinese (Cantonese)	jiāo dòu; lǜdòu; qīng xiǎodòu; zhí dòu
Chinese (Hakka)	liu̍k-theu

(*Continued*)

TABLE 16.7 (*Continued*)
Popular Names Denoting *Vigna radiata* (L.) R. Wilczek var. *radiata* in Some World Languages and Other Linguistic Taxa

Language/Taxon	Name
Chinese (Mandarin)	lǜdòu
Chinese (Wu)	lǜdòu
Croatian	zlatni grah
Czech	mungo fazole; vigna zlatá
Danish	Jerusalembønne; mungbønne; mung-bønne
Dhao	kabui iki
Dutch	mungboon
English	Burmese mung bean; Chinese mung bean; golden gram; golden-seeded mung bean; green gram; green-seeded mung bean; Jerusalem pea; moong bean; mung; mung bean; mung dahl; Tientsin green bean
English (Australia)	celera-bean
Esperanto	Jerusalemfabo; mungfabo; mungofabo
Estonian	munguba
Finnish	mungopapu
French	ambérique; ambérique verte; haricot crevost; haricot doré; haricot mung à grain vert; haricot mungo; haricot velu de la basse Nubie; soja vert
French (Canada)	ambérique; ambérique jaune
German	mungbohne; Mungobohne; Jerusalembohne Jerusalem-Bohne; Lunjabohne
Gujarati	maga
Hebrew	m'sh
Hindi	moong
Hungarian	mungóbab
Ilocano	balatong; munggo
Indonesian	arta ijo; kacang djong; kacang hijau
Italian	fagiolo aureo; fagiolo indiano verde; fagiolo mungo verde
Japanese	bundou; fundou; ryokutō; yaenari
Javanese	kacang ijo
Kannada	hesaru kāḷu
Kapampangan	balatung
Khmer	santek bay
Korean	nogdu
Kusunda	gitak
Lao	thwax khiaw; thwax ngok; thwax sadek
Latvian	mungo pupiņas; zeltainās pupiņas
Lithuanian	spindulinė pupuolė
Makasae	uta-mata
Malay	kacang hijau
Malayalam	ceṟupayar
Maldivian	nūmugu
Marathi	mūga
Min (Eastern)	liŏh-dâu

(*Continued*)

TABLE 16.7 (*Continued*)
Popular Names Denoting *Vigna radiata* (L.) R. Wilczek var. *radiata* in Some World Languages and Other Linguistic Taxa

Language/Taxon	Name
Nepali	mugi
Norwegian (Bokmål)	mungbønne
Occitan	mongeta mungó
Odia	mug
Pashto	mas
Persian	mash
Polish	fasola złota
Portuguese	feijão-da-China; feijão-mungo; feijão-rajado; feijão-soroco
Portuguese (Brazil)	feijão-mungo-verde
Punjabi (Eastern)	mūgī
Punjabi (Western)	moong
Russian	boby mung; fasol' zolotistaia; fasol' vidov; mash; vigna luchistaia
Sanskrit	mudga; mudgaḥ
Serbian	mungo; grah mungo; mungo; pasulj mungo; zlatni mungo; zlatni mungo pasulj
Sinhalese	bu me; mun; muṁ æṭa; mun eta
Spanish	frijol mungo; judía mungo; loctao; poroto chino; poroto mung; soja verde
Spanish (Latin America)	frijol mungo
Spanish (Peru)	frijolito chino; loctao
Sundanese	kacang héjo
Swahili	mung
Swedish	mungböna
Tagalog	monggo; munggo
Tajik	moş
Tamil	muṅku; pācip payaṟu
Telugu	pesalu
Tetum	fore mungu
Thai	T̄hạw k̄heīyw
Tibetan	sran ljang
Ukrainian	boby munh; mash zvychainyi; munh-dkhal; vigna promenysta; zolotysta kvasolia
Uzbek	mosh
Vietnamese	đậu xanh; đỗ xanh
Visayan	balatong
Waray	munggu
Welsh	ffeuen fwng
Zhuang	duhheu; duhyez

TABLE 16.8
Popular Names Denoting *Vigna subterranea* (L.) Verdc. in Some World Languages and Other Linguistic Taxa

Language/Taxon	Name
Arabic	lubia mtmur
Bambara	tiganingèlèn; tiganinkuru
Bariba	samboutourou
Catalan	bambara
Danish	angolaaert; jordaert
Dendi	densi
Dutch	afrikaanse aardnoot
Dyula	tigba
English	bambara bean; bambara groundnut; bambara nut; Congo goober; earth pea; ground bean; hog-peanut; jugo bean; Madagascar peanut
Esperanto	terpizo; voandzeo
Finnish	maapapu
Fon	azingokwin
French	pois bambara; pois de terre; voandzou
Gen	azingokwin
German	Angola-Erbse; Bambara-Erdnuss; Erderbse; Mandubi-Erbse
Igbo	opupa
Indonesian	kacang banten; kacang bogo; kacang bogor; kacang manila
Italian	pisello di terra
Japanese	banbaramame
Javanese	kacang banten; kacang bogor; kara pendem
Kimbundu	nguba
Kongo	mpinda; nguba
Lingala	ngúba
Malay	kacang bogor; kacang tanah
Mandinka	tige
Ndebele (Northern)	indlubu
Portuguese	jinguba-de-Cabambe
Serbian	bambara; grah bambara; pasulj bambara; podzemna vigna; vigna kikiriki
Shona	nyimo; nzungu
Spanish	bambarra; guandsú; guisante de tierra; maní africano; maní de bambarra
Sundanese	kacang bandung; kacang bogor; kacang génggé; kacang gondola; kacang jogo; acang manila
Swahili	mnjugu; mnjugu-mawe; njugumawe
Swedish	bambarajordnöt
Tswana	ditloo
Xhosa	indongomane
Yoruba	boro; ẹpa; ekpaboro
Zulu	iintongomane

TABLE 16.9
Popular Names Denoting *Vigna umbellata* (Thunb.) Ohwi & H. Ohashi in Some World Languages and Other Linguistic Taxa

Language/Taxon	Name
Arabic	lubia khimi
Chinese (Cantonese)	chìxiǎodòu
Chinese (Mandarin)	chì dòu; chìshān dòu chìxiǎodòu; dànbái dòu fàn dòu; hóng fàn dòu; mi dou; mi chi dou; xiao hong dou; ye mi dou
Danish	risbønne
English	climbing mountain-bean; mambi bean; oriental-bean; red bean; rice bean; ricebean; small red bean
Esperanto	rizvigno
Finnish	lehmänpapu
French	haricot grain de riz; haricot riz
German	Reisbohne
Indonesian	kacang uci
Japanese	shima tsuru azuki; raisu biin; take azuki; tsuru azuki
Khmer	santek riech mieh
Korean	deonggulpat
Lao	thov lang t'k; thov phi; thov sad't pa'
Malay	kacang sepalit
Myanmar	be nauk; be pwe; be sang; be te; be tyel; beli; ning krung shapre; pe nauk saung; pe yin
Nepali	masyāṇa
Portuguese	feijão-arroz
Spanish	frijol de arroz; frijol mambé; frijol rojo; frijolito rojo; judía arroz
Tagalog	anipai
Tetum	koto mean
Thai	ma pae; thua daeng; thua pae
Vietnamese	đậu nho nhe
Visayan	kapilan; pagsei

TABLE 16.10
Popular Names Denoting *Vigna unguiculata* (L.) Walp. subsp. *unguiculata* f. *biflora* in Some World Languages and Other Linguistic Taxa

Language/Taxon	Name
Arabic	lubiya baladi
Chinese	bai dou; duan jia jiang dou; fan dou; fan jiang dou; hung ch'iang tou; mei dou; yang dou jiao
Danish	vignabønne
Dutch	katjang pandjang
English	bombay cowpea; catjang; catjang pea; catjang cowpea; cylindric-shape-seeded cowpea; Hindi cowpea; Indian cowpea; sow-pea

(Continued)

TABLE 16.10 (*Continued*)
Popular Names Denoting *Vigna unguiculata* (L.) Walp. subsp. *unguiculata* f. *biflora* in Some World Languages and Other Linguistic Taxa

Language/Taxon	Name
French	dolique cajun; dolique catjan; dolique de Chine; dolique des vaches; dolique mongette
German	Catjangbohne
Indonesian	kacang merah; kacang peudjit; kacang tunggak
Italian	fagiolo del occhio
Japanese	hata sasage; yakko sasage
Khmer	sandaek khmau; sandaek kr'h'm; sandaek sa
Lao	thwo x si'nx
Malay	kacang merah; kacang peudjit; kacang tunggak
Portuguese	feijão-fradinho
Russian	korovii gorokh; korovii goroshek; vigna kitaiskaia
Spanish	judía catjang; caupi catjang
Thai	po thoh saa; thua khaao; thua rai
Vietnamese	đậu đỏ; đậu trắng

TABLE 16.11
Popular Names Denoting *Vigna unguiculata* (L.) Walp. subsp. *unguiculata* f. *sesquipedalis* in Some World Languages and Other Linguistic Taxa

Language/Taxon	Name
Arabic	lobiya; lobiya balad'
Chinese (Cantonese)	ch'euhng ch'eng dauh gok; ch'eung kong tau; chang qing dou jiao; ch'eung ts'ing tau kok
Chinese (Mandarin)	chang jiang dou; chang qing dou jiao; dou jiao
Danish	aspargesbønne; kaempeaspargesbønne; meterbønne
English	asparagus bean; Chinese long-bean; bodi bean; green asparagus bean; green-podded cowpea; long horn bean; long-podded cowpea; long-podded kidney bean; pea bean; snake bean; yardlong bean; yard-long cowpea
French	dolique asperge; dolique géante; haricot asperge; haricot kilomètre
German	Langbohne; Spargelbohne
Indonesian	kacang belut; kacang tolo
Italian	fagiolo asparago; fagiolo gigante
Japanese	juuroku sasage
Malay	kacang panjang; kachang panjang
Portuguese	dólico gigante; feijão-chicote; feijão-espargo

(*Continued*)

TABLE 16.11 (*Continued*)
Popular Names Denoting *Vigna unguiculata* (L.) Walp. subsp. *unguiculata* f. *sesquipedalis* in Some World Languages and Other Linguistic Taxa

Language/Taxon	Name
Russian	boby sparzhevye; metrovye boby; sparzhevaia fasol'; vigna v'jushchaiasia
Spanish	dólico espárrago; judía espárrago; poroto espárrago
Tagalog	sitao; sitaw
Tetum	fore naruk; fore sikote; fore xikote
Thai	tua fak yaow; tua phnom
Vietnamese	đậu đũa
Visayan	banor; hamtak

TABLE 16.12
Popular Names Denoting *Vigna unguiculata* (L.) Walp. subsp. *unguiculata* f. *unguiculata* in Some World Languages and Other Linguistic Taxa

Language/Taxon	Name
!Kung (Southeastern)	kàˤʔání
Afrikaans	akkerboon; boontjie; dopboontjie; koertjie; swartbekboon
Arabic	lubia zfri; mash
Arabic (Egypt)	lobia
Arabic (Jordan)	lobya
Arabic (Lebanon)	lobya
Arabic (Syria)	lobya
Bariba	swia
Bemba	nkainia
Bengali	bi'ulira ḍāla
Bulgarian	bebridza; papuda
Chinese (Cantonese)	hēi yǎn dòu
Chinese (Mandarin)	da jiao dou; dou jiao; jiāngdòu; mei dou
Danish	vignabønne; koaert
Dembo	mankundia
Dendi	dougouri
English	asparagus bean; bachapin bean; black-eyed bean; black-eyed cowpea; black-eyed dolichos; catjang; Chinese long bean; common cowpea; cowpea; crowder bean; crowder pea; kafir bean; marble pea; poona pea; Reeve's pea; southern pea; sow-pea; snake bean; southern pea; yardlong bean
English (Australia)	Reeve's pea; snake pea
English (North America)	black-eyed bean; black-eyed pea; goat pea
English (South Africa)	cowpea; bachapin bean; black-eye bean; black-eye pea; catjang; China pea; cowgram; southern pea

(*Continued*)

TABLE 16.12 (*Continued*)
Popular Names Denoting *Vigna unguiculata* (L.) Walp. subsp. *unguiculata* f. *unguiculata* in Some World Languages and Other Linguistic Taxa

Language/Taxon	Name
Esperanto	asparago vigno; katjango; nigraokula fabo; okulvigno
Faradje	tori koso
Finnish	lehmänpapu; pitkäpapu
Fon	ayikoun
French	catjang; cornille; dolique à œil noir; dolique asperge; dolique de Chine; niébé; niébé commun; haricot indigene; haricot kilomètre; pois à vache
French (Réunion)	voème
Gandajika	tshibungo lukundo
Garamba	abakpanvua
Gen	ayou
Georgian	dzadza
German	Augenbohne; Catjangbohne; Kuhbohne; Langbohne; Schwarzaugenbohne; Schlangenbohne; Spargelbohne
Greek	fasólia mavromatiká
Greek (Ancient)	fáselis; phásēlos
Gujarati	cōḷā
Hebrew	luv'vah
Hindi	chauli; lobiya
Icelandic	augnbaun
Indonesian	kacang bangkok; kacang perut ayam; kacang tolo; kacang tunggak
Italian	dolico dall'occhio nero; fagiolino dall'occhio; fagiolino piccolo; fagiolo con l'occhio; fagiolo dall'occhio; vigna cinese
Japanese	sasage
Japanese (Middle)	sasage
Japanese (Old)	sasage
Jul'hoan	kàˤʔání
Kannada	alasaṇḍe; alsande kalu
Kazakh	sïırburşaq; lobïya; qıtay sïırburşağı
Khmer	sandaek kang; sandaek engkuy
Kinyarwanda	ibishyimbo; igishyimbo; musambi
Korean	dongbu; tongpu
Lao	thwo do
Latin	fasēlus; faseolus; phasēlus; phasellus; Phaseolus
Lokutu	djembe
Luba	bikunde
Luba-Kasai	lusashi; montuba
Lukafu	lawanda
Lusanga	singa nzambi
Malagasy	lozy; mahalaindolo; voahimba; voanemba; voatsirokonangatra
Malay	kacang bol; kacang merah; kacang toonggak
Malayalam	vanpayar

(*Continued*)

TABLE 16.12 (*Continued*)
Popular Names Denoting *Vigna unguiculata* (L.) Walp. subsp. *unguiculata* f. *unguiculata* in Some World Languages and Other Linguistic Taxa

Language/Taxon	Name
Maldivian	lūbiyā; riha toḷi
Marathi	cavaḷī; chawli
Mbwasa	ndekona mongasa
Mulungu	mogandjale nkole
Mvuazi	tanga
Nepali	bōḍī
Ngbandi (Northern)	ngase
Oroko	impompo
Persian	lubiai tcshmblbli
Polish	wspięga wężowata
Portuguese	feijão-de-corda; feijão-frade; feijão-fradinho; feijão-macáçar; feijão-miúdo; feijão-peqeno
Portuguese (Brazil)	feijão-de-corda
Punjabi (Eastern)	lōbī'ā
Punjabi (Western)	lobia
Russian	korovyi gorokh; vigna kitaiskaia; vigna konoplevidnaia
Sango	gazi
Serbian	crni okasti grašak; crnookica; kravlji pasulj; mletački grašak; pasuljica; šarena vigna
Shangaan	dinaba; munaoa; tinyawa
Shona	nyemba
Sinhalese	kavupī; li me; me karal; mil me; wanduru me
Sotho (Northern)	dinawa; monawa; nawa
Spanish	caupí; carilla; chíchere; chícharo de vaca; chícharo salvaje; chícharo tropical; chiclayo; fríjol cabecita negra; fríjol chino; frijol de carita; frijol de costa; frijol de vaca; judía de careta; judía rabiza
Spanish (Bolivia)	cumandé; frijol camba
Spanish (Central America)	frijol de costa
Spanish (Ecuador)	tumbe
Spanish (Latin America)	caupí
Spanish (Mexico)	frijol ojo de cabra
Spanish (Nicaragua)	frijol de vaca; frijol de vara
Spanish (Paraguay)	cumandé
Spanish (Peru)	chiclayo; frijol Castilla
Spanish (Uruguay)	frijol Castilla; poroto tape
Spanish (Venezuela)	frijol
Swahili	kunde; mkobwe; mkude; mkunde-mwitu
Swedish	ögonböna; vignabøna
Tagalog	kibal; paayap
Tamil	kārāmaṇi
Telugu	alasanda

(*Continued*)

TABLE 16.12 (*Continued*)
Popular Names Denoting *Vigna unguiculata* (L.) Walp. subsp. *unguiculata* f. *unguiculata* in Some World Languages and Other Linguistic Taxa

Language/Taxon	Name
Tetum	fore talin
Thai	tua dam
Tonga	nzangui
Tswana	dinawa; nawa-ea-setswana
Turkish	börülce
Ukrainian	dovhi boby; korov'iachyi horokh; mash kytaiskyi; vigna kytais'ska
Urdu	lobia
Venda	munawa; nawa
Vietnamese	đậu dải; đậu dải trắng rốn nâu; đậu đen; đậu đũa
Visayan	batong; kibal; otong
Wolof	niébé
Yaka (Congo-Angola)	nkasa
Yoruba	ewa
Zande	abapu; aholo; bambalemba; gara; konde
Zulu	imbumba; indumba; isihlumaya

Usually rich in the roots connected with pulse crops, the vast Afroasiatic ethnolinguistic family could have only one or two attested, indirectly relating to *Vigna luteola*. It is either the Proto-Afroasiatic **mVr-*, designing something good, or the Proto-Afroasiatic **mar-*, resembling a property to help in recovering health, which gave the Proto-West-Chadic **mār-* (Stolbova 2006), with a slightly shifted meaning to recovering from poverty, such as in Hausa (Table 16.3).

Being aware of the geographical distribution of the *Vigna* species across Eurasia, it is not surprising that the only branch of the supposed Altaic macrofamily, which has genuine names relating to the *Vigna* species, is the Japonic one. The Japanese name for *Vigna angularis*, *azuki*, is largely adopted by many diverse languages (Table 16.2) and, in fact, represents a borrowing of the Mandarin Chinese *xiǎodòu*, literally meaning *small bean*. In Japanese, this name is a compound word, consisting of *ko*, meaning *small*, and *mame*, denoting a legume bean. The name of this species and its cultivated botanical variety in modern Japanese has its ultimate origin, through the Middle Japanese *màmè*, *màmé*, the Old Japanese *mame,* and the Proto-Japanese **mamai*, in the Proto-Altaic root **ńaŋo*, all of which refer to a kind of bean-like pulse crop (Starostin 2006b). It is noteworthy that *azuki* is present in the names for numerous other *Vigna* species, witnessing its significance, which may be the main reason why this word became a synonym for almost every species of this genus. Another term in Japanese relating to certain *Vigna* taxa (Tables 16.4, 16.5 and 16.10–16.13) was derived from the Proto-Japonic **sasa(n)kai*, via the Old Japanese *sasage* and the Middle Japanese *sasage* (Starostin 2006b), in which it denotes *Vigna unguiculata*. In its turn, this Proto-Japonic root stemmed directly from the Proto-Altaic root **zi̯ăbsa*,

TABLE 16.13
Popular Names Denoting *Vigna vexillata* (L.) A. Rich. in Some World Languages and Other Linguistic Taxa

Language/Taxon	Name
Amba	buligiangada
Bemba	kansimba simba
Chinese	yě jiāng dòu
English	wild cowpea; zombi-pea
French	pois poison; pois zombi
Japanese	aka-sasage; fuji sasage
Kela (Africa)	zanzizangi
Kikondo	igabagaba
Kilur	adjolo; adjura
Kinyarwanda	mutjaso; ngole; umushasuka
Kirundi	akanyayuchese; umuyambi
Kongo	zanzizangi
Lisala	edjokolo
Lombo	inaolo a kwakwa; kwakwa
Luki	madeso; madeso ya soto; masioto
Mandungu	mangasa
Mayumbe	zangi ya binimuba
Mbay	djoro
Nyagezi	tjikutuka
Swahili (Yangambi)	yonde
Talinga-Bwisi	buligiangada
Tiwi	wuliwirranga; wuluwirranga; wuluwurranga
Tshibinda	mihalula
Vietnamese	đậu cờ
Zande	mulelalia

referring equally to lentil (*Lens culinaris* Medik.) and pea (*Pisum sativum* L.), which, on a long-range historical scale, probably were the staple pulse crops for the ancestors of modern Altaic peoples (Starostin et al. 2003).

The Australian Aboriginal languages demonstrate a remarkable wealth of the names linked with *Vigna lanceolata* (Table 16.3), which demonstrates how heavy it may be to perform their thorough comparative linguistic analysis, attempting to assess the potential ties among all these terms and postulating a hypothetical attested common root or roots (McConvell and Bowern 2011). The elongated and relatively thin tubers of this species have an essential place in the nutrition of the local peoples, especially when an individual happens to be in the desert areas, since they are easily found, dug out, and cooked (Lawn and Holland 2003). *Vigna lanceolata* also has a prominent and genuine spiritual value in the indigenous Australian cultures as one of the sacred living things in the rituals of Dreaming, a kind of totemistic art closely associated with Dreamtime, a pivotal concept in the Australian indigenous animistic religion (Price-Williams 2016). The long and winding lines on the surface of the *Vigna lanceolata* tubers strongly resemble the visual perception of mythical beings and objects and were globally popularized by the esteemed twentieth century indigenous Australian artist, Emily Kame Kngwarreye (Butler 1997), remaining a constant inspiration for new generations of the Australian Aboriginal painters (McCulloch 2001).

Certainly there must be some lexical evidence that the Austronesian languages could have specific root-words associated with the *Vigna* species. However, it still remains unexplored and, quite understandable, in the shadow of much more essential and pivotal issues for this ethnolinguistic family (Blust 2014, Sagart 2016). Found in the wild flora of Hawaii, *Vigna marina* has many vernacular names in this Polynesian Oceanic Austronesian language (Table 16.5). Many of them are descriptive, such as those with the morpheme *-hili-*, referring to something akin to braid or plait, due to the stems longer than 3 m, and with the morpheme *lemu-*, indicating the species' slow growth (Lawn and Cottrell 2016). The others, such as *nanea*, are local female names, thus representing a compliment given by the Hawaiian people to this beautiful *Vigna* species (Native Plants Hawaii 2009). In addition to these names, exotic to the ears of a cool season and moderate climate reader, we may mention a well-known term *kacang* (see Chapter 13), which, in Indonesian and its linguistic relatives, is closely related to any kind of bean-like grain legume, entered Dutch as *katjang*, spread into other languages, and remained in many a name referring to *Vigna unguiculata* and its subspecies (Tables 16.10–16.12).

One of the attested Proto-Dravidian roots, **uẓ-untu-*, denotes *Vigna mungo* (Krishnamurti 2003) and is a linguistic testimony that this pulse crop was domesticated, cultivated, and used among the inhabitants of the Indian subcontinent before the settlement of the Indo-Aryan peoples (Fuller and Murphy 2014). This proto-word gave the contemporary names for the species in the Dravidian Kannada, Malayalam, Tamil, and Tulu and, through the Indo-European Sanskrit, found its place in the vocabulary of the close relatives of the latter, namely Gujarati, Hindi, Marathi, and Sinhalese. Later, the name entered other Indo-European languages, such as the Germanic Danish, English, and German, the Romance Asturian, French, Italian, Portuguese, and Spanish, and the Slavic Russian, Serbian, and Ukrainian, as well as the constructed Esperanto (Table 16.6). The common British colonial rule in both India and Papua New Guinea could bring the *Vigna* crop and its Dravidian name to the latter region, where it was adopted for human consumption and where the original word was morphologically modified, such as in the Trans-New Guinea Makasae (Table 16.7).

The Proto-Indo-European root **bhask(')-*, denoting (Nikolayev 2012) could give the Ancient Greek *fáselis*, *phásēlos* and the Latin *phasēlus*, all referring to cowpea and becoming the Linnean name for the genus *Phaseolus* (Linnaeus 1753, 1758), initially comprising a large number of American and Asian bean-like grain legume species (see Chapter 13).

The Proto-Indo-European root **mAis-*, denoting skin (Nikolayev 2012) and producing the names for lentil, pea, and vetches (*Vicia* spp.), as elaborated in more details in Chapters 11, 14, and 15, found two additional uses. The first one is referring to *Vigna aconitifolia*, with the names in the Indo-Aryan Gujarati, Hindi, and Marathi (Table 16.1) and borrowings into other Indo-European languages, such as the Iranian Pashto, Persian, and Tajik, the Germanic English and German, the Romance French, and the Slavic Serbian, Russian, and Ukrainian, as well as in the Afroasiatic Arabic and Hebrew, the Altaic Japanese and Uzbek, the Austronesian Malay, the Sino-Tibetan Myanmar, the Tai-Kadai Thai, and the constructed Esperanto. The second Proto-Indo-European root relating to the *Vigna* species, **mAis-*, is present in the names designating *Vigna mungo* in the Indo-Aryan Bengali, Hindi, both Eastern and Western Punjabi and Sanskrit, and the Slavic Russian and Serbian (Table 16.6). The Sanskrit word *mudga*, relating to *Vigna radiata*, evolved into the names with same meaning in the Indo-Aryan Gujarati, Hindi, Maldivian, Marathi, Nepali, Odia, Eastern and Western Punjabi,and Sinhalese, and was exported into other Indo-European languages, such as the Baltic Latvian, the Celtic Welsh, the Germanic Danish, Dutch, English, German, Norwegian and Swedish, the Romance Asturian, Catalan, Italian, Occitan, Portuguese, Brazilian Portuguese and Spanish and the Slavic Czech, Russian and Serbian, as well as the Austronesian Ilocano, Tagalog and Waray, the Niger-Congo Swahili, the Uralic Estonian, Finnish and Hungarian, and the constructed Esperanto (Table 16.7). The origin of the said Sanskrit name for *Vigna radiata* and its potential link with the Proto-Indo-European **mAis-* could remain, together with the confusion between the taxonomic attribute *mungo* and the vernacular noun *mung*, a kind of challenge for both botanists and linguists.

It is by no means common to encounter the names for exact plant species, especially those we consider crops, in the persisting hunting and gathering societies. There is an attested proto-word in the language of such society: *kàˤʔání*, denoting *Vigna unguiculata* in Ju|'hoan, a dialect of the Southeastern !Kung language (Table 16.12), stemming out of the Khoisan Proto-Juu *√kaˤʔani*, with the same meaning (Starostin 2008). All this represents a rather nice example of indirect testimony regarding the centers of origin and diversity of the African *Vigna* species, such as *Vigna unguiculata*, and their presence in numerous wild floras (Molosiwa et al. 2016), which the !Kung people and the others with similar culture obtain for food solely by foraging in the Kalahari Desert and similar environments (Smith 2017). In addition, the survival and the evolution of the aforementioned name for *Vigna unguiculata* may only testify how the peoples using the grain of some plant in a relatively short time span, that is, from collecting it in wild flora to cooking it and leaving some reserve for later, do learn to distinguish various species and manage to pass their accumulated knowledge from one generation to another in an exclusively oral form (Wiessner 2014).

The terms relating to the various *Vigna* species, native to Africa and belonging to the Nilo-Saharan ethnolinguistic family are extraordinarily scarce and quite difficult to accurately and precisely define, being just one of numerous demonstrations that it has been overshadowed by the neighboring Niger-Congo family (Ramesar 2015). However, three names in three more or less geographically close languages, namely Dendi, Lendu, and Mbay, and denoting three different species, *Vigna unguiculata*, *Vigna macrodon*, and *Vigna vexillata*, respectively, may show certain morphological resemblance (16.12 and 16.13. Thus, we take liberty to suggest that the Dendi *dougouri*, the Lendu *dedjalo-*, and the Mbay *djoro* may be derived from some still unidentified Proto-Nilo-Saharan root, which could contain few morphemes, such as *d-*, *-e-/-o-*, and *-l-/-r-*, although we are also fully aware that any further and properly carried out comparative analysis would be rather demanding, especially since the fact these three languages belong to diverse subgroups within the Nilo-Saharan family, with Dendi to

Southern Songhay, Lendu to Eastern Central Sudanic, and Mbay to Bongo-Bagirmi Central Sudanic (Starostin 2016, 2017). At any rate, however weakly based, such hypothesis could give at least the slightest contribution to solving this curious botanical and linguistic issue, which has already been the focus in some other native African crops (Blench 2016a, b).

If we compare the names relating to *Vigna* and other pulse crops, such as pea (see Chapter 14), in the languages of the complex Niger-Congo ethnolinguistic family, we shall easily discern that those denoting the *Vigna* species of the African origin are rather numerous, especially in the cases of *Vigna subterranea* (Table 16.8) and *Vigna unguiculata* (Tables 16.10–16.12. There could be one or two still unattested Proto-Niger-Congo roots relating primarily to *Vigna subterranea* and *Vigna unguiculata*, depending on the specific region, and, secondarily, to peanut (*Arachis hypogaea* L.), the introduced grain legume from South America, which has a similar growth habit to *Vigna subterranea* and which replaced it in many a place in its new transatlantic homeland, namely taking its name as well (see Chapter 3). If we make a brief linguistic analysis, based upon morphology, and choose to accept two somewhat mutually resembling root-words, but not necessarily related to each other in an etymological sense, it could result in the following form:

- The first group of names is conditionally gathered around the Kongo *mpinda*, referring to *Vigna subterranea* (Table 16.8), with the exports to the Indo-European Dutch, U.S. English, and Puerto Rican Spanish, and the creole Papiamento and the shift of meaning in all of them to *Arachis hypogaea* (Krampner 2014); the words, possibly stemming from the same source like this name in Kongo, may be found in several other Niger-Congo languages, where they denote several other *Vigna* species, such as *Vigna ambacensis* Welw. ex Baker in Lokutu, *Vigna membranaceoides* Robyns & Boutique in Kinyarwanda, and *Vigna multiflora* in Oroko and *Vigna unguiculata* in Bariba (Table 16.12). There may be a very slight possibility that the said Lokutu word is associated with the names designating the *Vigna* species in the Nilo-Saharan languages presented in the previous paragraph, in some still insufficiently assessed way and in accordance to the confirmed relationship between these two African ethnolinguistic families (Westengen et al. 2014, Dobon et al. 2015);
- The second set of terms in the Niger-Congo languages, much more abundant than the previous one and referring to only three *Vigna* species, is loosely based upon the word *nguba*, which literally denotes something kidney-like and is the name for both *Vigna subterranea* in the Zone C Bantu Lingala and the Zone H Bantu Kimbundu and Kongo; we take liberty to find some resemblance between these and the names associated with (1) *Vigna multiflora* in Lingala, (2) *Vigna unguiculata* in the Luban Bantu, with Luba-Kasai, the Great Lakes Bantu, with Kinyarwanda, the Northeast Coast Bantu, with Mulungu and Swahili, the Sabi Bantu, with Bemba, the Southern Bantu, with Northern Sotho, Tswana, and Venda, the Zone C Bantu, with Kela, the Zone H Bantu, with Katana, Lusanga, Mandungu, Mayumbe, and Yaka, the Mbam Mbwasa, the Southern Bantoid Shona, the Savanna Ngbandi and its creolized Sango, the Senegambian Wolof, and the isolate Zande (Tables 16.10–16.12) and (3) *Vigna subterranea* in Swahili, the Southern Bantu, with Northern

> Ndebele, Xhosa, and Zulu, the Volta-Niger Fon, Gen, Igbo, and Yoruba, and the Mande Bambara, Dyula, and Mandinka (Table 16.8), with a morpheme **(n)gub-* as morphologically similar as possible to the hypothetical Proto-Niger-Congo root associated with the *Vigna* species (Figure 16.1); in the end of this rather extensive paragraph, we shall just add that *nguba*, through the creole Gullah and in the form of a crop and its secondary name, made a return journey over the ocean and established the history of *Arachis hypogaea* in North America (see Chapter 3).

As in the case of nearly all the other pulse crops, the introduced or, especially in recent times, simply imported as declared and properly labeled bags with their grains, the *Vigna* species from all the continents were considered somewhat similar

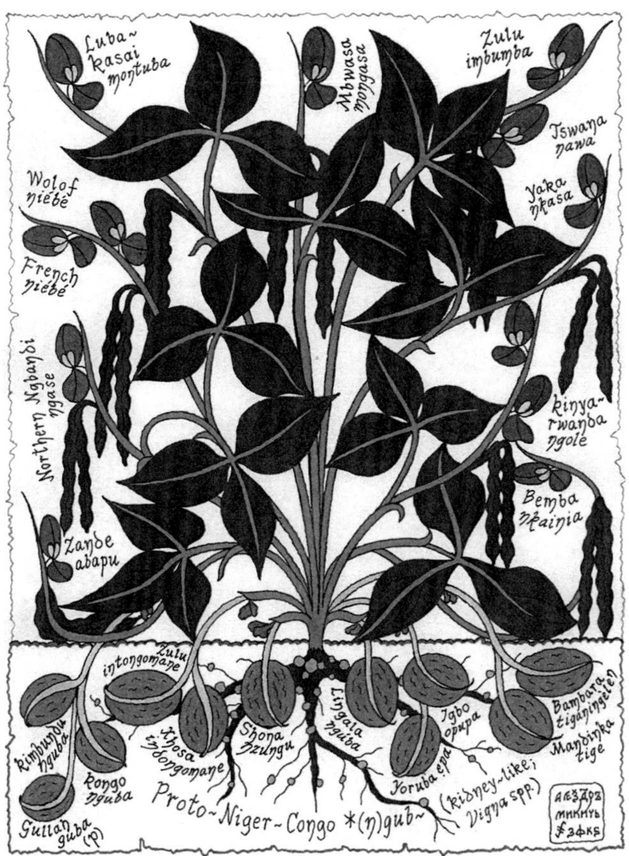

FIGURE 16.1 (See color insert.) One of the possible evolutions of the hypothetical Proto-Niger-Congo morpheme denoting something kidney-like and, possibly, grain of the *Vigna* species, into its contemporary descendants and with borrowing into French; the words associated with *Vigna subterranea* are represented as pods, while those referring to *Vigna unguiculata* are portrayed as flowers.

to diverse traditionally cultivated legumes and thus received their local names, such as (Tables 16.1–16.13):

- Peanut in the Indo-European languages, such as the Germanic, with Afrikaans, Danish, English, German, and Swedish, and the Romance, with Spanish (see Chapter 3);
- Pigeonpea (*Cajanus cajan* [L.] Huth), in Spanish (see Chapter 4);
- Soybean (*Glycine max* [L.] Merr.), in the Indo-European Romance, with Asturian, French, and Spanish, and Slavic, with Polish, and the constructed languages, with Esperanto (see Chapter 8);
- Hyacinth bean (*Lablab purpureus* [L.] Sweet, syn. *Dolichos lablab* L.), in French (see Chapter 9);
- *Lathurus cicera* L., in Spanish (see Chapter 10);
- Lentil, in the Indo-European Germanic, with English, Romance, with Asturian, Catalan, Brazilian Portuguese, and Spanish, and Slavic, with Ukrainian (see Chapter 11);
- *Phaseolus* beans, in the Altaic Arabic and Hebrew and with exports into the Indo-European Maldivian, Persian, Eastern and Western Punjabi, and Urdu, and in the Indo-European Baltic, with Lithuanian, Romance, with Asturian, French, Italian, Portuguese, Spanish and Latin American Spanish, Slavic, with Croatian, Czech, Macedonian, Polish, Russian, Serbian, Slovak, and Ukrainian (see Chapter 13);
- Pea, in the Altaic Korean, in the Indo-European Germanic, with Danish, Dutch, English, and German, Romance, with French, Italian, and Spanish, and Slavic, with Russian, Serbian, and Ukrainian, and in the constructed Esperanto (see Chapter 14);
- Bitter vetch (*Vicia ervilia* [L.] Willd.), in Spanish (see Chapter 6);
- Faba bean (*Vicia faba* L.), in the Afroasiatic Arabic and with borrowing into the Indo-European Persian, in the Indo-European Slavic, with Russian and Ukrainian and imports by the Uralic Finnic and Hungarian, and in the constructed Esperanto (see Chapter 7);
- A kind of mangetout, in the Indo-European Romance, with Catalan, French, and Occitan (see Chapters 13 and 14);
- Either *Phaseolus* or faba bean or both, in the Indo-European Germanic, with Afrikaans, Danish, Dutch, English, German, and Swedish, and borrowings by the Uralic Finnish;
- A kind of bean-like grain legume, in the Altaic Japanese, the Austroasiatic Khmer and Vietnamese, the Austronesian Indonesian and Malay, the Sino-Tibetan Cantonese and Mandarin Chinese and Eastern Min, the Tai-Kadai Lao, Thai, and Zhuang;
- A kind of plant grain, in the Austronesian Malagasy, the Indo-European English, and the creole Réunion French.

It may be curious how, on the one hand, the most renowned name referring to *Vigna subterranea* in many languages, with an emphasis on those spoken in the former chief West European colonial powers, is, in fact, the name of both the people who had been cultivating it for thousands of years and its language and how, on the other hand, the names denoting this crop in that very language, Bambara, became associated at one moment during the last five centuries with *Arachis hypogaea*: *tiganingèlèn* designs small and hard *peanut* and *tiganinkuru* stands for small and round *peanut*. Nothing surprising, one may say, especially in light of the aforementioned odysseys, introductions, and homecomings of the two crops, too similar in their appearance, cultivation, and use to the eye of the farmers from both Africa and the Americas.

Afterword

The sky spoke of rain to come; but the light was broadening quickly, and the red flowers on the beans began to glow against the wet green leaves.

John Ronald Reuel Tolkien, *The Fellowship of the Ring*

* * *

There was a migration and such there will ever be, just as there will be births. There are migrations. There is no death.

Miloš Crnjanski, *The Second Book of Migrations* (translated by Mladen and Vanessa Pupavac)

* * *

- *You memorized the names of all the stars—how many are there?*
- *What difference does it make? I know their names. I don't know how many there are. It's their names that matter.*

Madeleine L'Engle, *A Wind in the Door*

* * *

Is it true, prince, that you once declared that "beauty would save the world"? Great Heaven! The prince says that beauty saves the world!

Fyodor Mikhailovich Dostoyevsky, *The Idiot* (translated by Eva Martin)

References

Adanson, M. 1763. *Familles naturelles des plantes*. Paris, France: Chez Vincent.

Ainouche, A. K., and R. J. Bayer. 1999. Phylogenetic relationships in *Lupinus* (Fabaceae—Papilionoideae) based on internal transcribed spacer sequences (ITS) of nuclear ribosomal DNA. *American Journal of Botany* 86:590–607.

Alasalvar, C., and F. Shahidi. 2008. Tree nuts: Composition, phytochemicals, and health effects: An overview. In *Tree Nuts: Composition, Phytochemicals, and Health Effects*, C. Alasalvar and F. Shahidi (Eds.), pp. 1–10. Boca Raton, FL: CRC Press.

Allkin, R., D. J. Goyder, F. A. Bisby, and R. J. White. 1986. *Names and Synonyms of Species and Subspecies in the Vicieae: Issue 3* (Vicieae Datab. Proj.). Southampton, UK: Biology Department, University of Southampton, Vol. 7, p. 46.

Anderson, R. G., S. Bale, and W. Jia. 1996. Hyacinth bean: Stems for the cut flower market. In *Progress in New Crops*, J. Janick (Ed.), pp. 540–542. Arlington, TX: ASHS Press.

Anderson, S. 2012. *Languages: A Very Short Introduction*. Oxford, UK: Oxford University Press.

Andronov, M. S. 2003. *A Comparative Grammar of the Dravidian Languages*. Wiesbaden, Germany: Harrassowitz Verlag.

Angioi, S. A., D. Rau, G. Attene et al. 2010. Beans in Europe: Origin and structure of the European landraces of *Phaseolus vulgaris* L. *Theoretical and Applied Genetics* 121:829–843.

Annicchiarico, P., N. Harzic, and A. M. Carroni. 2010. Adaptation, diversity, and exploitation of global white lupin (*Lupinus albus* L.) landrace genetic resources. *Field Crops Research* 119:114–124.

Antanasović, S., A. Mikić, B. Ćupina et al. 2011. Some agronomic aspects of the intercrops of semi-leafless and normal-leafed dry pea cultivars. *Pisum Genetics* 43:25–28.

Anthony, D. W. 2007. *The Horse, the Wheel, and Language: How Bronze-Age Riders from the Eurasian Steppes Shaped the Modern World*. Princeton, NJ: Princeton University Press.

Arnoldi, A., G. Boschin, C. Zanoni, and C. Lammi. 2015. The health benefits of sweet lupin seed flours and isolated proteins. *Journal of Functional Foods* 18:550–563.

Asmussen, C., and A. Liston. 1998. Chloroplast DNA characters, phylogeny, and classification of Lathyrus (Fabaceae). *American Journal of Botany* 85:387–401.

Atchison, G. W., B. Nevado, R. J. Eastwood et al. 2016. Lost crops of the Incas: Origins of domestication of the Andean pulse crop tarwi, *Lupinus mutabilis*. *American Journal of Botany* 103:1592–1606.

Aura, J. E., Y. Carrión, E. Estrelles, and G. P. Jordà. 2005. Plant economy of hunter-gatherer groups at the end of the last Ice Age: Plant macroremains from the cave of Santa Maira (Alacant, Spain) ca. 12000–9000 BP. *Vegetation History and Archaeobotany* 14:542–550.

AuSIL. 2017. Dictionaries. The Australian Society for Indigenous Languages, Darwin—Alice Springs—Katherine—Kupang. http://www.ausil.org.au/ (accessed November 25, 2017).

Bailey, L. H., and E. Z. Bailey. 1976. *Hortus Third: A Concise Dictionary of Plants Cultivated in the United States and Canada*. New York: Macmillan.

Beckman, G. 2011. The Hittite language: Recovery and grammatical sketch. In *The Oxford Handbook of Ancient Anatolia*, S. R. Steadman and G. McMahon (Eds.), pp. 517–533. New York: Oxford University Press.

Bedoussac, L., E. P. Journet, H. Hauggaard-Nielsen et al. 2015. Ecological principles underlying the increase of productivity achieved by cereal-grain legume intercrops in organic farming. A review. *Agronomy for Sustainable Development* 35:911–935.

Beekes, R. S. P., and L. van Beek. 2010. φάσηλος. In *Etymological Dictionary of Greek* (Leiden Indo-European Etymological Dictionary Series; 10), Vol. II, p. 1556, Leiden, the Netherlands: Brill.

Bellwood, P. S. 1997. *Prehistory of the Indo-Malaysian Archipelago.* Honolulu, HI: University of Hawaii Press.

Bengtson, J. 1997. Ein vergleich von buruschaski und nordkaukasisch. *Georgica* 20:88–94.

Bengtson, J. 1998. Caucasian and Sino-Tibetan: A hypothesis of S.A. Starostin. *General Linguistics* 36:33–49.

Bengtson, J. 2015. Basque etymology. The Tower of Babel, an International Etymological Database Project. http://starling.rinet.ru (accessed November 25, 2017).

Bengtson, J. D., and V. Blažek. 2011. On the Burushaski-Indo-European hypothesis by I. Čašule. *Journal of Language Relationship* 6:25–63.

Berger, J. D., L. D. Robertson, and P. S. Cocks. 2003. Agricultural potential of Mediterranean grain and forage legumes: 2) Anti-nutritional factor concentrations in the genus Vicia. *Genetic Resources and Crop Evolution* 50:201–212.

Beyra, A., and G. Reyes Artiles. 2004. Revisión taxonómica de los géneros *Phaseolus* y *Vigna* (Leguminosae—Papilionoideae) en Cuba. *Anales del Jardín Botánico de Madrid* 61:145–146.

Bitocchi, E., D. Rau, E. Bellucci et al. 2017. Beans (*Phaseolus* ssp.) as a model for understanding crop evolution. *Frontiers in Plant Science* 8:722.

Blažek, V. 2006. Current progress in Altaic etymology. *Linguist Online* 1:1–9.

Blench, R. 2006. *Archaeology, Language, and the African Past.* Lanham, MD: AltaMira Press.

Blench, R. 2016a. Finger millet: The contribution of vernacular names towards its prehistory. *Archaeological and Anthropological Sciences* 8:79–88.

Blench, R. M. 2016b. Vernacular names for African millets and other minor cereals and their significance for agricultural history. *Archaeological and Anthropological Sciences* 8:1–8.

Blundell, D. 2006. Revisiting cultural heritage in Sri Lanka: The Vedda (Vanniyaletto). *Bulletin of the Indo-Pacific Prehistory Association, Australian National University* 26:163–167.

Blust, R. 2014. Some recent proposals concerning the classification of the Austronesian languages. *Oceanic Linguistics* 53:300–391.

Boas, F. 1911. *Handbook of American Indian Languages* (Vol. 1). Bureau of American Ethnology, Bulletin 40. Washington, DC: Government Print Office.

Boeder, W. 2005. The South Caucasian languages. *Lingua* 115:5–89.

Botanic Garden Meise. 2013. The digital flora of Central Africa. Democratic Republic of the Congo, Rwanda & Burundi. http://www.botanicgarden.be/index.php (accessed November 25, 2017).

Bouckaert, R., P. Lemey, M. Dunn et al. 2012. Mapping the origins and expansion of the Indo-European language family. *Science* 337:957–960.

Bowern, C., and Q. Atkinson. 2012. Computational phylogenetics and the internal structure of Pama-Nyungan. *Language* 88:817–845.

Broué, P., D. R. Marshall, and W. J. Muller. 1977. Biosystematics of subgenus Glycine (Verdc.): Isoenzymatic data. *Australian Journal of Botany* 25:555–566.

Brown, C. H., C. R. Clement, P. Epps, E. Luedeling, and S. Wichmann. 2014. The paleobiolinguistics of the common bean (*Phaseolus vulgaris* L.). *Ethnobiology Letters* 5:104–115.

Bryant, J. A., and S. G. Hughes. 2011. Vicia. In *Wild Crop Relatives: Genomic and Breeding Resources, Legume Crops and Forages*, C. Kole (Ed.), pp. 273–289. Berlin, Germany: Springer.

Budanova, V. I. 1982. Novoe i sistematike rodov *Phaseolus* L. i *Vigna* Savi. (Some new data on the systematics of the genera *Phaseolus* L. and *Vigna* Savi.) *Trudy po Prikladnoj Botanike Genetike i Selekcii* 72:16–20.

Burrow, T., and M. B. Emeneau. 1998. *A Dravidian Etymological Dictionary*. New Delhi, India: Munshiram Manoharlal Publishers.

Bussmann, R. W., N. Y. Paniagua Zambrana, S. Sikharulidze et al. 2016. A comparative ethnobotany of Khevsureti, Samtskhe-Javakheti, Tusheti, Svaneti, and Racha-Lechkhumi, Republic of Georgia (Sakartvelo), Caucasus. *Journal of Ethnobiology and Ethnomedicine* 12:43.

Butler, R. 1997. The impossible painter. *Australian Art Collector* 1:42–45.

Cacan, E., K. Kokten, H. Inci, A. Das, and A. Y. Sengul. 2016. Fatty acid composition of the seeds of some vicia species. *Chemistry of Natural Compounds* 52:1084–1086.

Campbell, L. 2000. *American Indian Languages: The Historical Linguistics of Native America*. Oxford, UK: Oxford University Press.

Campbell, L., and W. J. Poser. 2008. *Language Classification: History and Method*. Cambridge, UK: Cambridge University Press.

Campbell, M. C., and S. A. Tishkoff. 2010. The evolution of human genetic and phenotypic variation in Africa. *Current Biolology* 20:166–173.

Cappers, R. T. J., and R. Hamdy. 2007. Ancient Egyptian plant remains in the agricultural museum (Dokki, Cairo). In *Fields of Change: Progress in African Archaeobotany*, R. Cappers (Ed.), pp. 165–214. Groningen, the Netherlands: Barkhuis/Groningen University Library.

Caracuta, V., J. Vardi, Y. Paz, and E. Boaretto. 2017. Farming legumes in the pre-pottery Neolithic: New discoveries from the site of Ahihud (Israel). *PLoS One* 12:e0177859.

Caracuta, V., M. Weinstein-Evron, D. Kaufman, R. Yeshurun, J. Silvent, and E. Boaretto. 2016. 14,000-year-old seeds indicate the Levantine origin of the lost progenitor of faba bean. *Scientific Reports* 6:37399.

Carrouée, B. 1993. Different types of peas: To clarify a complex status. *Grain Legumes* 3:26–27.

Castetter, E. F. and R. M. Underhill. 1935. *The Ethnobiology of the Papago Indians*. Albuquerque, NM: University of New Mexico.

Čašule, I. 2009. Burushaski shepherd vocabulary of Indo-European origin. *Acta Orientalia* 70:147.

Cavalli-Sforza, L. L., and M. Seielstad. 2001. *Genes, Peoples and Languages*. London, UK: Penguin Press.

Ceccon, G., L. A. Staut, E. Sagrilo, L. A. Z. Machado, D. P. Nunes, and V. B. Alves. 2013. Legumes and forage species sole or intercropped with corn in soybean-corn succession in midwestern Brazil. *Revista Brasileira de Ciência do Solo* 37:204–212.

Çekal, N., D. Derin, and M. Akman. 2012. Evaluation of Turkish cuisine in terms of nutrition and health. *European Journal of Social Sciences* 34:5–10.

Cergna, S. 2015. *Vocabolario del dialetto di Valle d'Istria*. Rovigno, Croatia: Centro Ricerche Storiche.

Chalup, L., M. Grabiele, V. S. Neffa, and G. Seijo. 2014. DNA content in South American endemic species of Lathyrus. *Journal of Plant Research* 127:469–480.

Chamoux, M. N. 1997. La cuisine de la Toussaint chez les Aztèques de la Sierra de Puebla (Mexique). *Babel. Internationales de l'imaginaire* 7:85–99.

Chantraine, P. 1968. *Dictionnaire Étymologique de la langue grecque*. Paris, France: Klincksieck.

Chomsky, N. 2000. *The Architecture of Language*. Oxford, UK: Oxford University Press.

Christenhusz, M. J. M., and J. W. Byng. 2016. The number of known plants species in the world and its annual increase. *Phytotaxa* 261:201–217.

Clark, A. L., C. L. King, H. R. Buckley et al. 2017. Biological anthropology in the Indo-Pacific Region: New approaches to age-old questions. *Journal of Indo-Pacific Archaeology* 41:78–94.

Collins Dictionaries. 2014. *Collins English Dictionary—Complete and Unabridged*, 12th ed. Glasgow, Scotland: HarperCollins.

Cooper, M. 1976. Review: The Nippo Jisho. *Monumenta Nipponica* 31:417–430.

Corinto, G. L. 2014. Nikolai vavilov's centers of origin of cultivated plants with a view to conserving agricultural biodiversity. *Human Evolution* 29:285–301.

Crystal, D. 2000. *Language Death*. Cambridge, UK: Cambridge University Press.

Ćupina, B., B. Zlatković, P. Smýkal et al. 2011. In situ evaluation of a *Pisum sativum* subsp. *elatius* population from the valley of the river Pčinja in southeast Serbia. *Pisum Genetics* 43:20–24.

Cvijetić, R. 2014. *Rečnik Užičkog govora*. Belgrade, Sebia: Službeni glasnik, Užice: Faculty of Pedagogy.

D'Anastasio, R., S. Wroe, C. Tuniz et al. 2013. Micro-biomechanics of the Kebara 2 hyoid and its implications for speech in Neanderthals. *PLoS One* 8:e82261.

Davies, D., and A. Jones. 1995. *Welsh Names of Plants*. Cardiff, Wales: National Museum of Wales.

Davies, P. H. 1970. *Vavilovia* A. Fed. In *Flora of Turkey and East Aegean Islands 3*, P. H. Davies (Ed.), pp. 44–45. Edinburgh, Scotland: University of Edinburgh.

de Filippo, C., C. Barbieri, M. Whitten et al. 2010. Y-chromosomal variation in sub-Saharan Africa: Insights into the history of Niger-Congo groups. *Molecular Biology and Evolution* 28:1255–1269.

Debouck, D. G. 1999. Diversity in *Phaseolus* species in relation to the common bean. In *Common Bean Improvement in the Twenty-first Century. Developments in Plant Breeding*, vol. 7, S. P. Singh (Ed.), pp. 25–52. Dordrecht, the Netherlands: Springer.

Delgado-Salinas, A., M. Thulin, R. Pasquet, N. Weeden, and M. Lavin. 2011. *Vigna* (Leguminosae) sensu lato: The names and identities of the American segregate genera. *American Journal of Botany* 98:1694–1715.

Delgado-Salinas, A., R. Bibler, and M. Lavin. 2006. Phylogeny of the genus *Phaseolus* (Leguminosae): A recent diversification in an ancient landscape. *Systematic Botany* 31:779–791.

Delgado-Salinas, A., T. Turley, A. Richman, and M. Lavin. 1999. Phylogenetic analysis of the cultivated and wild species of *Phaseolus* (Fabaceae). *Systematic Botany* 31:438–460.

Diakonoff, I. M. 1988. *Afrasian Languages*. Moscow, Russia: Nauka.

Diamond, J. 1997. *Guns, Germs, and Steel: The Fates of Human Societies*. New York: Norton & Co.

Dillehay, T. D., J. Rossen, T. C. Andres, and D. E. Williams. 2007. Preceramic adoption of peanut, squash, and cotton in northern Peru. *Science* 316:1890–1893.

Diller A., J. Edmondson, and Y. Luo. 2008. *The Tai-Kadai Languages*. London, UK: Routledge.

Dixon, R. M. W. 2002. *Australian Languages: Their Nature and Development*. Cambridge, UK: Cambridge University Press.

Dixon, R. M., and R. M. W. Dixon. 2011. *The Languages of Australia*. Cambridge, UK: Cambridge University Press.

Dobon, B., H. Y. Hassan, H. Laayouni et al. 2015. The genetics of East African populations: A Nilo-Saharan component in the African genetic landscape. *Scientific Reports* 5:9996.

Doi, K., A. Kaga, N. Tomooka, and D. A. Vaughan. 2002. Molecular phylogeny of genus Vigna subgenus Ceratotropis based on rDNA ITS and atpB-rbcL intergenic spacer of cpDNA sequences. *Genetica* 114:129–145.

Domonoske, C. 2014. A legume with many names: The story of "goober." Code switch, NPR. https://www.npr.org/sections/codeswitch/2014/04/20/304585019/a-legume-with-many-names-the-story-of-goober (accessed November 25, 2017).

Doyle, J. J., and M. A. Luckow. 2003. The rest of the iceberg. Legume diversity and evolution in a phylogenetic context. *Plant Physiology* 131:900–910.

Drummond, C. S., R. J. Eastwood, S. T. Miotto, and C. E. Hughes. 2012. Multiple continental radiations and correlates of diversification in *Lupinus* (Leguminosae): Testing for key innovation with incomplete taxon sampling *Systematic Botany* 61:443–460.

du Cange, C. 1883. *Glossarium Mediæ et Infimæ Latinitatis* (in Latin), G. A. L. Henschel, P. Carpentier, and L. Favre (Eds.), Niort, France: L. Favre.

Duke, J. A. 1981. *Handbook of Legumes of World Economic Importance.* New York: Plenum Press.

Dunaway, W. F. 1913. *Reminiscences of a Rebel.* New York: The Neale Publishing Company.

Dworkin, S. N. 2012. *A History of the Spanish Lexicon: A Linguistic Perspective.* Oxford, UK: Oxford University Press on Demand.

Dybo, A. 2006a. Tungus etymology. The Tower of Babel, an International Etymological Database Project. http://starling.rinet.ru (accessed November 25, 2017).

Dybo, A. 2006b. Turkic etymology. The Tower of Babel, an International Etymological Database Project. http://starling.rinet.ru/ (accessed November 25, 2017).

Eco, U. 1995. *The Search for the Perfect Language.* Oxford, UK: Blackwell.

Ecocrop 2007. http://ecocrop.fao.org/ecocrop/srv/en/home (accessed November 25, 2017).

Ehret, C. 1995. *Reconstructing Proto-Afroasiatic (Proto-Afrasian): Vowels, Tone, Consonants, and Vocabulary.* Berkeley, CA: University of California Press.

Elliott, J. 2008. Tex-Arcana: How did bluebonnets become state flower? Decision blossomed into a 70-year fight. Houston Cronicle March 23. http://www.chron.com/news/houston-texas/article/Tex-Arcana-How-bluebonnets-became-state-flower-1792133.php (accessed November 25, 2017).

Ellis, T. H. N. 2011. Pisum. In *Wild Crop Relatives: Genomic and Breeding Resources, Legume Crops and Forages*, C. Kole (Ed.), pp. 237–247. Berlin, Germany: Springer.

Ellis, T. H. N., J. M. I. Hofer, G. M. Timmerman-Vaughan, C. J. Coyne, and R. P. Hellens. 2011. Mendel, 150 years on. *Trends in Plant Science* 16:590–596.

Elzaki, O. T., T. O. Khider, and S. H. Omer. 2012. Pulp and papermaking characteristics of Cajanus cajan stems from Sudan. *Cellulose* 46:47–31.

Endo, Y., B. H. Choi, H. Ohashi, and A. Delgado-Salinas. 2008. Phylogenetic relationships of New World *Vicia* (Leguminosae) inferred from nrDNA Internal Transcribed Spacer sequences and floral characters. *Systematic Botany* 33:356–363.

EPPO 2017. European and Mediterranean Plant Protection Organization (EPPO), Paris. http://www.eppo.int/ (accessed November 25, 2017).

Erskine, W., S. Chandra, M. Chaudhry et al. 1998. A bottleneck in lentil: Widening its genetic base in South Asia. *Euphytica* 101:207–211.

Ethnologue 2017. *Ethnologue: Languages of the World.* Dallas, TX: SIL International. https://www.ethnologue.com/ (accessed November 25, 2017).

Fabre, A. 2005. Diccionario etnolingüìstico y güía bibliográfica de los pueblos indígenas sudamericanos. http://www.ling.fi/Diccionario%20etnoling.htm (accessed November 25, 2017).

Fabre, A. 2016. Towards a data base for typological studies of South American and Lower Central American indigenous languages (morphology and syntax). http://www.ling.fi/typologydatabase.html (accessed November 25, 2017).

Fairbairn, A., D. Martinoli, A. Butler, and G. Hillman. 2007. Wild plant seed storage at Neolithic Çatalhöyük East, Turkey. *Vegetation History and Archaeobotany* 16:467–479.

FAOSTAT 2017. *Food and Agriculture Data* (*FAOSTAT*). Food and Agriculture Organization (FAO) of the United Nations, Rome, Italy. http://www.fao.org/faostat/ (accessed November 25, 2017).

Felger, R. S., and M. B. Moser. 1985. *People of the Desert and Sea: Ethnobotany of the Seri Indians.* Tucson, AZ: University of Arizona Press.

Ferguson, M. E., N. Maxted, M. van Slageren, and L. D. Robertson. 2000. A re-assessment of the taxonomy of Lens Mill. (Leguminosae, Papilionoideae, Vicieae). *Botanical Journal of the Linnean Society* 133:41–59.

Fernald, M. L. 1950. *Gray's Manual of Botany*, 8th ed. New York: American Book Company.

Fortescue, M. 2005. *Comparative Chukotko-Kamchatkan Dictionary*. Berlin, Germany: Mouton de Gruyter.

Frediani, M., F. Maggini, M. T. Gelati, and R. Cremonini. 2004. Repetitive DNA sequences as probes for phylogenetic analysis in *Vicia* genus. *Caryologia* 57:379–386.

Freytag, G. F., and D. G. Debouck 2002. *Taxonomy, Distribution, and Ecology of the Genus Phaseolus (Leguminosae—Papilionoideae) in North America, Mexico and Central America*. Fort Worth, TX: Botanical Research Institute of Texas.

Fuller, D. Q., and C. Murphy. 2014. Overlooked but not forgotten: India as a center for agricultural domestication. *General Anthropology* 21:1–8.

Fuller, D. Q., and E. L. Harvey. 2006. The archaeobotany of Indian pulses: Identification, processing and evidence for cultivation. *Environmental Archaeology* 11:219–246.

Fuller, D. Q., G. Willcox, and R. G. Allaby. 2012. Early agricultural pathways: Moving outside the "core area" hypothesis in Southwest Asia. *Journal of Experimental Botany* 63:617–633.

Fuller, D., R. Korisettar, P. C. Venkatasubbaiah, and M. K. Jones. 2004. Early plant domestications in southern India: Some preliminary archaeobotanical results. *Vegetation History and Archaeobotany* 13:115–129.

García-Granero, J. J., C. Lancelotti, and M. Madella. 2017. A methodological approach to the study of microbotanical remains from grinding stones: A case study in northern Gujarat (India). *Vegetation History and Archaeobotany* 26:43–57.

Gell-Mann, M., and M. Ruhlen. 2011. The origin and evolution of word order. *Proceedings of the National Academy of Sciences USA*. 108:17290–17295.

Georg, S., P. A. Michalove, A. M. Ramer, and P. J. Sidwell. 1999. Telling general linguists about Altaic. *Journal of Linguistics* 35:65–98.

Georgiev, V. I. 1981. *Introduction to the History of the Indo-European Languages*. Sofia, Bulgaria: Bulgarian Academy of Science.

Gepts, P., and F. A. Bliss. 1988. Dissemination pathways of common bean (*Phaseolus vulgaris*, Fabaceae) deduced from phaseolin electrophoretic variability. II. Europe and Africa. *Economic Botany* 42:86–104.

Gerard, J. 1597. *Great Herball, or, Generall Historie of Plantes*. London, UK: John Norton.

Gianfranco, V., C. Ravalli, and R. Cremonini. 2008. The karyotype as a tool to identify plant species: *Vicia* species belonging to *Vicia* subgenus. *Caryologia* 61:300–319.

Gimbutas, M., M. R. Dexter, and K. Jones-Bley. 1997. *The Kurgan Culture and the Indo-Europeanization of Europe: Selected Articles from 1952 to 1993*. Washington, DC: Institute for the Study of Man.

Giorgadze, M., and N. Inaishvili. 2016. The colchis black sea littoral in the archaic and classical periods. *HISTORIKA Studi di storia greca e romana* 5:151–165.

Gladstones, J. S. 1974. Lupinus of the Mediterranean region and Africa. *Technical bulletin. Western Australia. Department of Agriculture* 26:48.

Gladstones, J. S. 1976. The Mediterranean white lupin. *Journal of Agriculture of Western Australia* 17:70–74.

Gledhill, D. 2008. *The Names of Plants*. Cambridge, UK: Cambridge University Press.

Gligić, V. 1954. *Etimološki botanički rečnik*. Sarajevo, Bosnia and Herzegovina: Veselin Masleša.

Gombocz, Z. 1912. *Die bulgarisch-türkische Lehnwörter in der ungarischen Sprache*. Helsinki, Finland: Mémoires de la Société Finno-Ougrienne (MSFOu 30).

González, A. M., O. García, J. M. Larruga, and V. M. Cabrera. 2006. The mitochondrial lineage U8a reveals a Paleolithic settlement in the Basque country. *BMC Genomics* 7:124.

Gorim, L. Y., and A. Vandenberg. 2017. Evaluation of wild lentil species as genetic resources to improve drought tolerance in cultivated lentil. *Frontiers in Plant Science* 8:1129.

Greenberg, J. H. 1971. The Indo-Pacific hypothesis. In *Current Trends in Linguistics, Vol. 8: Linguistics in Oceania*, T. A. Sebeok (Ed.), pp. 808–871. The Hague, the Netherlands: Mouton.

Greenberg, J. H. 1972. Linguistic evidence regarding Bantu origins. *The Journal of African History* 13:189–216.

Greenberg, J. H. 1987. *Language in the Americas*. Stanford, CA: Stanford University Press.

Greenberg, J. H. 1996. In defense of Amerind. *International Journal of American Linguistics* 62:131–164.

Greenberg, J. H. 2000. *Indo-European and Its Closest Relatives: The Eurasiatic Language Family. Volume 1, Grammar.* Stanford, CA: Stanford University Press.

Greenberg, J. H. 2002. *Indo-European and Its Closest Relatives: The Eurasiatic Language Family. Volume 2, Lexicon*. Stanford, CA: Stanford University Press.

Greenberg, J. H., and M. Ruhlen. 1992. Linguistic origins of native Americans. *Scientific American* 267:94–99.

Greenberg, J. H., and M. Ruhlen. 2007. *An Amerind Etymological Dictionary*. Stanford, CA: Stanford University Press.

Hage, P. 2001. The evolution of Dravidian kinship systems in Oceania: Linguistic evidence. *Journal of the Royal Anthropological Institute* 7:487–508.

Hajdú, P. 1969. Finno-ugrische Urheimatforschung. *Ural-Altaishe Jahrbücher* 41:252–264.

Halliday, M. A. K. 2003. *On Language and Linguistics*. London, UK: Continuum.

Hammer, K., K. Khoshbakht, V. Montesano, and G. Laghetti 2015. *The Big Five—A Domestication Assessment of the Five Largest Plant Families*. Rotondella, Italy: Archivia.

Hammer, M. F., T. M. Karafet, A. J. Redd et al. 2001. Hierarchical patterns of global human Y-chromosome diversity. *Molecular Biology and Evolution* 18:1189–1203.

Handel, Z. 2008. What is Sino-Tibetan? Snapshot of a field and a language family in flux. *Language and Linguistics Compass* 2:422–441.

Hannan, R., N. Açikgöz, and L. D. Robertson. 2013. Chickpeas (*Cicer* L.). In *Plant Genetic Resources of Legumes in the Mediterranean*, N. Maxted and S. J. Bennett (Eds.), pp. 115–124. Dordrecht, the Netherlands: Springer.

Hechenleitner, V. P. 2015. *Biogeography and Systematics of South American Vicia* (*Leguminosae*). Doctoral dissertation, Aberdeen, Scotland: University of Aberdeen.

Henry, A. G., A. S. Brooks, and D. R. Piperno. 2011. Microfossils in calculus demonstrate consumption of plants and cooked foods in Neanderthal diets (Shanidar III, Iraq; Spy I and II, Belgium). *Proceedings of the National Academy of Sciences of the USA* 108:486–491.

Hermann, F. J. 1960. *Vetches of the United States—Native, Naturalized and Cultivated. Agriculture Handbook 168*. Washington, DC: U.S. Department of Agriculture.

Holst, J. H. 2014. *Advances in Burushaski Linguistics*. Tübingen, Germany: Narr Verlag.

Honkola, T., O. Vesakoski, K. Korhonen, J. Lehtinen, K. Syrjänen, and N. Wahlberg. 2013. Cultural and climatic changes shape the evolutionary history of the Uralic languages. *Journal of Evolutionary Biology* 26:1244–1253.

Hostetter, C. F. 2007a. I Lam na Ngoldathon: The grammar and Lexicon of the Gnomish tongue. In *J. R. R. Tolkien Encyclopedia: Scholarship and Critical Assessment*, M. D. C. Drout (Ed.), pp. 155–159. New York: Routledge.

Hostetter, C. F. 2007b. Qenyaqetsa: The qenya phonology and Lexicon. In *J. R. R. Tolkien Encyclopedia: Scholarship and Critical Assessment*, M. D. C. Drout (Ed.), pp. 551–552. New York: Routledge.

Huehnergard, J. 2004. Afro-Asiatic. In *The Cambridge Encyclopedia of the World's Ancient Languages*, R. D. Woodard (Ed.), pp. 138–159. Cambridge, UK: Cambridge University Press.

Hyman, L. M. 2011. The macro-Sudan belt and Niger-Congo reconstruction. *Language Dynamics and Change* 1:3–49.

Hymowitz, T., and C. A. Newell. 1981. Taxonomy of the genus *Glycine*, domestication and uses of soybeans. *Economic Botany* 35:272–288.

Hymowitz, T., R. J. Singh, and K. P. Kollipara. 1998. The genomes of the Glycine. *Plant Breeding Reviews* 16:289–318.

IBIS. 2017. IBIS Integrated Botanical Information System (IBIS), Australian plant common name database. https://www.anbg.gov.au/ibis/ (accessed November 25, 2017).

ILDIS. 2017. *International Legume Database & Information Service World Database of Legumes*. University of Reading, Reading. http://www.ildis.org/ (accessed November 25, 2017).

ISTA. 1982. International Seed Testing Association (ISTA). *A Multilingual Glossary of Common Plant-Names: Field Crops, Grasses and Vegetables*. H. Koster (Ed.) Wageningen, the Netherlands: ISTA.

Itkonen, E., and A. J. Joki. 1978. *Suomen kielen etymologinen sanakirja. 6. [vatrata-ööttää]*. Helsinki, Finland: Suomalais-ugrilainen seura.

Jacobsen, S. E., and A. Mujica. 2008. Geographical distribution of the Andean lupin (*Lupinus mutabilis* Sweet). *Plant Genetic Resources Newsletter* 155:1–8.

Janhunen, J. 1998. Samoyedic. In *The Uralic Languages*, D. Abondolo (Ed.), pp. 457–479. London, UK: Routledge.

Janhunen, J. 2009. Proto-Uralic—What, where, and when? *Mémoires de la Société Finno-Ugrian Society* 258:57–78.

Janhunen, J. 2010. Enclitic zero verbs in some Eurasian languages. In *Transeurasian Verbal Morphology in a Comparative Perspective: Genealogy, Contact, Chance (Turcogica 78)*, L. Johanson and M. Robbeets (Eds.), pp. 165–180. Wiesbaden, Germany: Harrassowitz Verlag.

Jantz, N., and H. Behling. 2012. A Holocene environmental record reflecting vegetation, climate, and fire variability at the Páramo of Quimsacocha, southwestern Ecuadorian Andes. *Vegetation History and Archaeobotany* 21:169–185.

Javadi, F., Y. T. Tun, M. Kawase, K. Guan, and H. Yamaguchi. 2011. Molecular phylogeny of the subgenus *Ceratotropis* (genus *Vigna*, Leguminosae) reveals three eco-geographical groups and Late Pliocene-Pleisticene diversification: Evidence from four plastid DNA region sequences. *Annals of Botany* 108:367–380.

Jing, R., A. Vershinin, J. Grzebyta et al. 2010. The genetic diversity and evolution of field pea (*Pisum*) studied by high throughput retrotransposon based insertion polymorphism (RBIP) marker analysis. *BMC Evolutionary Biology* 10:44.

Jing, R., M. A. Ambrose, M. R. Knox et al. 2012. Genetic diversity in European *Pisum* germplasm collections. *Theoretical and Applied Genetics* 125:367–380.

Jovanović, Ž., N. Stanisavljević, A. Nikolić et al. 2011. *Pisum & ervilia Tetovac*—made in early iron age Leskovac. Part two. Extraction of the ancient DNA from charred seeds from the site of Hissar in South Serbia. *Ratarstvo i Povrtarstvo* 48:227–232.

Kalaiselvan, V., M. Kalaivani, A. Vijayakumar, K. Sureshkumar, and K. Venkateskumar. 2010. Current knowledge and future direction of research on soy isoflavones as a therapeutic agents. *Pharmacognosy Reviews* 4:111–117.

Kassa, M. T., L. J. G. van der Maesen, C. Krieg, E. and J. B. von Wettberg. 2016. Historical and phylogenetic perspectives of pigeonpea. *Legume Perspectives* 11:5–7.

Kassa, M. T., R. V. Penmetsa, N. Carrasquilla-Garcia et al. 2012. Genetic patterns of domestication in pigeonpea (*Cajanus cajan* (L.) Millsp.) and wild Cajanus relatives. *PLoS One* 7:e39563.

Kaufman, T., and J. Justeson. 2006. History of the word for "Cacao" and related terms in ancient meso-America. In *Chocolate in Mesoamerica: A Cultural History of Cacao*, C. L. McNeil (Ed.), pp. 118–139. Gainesville, FL: University Press of Florida.

Kenicer, G. 2008. An introduction to the genus *Lathyrus* L. *Curtis's Botanical Magazine* 25:286–295.

Kenicer, G. J. 2007. *Systematics and Biogeography of Lathyrus L.* (Leguminosae—Papilionoideae), Doctoral dissertation, Edinburgh, Scotland: University of Edinburgh.

Kenicer, G. J., T. Kajita, R. T. Pennington, and J. Murata. 2005. Systematics and biogeography of Lathyrus (Leguminosae) based on internal transcribed spacer and cpDNA sequence data. *American Journal of Botany* 92:1199–1209.

Kew Science. 2017. Lupinus L. *The International Plant Names Index and World Checklist of Selected Plant Families.* Kew Science, Plants of the World Online, London, UK: Royal Botanical Gardens. http://powo.science.kew.org/ (accessed November 25, 2017).

Kislev, M. E., and O. Bar-Yosef. 1988. The legumes: The earliest domesticated plants in the Near East? *Current Anthropology* 29:175–179.

Klimov, G. 1998. *Languages of the World: Caucasian Languages.* Moscow, Russia: Academia.

Koerner, E. F. K. 1999. *Linguistic Historiography: Projects & Prospects.* Amsterdam, the Netherland: John Benjamins.

Kosterin, O. E., and V. S. Bogdanova. 2008. Relationship of wild and cultivated forms of *Pisum* L. as inferred from an analysis of three markers, of the plastid, mitochondrial and nuclear genomes. *Genetic Resources and Crop Evolution* 55:735–755.

Kosterin, O. E., O. O. Zaytseva, V. S. Bogdanova, and M. J. Ambrose. 2010. New data on three molecular markers from different cellular genomes in Mediterranean accessions reveal new insights into phylogeography of *Pisum sativum* L. subsp. *elatius* (Bieb.) Schmalh. *Genetic Resources and Crop Evolution* 57:733–739.

Krampner, J. 2014. *Creamy and Crunchy: An Informal History of Peanut Butter, the All-American Food.* New York: Columbia University Press.

Krapovickas, A., and W. C. Gregory. 1994. Taxonomía del género *Arachis* (Leguminosae). *Bonplandia* 8:26–27.

Krishna, K. A., and K. D. Morrison. 2009. History of South Indian agriculture and agroecosystems. *South Indian Agroecosystems: Nutrient Dynamics and Productivity* 1–51.

Krishnamurti, B. 2003. *The Dravidian Languages.* Cambridge, UK: Cambridge University Press.

Kurlovich, B. S. 2002. *Lupins: Geography, Classification, Genetic Resources and Breeding.* St. Petersburg, Russia: Intan.

Kuz'mina, E. E. 2007. *The Origin of the Indo-Iranians.* Leiden, the Netherlands: Brill.

Ladizinsky, G., and S. Abbo. 2015. *The Search for Wild Relatives of Cool Season Legumes.* Cham, Switzerland: Springer.

Lawn, R. J., and A. Cottrell. 2016. Seeds of "Vigna Marina" (burm.) Merrill survive up to 25 years flotation in salt water. *Queensland Naturalist* 54:3–13.

Lawn, R. J., and A. E. Holland. 2003. Variation in the *Vigna lanceolata* complex for traits of taxonomic, adaptive or agronomic interest. *Australian Journal of Botany* 51:295–307.

Lee, G. A., G. W. Crawford, L. Liu, Y. Sasaki, and X. Chen. 2011. Archaeological soybean (Glycine max) in East Asia: Does size matter? *PLoS One* 6:e26720.

Lee, M. J., J. S. Pate, D. J. Harris, and C. A. Atkins. 2006. Synthesis, transport and accumulation of quinolizidine alkaloids in *Lupinus albus* L. and *L. angustifolius* L. *Journal of Experimental Botany* 58:935–946.

Leht, M. 2005. Cladistic and phenetic analyses of relationships in *Vicia* subgenus *Cracca* (Fabaceae) based on morphological data. *Taxon* 54:1023–1032.

Leht, M. 2009. Phylogeny of old world *Lathyrus L.* (Fabaceae) based on morphological data. *Feddes Repertorium* 120:59–74.

Lewis, C. T. 1890. *An Elementary Latin Dictionary.* New York: American Book Company.

Lewis, C. T., and C. Short. 1879. *A Latin Dictionary.* Oxford, UK: Clarendon Press.

Liddell, H. G., R. Scott, H. S. Jones, and R. McKenzie. 1940. *A Greek-English Lexicon*, Vol. 2. Oxford, UK: Clarendon Press.

Lim, T. K. 2012. *Edible Medicinal and Non-Medicinal Plants: Volume 2, Fruits*. New York: Springer.

Lin, C. C., S. J. Wu, J. S. Wang, J. J. Yang, and C. H. Chang. 2001. Evaluation of the antioxidant activity of legumes. *Pharmaceutical Biology* 39:300–304.

Linnaeus, C. 1753. *Species plantarum exhibentes plantas rite cognitas, ad genera relatas, cum differentiis specificis, nominibus trivialibus, synonymis selectis, locis natalibus, secundum systema sexuale digestas, I-II.* Stockholm, Sweden: Impensis Laurentii Salvii.

Linnaeus, C. 1758. *Systema naturae per regna tria naturae: Secundum classes, ordines, genera, species, cum characteribus, differentiis, synonymis, locis.* Stockholm, Sweden: Impensis Laurentii Salvii.

Littré, É. 1863–1872. *Dictionnaire de la langue française.* Paris, France: Hachette Livre.

Lizarazo, C. I., A. M. Lampi, J. Liu, T. Sontag-Strohm, V. Piironen, and F. L. Stoddard. 2015. Nutritive quality and protein production from grain legumes in a boreal climate. *Journal of the Science of Food and Agriculture* 95:2053–2064.

Lockwood, W. B. 1977. *Indo-European Philology, Historical and Comparative.* London, UK: Hutchinson.

Logos. 2017. *Logos—Multilingual Translation Portal.* Modena, Italy: Logos. http://logos.it (accessed November 25, 2017).

Lytkin, V. I., and E. S. Gulyaev. 1970. *Kratkii etimologicheskii slovar' komi yazyka.* Moscow, Russia: Nauka.

Maass, B. L. 2016. Origin, domestication and global dispersal of Lablab purpureus (L.) Sweet (Fabaceae): Current understanding. *Legume Perspectives* 13:5–8.

Maass, B. L., M. R. Knox, S. C. Venkatesha, T. A. Tefera, S. Ramme, and B. C. Pengelly. 2010. Lablab purpureus—A crop lost for Africa?. *Tropical Plant Biology* 3:123–135.

Maass, B. L., R. H. Jamnadass, J. Hanson, and B. C. Pengelly. 2005. Determining sources of diversity in cultivated and wild *Lablab purpureus* related to provenance of germplasm by using amplified fragment length polymorphism. *Genetic Resources and Crop Evolution* 52:683–695.

Macbain, A. 1911. *Etymological Dictionary of the Gaelic Language.* Stirling, Scotland: Eneas Mackay.

Makasheva, R. K. H. 1979. Gorokh. In *Kul'turnaia Flora SSSR. Zernovye bobovye kul'tury, chast'* 1, D. D. Brezhnev (Ed.), pp. 45–49. Leningrad, Russia: Kolos.

Mallory, J. P. 1989. *In Search of the Indo-Europeans.* London, UK: Thames and Hudson.

Manderson, L. 1981. Traditional food classifications and humoral medical theory in Peninsular Malaysia. *Ecology of Food and Nutrition* 11:81–92.

Maréchal, R., J. M. Mascherpa, and F. Stainier. 1978. Étude taxonomique d'un groupe complexe d'espèces des genres Phaseolus et Vigna (Papilionaceae) sur la base données morphologiques et polliniques, traitées par l'analyse inf.tique. *Boissiera* 28:1–273.

Marin, P., and B. Tatić. 2004. *Etimološki rečnik naziva rodova i vrsta vaskularne flore Evrope.* Belgrade, Serbia: NNK internacional.

Marjanović-Jeromela, A., A. M. Mikić, S. Vujić et al. 2017. Potential of legume-brassica intercrops for forage production and green manure: Encouragements from a temperate Southeast European environment. *Frontiers in Plant Science* 8:312.

Maxted, N. 1995. An ecogeographical study of Vicia subgenus Vicia. *Systematic and Ecogeographic Studies on Crop Genepools 8*, pp. 1–184. Rome, Italy: International Board for Plant Genetic Resources.

Maxted, N., and M. Ambrose. 2001. Peas (Pisum L.). In *Plant Genetic Resources of Legumes in the Mediterranean*, N. Maxted and S. J. Bennett (Eds.), pp. 181–190. Dordrecht, the Netherlands: Kluwer.

Maxted, N., B. V. Ford-Lloyd, S. Jury, S. Kell, and M. Scholten. 2006. Towards a definition of a crop wild relative. *Biodiversity and Conservation* 15:2673–2685.

Maxted, N., P. Mabuza-Dlamini, H. Moss, S. Padulosi, A. Jarvis, and L. Guarino 2004. *An Ecogeographic Survey: African Vigna. Systematic and Ecogeographic Studies of Crop Genepools 10*, pp. 1–184. Rome, Italy: International Board for Plant Genetic Resources.

Mažiulis, V. 2004. *Prūsų kalbos istorinė gramatika*. Vilnius, Lithuania: Vilniaus universiteto leidykla.

McConvell, P., and C. Bowern. 2011. The prehistory and internal relationships of Australian languages. *Language and Linguistics Compass* 5:19–32.

McCulloch, S. 2001. *Contemporary Aboriginal Art: A Guide to the Rebirth of an Ancient Culture*. Crows Nest, Australia: Allen & Unwin.

Meakins, F. 2013. Mixed languages. In *Contact Languages: A Comprehensive Guide*, P. Bakker and Y. Matras (Eds.), pp. 159–228. Berlin, Germany: Mouton de Gruyter.

Medović, A., A. Mikić, B. Ćupina et al. 2011. Pisum & Ervilia tetovac—Made in early iron age Leskovac. Part one. Two charred pulse crop storages of the fortified hill fort settlement Hissar in Leskovac, South Serbia. *Ratarstvo i povrtatstvo* 48:219–226.

Medović, A., Ž Jovanović, N. Stanisavljević et al. 2010. An archaeobotanical and molecular fairy tale about the early Iron Age Balkan princess and the charred pea. *Pisum Genetics* 42:35–38.

Melamed, Y., U. Plitmann, and M. E. Kislev. 2008. Vicia peregrina: An edible early Neolithic legume. *Vegetation History and Archaeobotany* 17:29–34.

Mendel, G. 1866. Versuche über Pflanzen-Hybriden. *Verhandlungen des naturforschenden Vereines in Brünn* 4:3–47.

Michalove, P. A. 2002. The classification of the Uralic languages: Lexical evidence from Finno-Ugric. *Finnisch-Ugrische Forschungen* 57:58–67.

Midant-Reynes, B. 1999. *The Prehistory of Egypt: From the First Egyptians to the First Pharaohs*. Oxford, UK: Blackwell.

Mihailović, V., A. Mikić, and B. Ćupina 2004. Botanical and agronomic classification of fodder pea (Pisum sativum L.). *Acta Agriculturae Serbica* 17:61–65.

Mihailović, V., A. Mikić, M. Ćeran et al. 2016. Some aspects of biodiversity, applied genetics and agronomy in hyacinth bean (Lablab purpureus) research. *Legume Perspectives* 13:9–15.

Mikić, A. 2009. Words denoting pea (Pisum sativum) in European languages. *Pisum Genetics* 41:29–33.

Mikić, A. 2010. Words denoting lentil (Lens culinaris) in European languages. *Journal of Lentil Research* 4:15–19.

Mikić, A. 2011a. A note on some Proto-Indo-European roots related to grain legumes. *Indogermanische Forschungen* 116:60–71.

Mikić, A. 2011b. Words for ancient Eurasian food legumes in the languages of the Dene-Caucasian macrofamily. *Mother Tongue* 16:113–124.

Mikić, A. 2011c. Can we reconstruct the most ancient words for pea (Pisum sativum)? *Pisum Genetics* 43:36–42.

Mikić, A. 2011d. Words denoting faba bean (Vicia faba) in European languages. *Ratarstvo i povrtarstvo* 48:233–238.

Mikić, A. 2012. Origin of the words denoting some of the most ancient Old World pulse crops and their diversity in modern European languages. *PLoS One* 7:e44512.

Mikić, A. 2013a. Of orivaine, kamilot, and their kindred: Words for legumes in Tolkien's invented languages. *Arda Philology* 4:54–61.

Mikić, A. 2013b. Origin and diversity of the words for ancient Eurasian grain legumes in Slavic languages. *Onomázein* 28:280–287.

Mikić, A. 2014a. Grain legume crop history among Slavic nations traced using linguistic evidence. *Czech Journal of Genetics and Plant Breeding* 50:65–68.

Mikić, A. 2014b. A note on the words in the Baltic languages for some of the most ancient European grain legume crops. *Dialectologia et Geolinguistica* 22:39–45.

Mikić, A. 2014c. A note on the etymology and lexicology relating to traditional European pulses in the Celtic languages. *Dialectologia et Geolinguistica* 22:123–130.

Mikić, A. 2014d. Diversity of the words denoting chickpea in modern European languages. *Legume Perspectives* 3:7.

Mikić, A. 2015a. All legumes are beautiful, but some legumes are more beautiful than others. *Legume Perspectives* 8:26–27.

Mikić, A. 2015b. Around the world in two centuries or why French serradella (Ornithopus sativus) should return to its Mediterranean homeland. *Legume Perspectives* 10:20–21.

Mikić, A. 2016. The most ancient words relating to pigeonpea (Cajanus cajan). *Legume Perspectives* 11:52–53.

Mikić, A., B. Ćupina, D. Rubiales et al. 2015a. Models, developments, and perspectives of mutual legume intercropping. *Advances in Agronomy* 130:337–419.

Mikić, A., V. Đorđević, and V. Perić. 2013a. Origin of the word soy. *Legume Perspectives* 1:5–6.

Mikić, A., A. Ignjatović-Ćupina, and B. Ćupina. 2010. Words denoting pea (Pisum sativum) in constructed languages. *Pisum Genetics* 42:39–40.

Mikić, A., A. Medović, Ž. Jovanović, and N. Stanisavljević. 2014a. Integrating archaeobotany, paleogenetics and historical linguistics may cast more light onto crop domestication: The case of pea (Pisum sativum). *Genetic Resources and Crop Evolution* 61:887–892.

Mikić, A., A. Medović, Ž. Jovanović, and N. Stanisavljević. 2015b. A note on the earliest distribution, cultivation and genetic changes in bitter vetch (Vicia ervilia) in ancient Europe. *Genetika Beograd* 47:1–11.

Mikić, A., and V. Mihailović. 2014a. Significance of genetic resources of cool season annual legumes. I. Crop wild relatives. *Ratarstvo i Povrtarstvo* 51:62–82.

Mikić, A., and V. Mihailović. 2014b. Significance of genetic resources of cool season annual legumes. II. Neglected and underutilised crops. *Ratarstvo i Povrtarstvo* 51:127–144.

Mikić, A., V. Mihailović, B. Ćupina et al. 2011. Achievements in breeding autumn-sown annual legumes for temperate regions with emphasis on the continental Balkans. *Euphytica* 180:57–67.

Mikić, A., V. Mihailović, B. Ćupina et al. 2014b. Forage yield components and classification of common vetch (Vicia sativa L.) cultivars of diverse geographic origin. *Grass and Forage Science* 69:315–322.

Mikić, A., V. Mihailović, M. Dimitrijević et al. 2013b. Evaluation of seed yield and seed yield components in red-yellow (Pisum fulvum) and Ethiopian (Pisum abyssinicum) peas. *Genetic Resources and Crop Evolution* 60:629–638.

Mikić, A., V. Mihailović, S. Vasiljević, S. Katanski, B. Milošević, and D. Živanov. 2016. Potential of Noë's vetch (Vicia noeana) for forage production. *Legumes and Groat Crops—Zernobobovye i krupyanye kul'tury* 1:52–56.

Mikić, A., and V. Perić. 2011. An etymological and lexicological note on the words for some ancient Eurasian grain legume crops in Turkic languages. *Turkish Journal of Field Crops* 16:179–182.

Mikić, A., and V. Perić. 2012. Origin and diversity of the words denoting some traditional Eurasian pulse crops in Mongolic and Tungusic. *Dialectologia et Geolinguistica* 20:63–70.

Mikić, A., and V. Perić. 2016. Origin of some scientific and popular names designating hyacinth bean (Lablab purpureus). *Legume Perspectives* 13:39–41.

Mikić, A., P. Smýkal, G. Kenicer et al. 2014c. Beauty will save the world, but will the world save beauty? The case of the highly endangered Vavilovia formosa (Stev.) Fed. *Planta* 240:1139–1146.

Mikić, A., and F. L. Stoddard. 2009. Legumes in the work of J. R. R. Tolkien. *Grain Legumes* 51:34.

Mikić, A., and F. L. Stoddard. 2013. Words for traditional Eurasian grain legumes in Uralic languages. *Dialectologia et Geolinguistica* 21:123–131.

Mikić, A., and M. Vishnyakova. 2012. A short survey on the lexicological continuum related to the most ancient Eurasian pulses in the North Caucasian languages. *Iran and the Caucasus* 16:217–223.

Mikić, A., L. Zorić, and B. Zlatković. 2015c. Origin of the binomial Linneaus nomenclature used to name some Old World legume species. Iheringia. *Série Botânica* 70:173–176.

Mikić, A. M. 2015c. The first attested extraction of ancient DNA in legumes (Fabaceae). *Frontiers in Plant Science* 6:1006.

Mikić-Vragolić, M., A. Mikić, B. Ćupina et al. 2007. Words related to some annual legumes in Slavic and other Indo-European languages. *Ratarstvo i povrtarstvo* 44:91–96.

Militarev, A. 2002. The prehistory of a dispersal: The Proto-Afrasian (Afroasiatic) farming Lexicon. In *Examining the Farming/Language Dispersal Hypothesis*, P. Bellwood and C. Renfrew (Eds.), pp. 135–150. Cambridge, UK: McDonald Institute for Archaeological Research.

Militarev, A., and O. Stolbova. 2007. *Afroasiatic Etymology*. The Tower of Babel, an International Etymological Database Project. http://starling.rinet.ru/ (accessed November 25, 2017).

Miller, R. A. 1996. *Languages and History: Japanese, Korean and Altaic*. Oslo, Norway: Institute for Comparative Research in Human Culture.

Mintz, S. W., and C. B. Tan. 2001. Bean-curd consumption in Hong Kong. *Ethnology* 40:113–128.

Molosiwa, O. O., C. Gwafila, J. Makore, and S. M. Chite. 2016. Phenotypic variation in cowpea (Vigna unguiculata [L.] Walp.) germplasm collection from Botswana. *International Journal of Biodiversity and Conservation* 8:153–163.

Morris, A. 1999. The agricultural base of the pre-Incan Andean civilizations. *Geographical Journal* 165:286–295.

Moseley, C. 2010. *Atlas of the World's Languages in Danger. Memory of Peoples*. Paris, France: UNESCO Publishing.

Mudaraddi, B., K. B. Saxena, R. K. Saxena, and R. K. Varshney. 2013. Molecular diversity among wild relatives of Cajanus cajan (L.) Millsp. *African Journal of Biotechnology* 12:3797–3801.

Mudrak, O. 2005. *Eskimo Etymology*. The Tower of Babel, an International Etymological Database Project. http://starling.rinet.ru/ (accessed November 25, 2017).

Mudrak, O. 2006. *Mongolian Etymology*. The Tower of Babel, an International Etymological Database Project. http://starling.rinet.ru/ (accessed November 25, 2017).

Myers, J. R., J. R. Baggett, and C. Lamborn. 2001. Origin, history, and genetic improvement of the snap pea. In *Plant Breeding Reviews 21*, J. Janick (Ed.), pp. 93–138. Toronto, Canada: John Wiley & Sons.

Native Plants Hawaii. 2009. *Vigna Marina*. Honolulu, HI: University of Hawaii. http://nativeplants.hawaii.edu/plant/view/Vigna_marina (accessed November 25, 2017).

Ngelenzi, M. J., S. Mwanarusi, and O. J. Otieno. 2016. Improving french bean (Phaseolus vulgaris L.) pod yield and auality through the use of different coloured agronet covers. *Sustainable Agriculture Research* 6:62–72.

Nikolayev, S. L. 2012. *Indo-European Etymology*. The Tower of Babel, an International Etymological Database Project. http://starling.rinet.ru (accessed November 25, 2017).

Nikolayev, S. L., and S. A. Starostin. 1994. *A North Caucasian Etymological Dictionary*. Moscow, Russia: Asterisk.

Nişanyan, S. 2017. *Sözlerin Soyağaci: Çağdaş Türkçenin Etimolojik Sözlüğü*. Istanbul, Turkey: Adam yay.

NPGS 2017. National Plant Germplasm System (NPGS). *Germplasm Resources Information Network*. Washington, DC: Agricultural Research Service, United States Department of Agriculture. http://www.ars-grin.gov/npgs/index.html (accessed November 25, 2017).

Oas, S. E., A. C. D'Andrea, and D. J. Watson. 2015. 10,000 year history of plant use at Bosumpra Cave, Ghana. *Vegetation History and Archaeobotany* 24:635–653.

Olson, K. S. 2004. An evaluation of Niger-Congo classification. Summer Institute of Linguistics (SIL) Electronic Working Papers 1–27.

Pagani, L., T. Kivisild, A. Tarekegn et al. 2012. Ethiopian genetic diversity reveals linguistic stratification and complex influences on the Ethiopian gene pool. *The American Journal of Human Genetics* 91:83–96.

Pagel, M., Q. D. Atkinson, A. S. Calude, and A. Meade. 2013. Ultraconserved words point to deep language ancestry across Eurasia. *Proceedings of the National Academy of Sciences USA*. 110:8471–8476.

Pasquet, R. S. 1999. Genetic relationships among subspecies of Vigna unguiculata (L.) Walp. *Theoretical and Applied Genetics* 98:1104–1119.

Patel, K. B. 2016. Studies on control of bird pests to damage millet crops in semiarid zone of northern Gujarat, India. *Journal of Environmental Research and Development* 11:132–141.

Pawley, A. 2012. How reconstructable is Proto Trans New Guinea? Problems, progress, prospects. *Language and Linguistics in Melanesia* 1:88–164.

Peiros, I. 2004. *A Genetic Classification of Austroasiatic Languages*. Moscow, Russia: Russian State University for the Humanities.

Peiros, I. 2005. *Austro-Asiatic Etymology*. The Tower of Babel, an International Etymological Database Project. http://starling.rinet.ru (accessed November 25, 2017).

Peiros, I. 2009. *Tai-Kadai Etymology*. The Tower of Babel, an International Etymological Database Project. http://starling.rinet.ru (accessed November 25, 2017).

Peiros, I., and S. Starostin. 2005. *Austric Etymology*. The Tower of Babel, an International Etymological Database Project. http://starling.rinet.ru (accessed November 25, 2017).

Pengelly, B. C., and B. L. Maass. 2001. Lablab purpureus (L.) Sweet-diversity, potential use and determination of a core collection of this multi-purpose tropical legume. *Genetic Resources and Crop Evolution* 48:261–272.

Pennington, C. W. 1981. *Arte y vocabulario de la lengua dohema, heve o eudeva*. Mexico City, Mexico: Universidad Nacional Autónoma de México.

Piperno, D. R., and T. D. Dillehay. 2008. Starch grains on human teeth reveal early broad crop diet in northern Peru. *Proceedings of the National Academy of Sciences of the USA* 105:19622–19627.

Pokorny, J. 1959. *Indogermanisches Etymologisches Wörterbuch 1*. Bern, Switzerland: Francke.

Polese, J. M. 2006. *La culture des haricots et des pois*. Chamalières, France: Editions Artemis.

Porcher, M. H. 2008. *Multilingual Multiscript Plant Name Database*. Melbourne, Australia: The University of Melbourne.

Potokina, E., N. Tomooka, D. A. Vaughan, T. Alexandrova, and R. Q. Xu. 1999. Phylogeny of Vicia subgenus Vicia (Fabaceae) based on analysis of RAPDs and RFLP of PCR-amplified chloroplast genes. *Genetic Resources and Crop Evolution* 46:149–161.

Powell, J. W. 1891. *Indian Linguistic Families of America North of Mexico, Seventh Annual Report*. Washington, DC: Bureau of American Ethnology, Government Printing Office.

Price-Williams, D. 2016. Altjira, dream and god. In *Religion and Non-religion Among Australian Aboriginal Peoples*, J. L. Cox and A. Possamai (Eds.), pp. 85–108. New York: Routledge.

Quattrocchi, U. 2012. CRC *World Dictionary of Medicinal and Poisonous Plants: Common Names, Scientific Names, Eponyms, Synonyms, and Etymology*. Boca Raton, FL: CRC Press.

Quiroga, R., L. Meneses, and R. W. Bussmann. 2012. Medicinal ethnobotany in Huacareta (Chuquisaca, Bolivia). *Journal of Ethnobiology and Ethnomedicine* 8:29.

Raes, N., L. G. Saw, P. C. van Welzen, and T. Yahara. 2013. Legume diversity as indicator for botanical diversity on Sundaland, South East Asia. *South African Journal of Botany* 89:265–272.

Rahmianna, A. A., and B. S. Radjit. 2000. Common beans: Their potential and opportunities in Indonesia. *Edisi khusus Balitkabi* 16.: 160–168.

Ramesar, R. 2015. Genomics: African dawn. *Nature* 517:276–277.

Ratliff, M. S. 2010. *Hmong-Mien Language History 613*. Canberra, Australia: Pacific Linguistics.

Ray, J. 1686. *Historia Plantarum 1*. London, UK: Mary Clark.

Redden, R. J., and J. D. Berger. 2007. History and origin of chickpea. In *Chickpea Breeding and Management*, S. S. Yadav, R. J. Redden, W. Chen, and B. Sharma (Eds.), pp. 1–13. Wallingford, CT: CAB International.

Reddy, M. R. K., R. Rathour, N. Kumar, P. Katoch, and T. R. Sharma. 2010. Cross-genera legume SSR markers for analysis of genetic diversity in Lens species. *Plant Breeding* 129:514–518.

Rehm, S. 1994. *Multilingual Dictionary of Agronomic Plants*. Dordrecht, the Netherlands: Springer.

Ringe, D. 2006. *From Proto-Indo-European to Proto-Germanic*. Oxford, UK: Oxford University Press.

Ringe, D., A. Warnow, and A. Taylor. 2002. Indo-European and computational cladistics. *Transactions of the Philological Society* 100:59–129.

Rippke, U., J. Ramirez-Villegas, A. Jarvis et al. 2016. Timescales of transformational climate change adaptation in sub-Saharan African agriculture. *Nature Climate Change* 6:605–609.

Ross, M. 2009. Proto Austronesian verbal morphology: A reappraisal. In *Austronesian Historical Linguistics and Culture History: A Festschrift for Robert Blust*, A. Adelaar and A. Pawley (Eds.), pp. 295–326. Canberra, Australia: Pacific Linguistics.

Rubiales, D., and A. Mikić. 2015. Introduction: Legumes in sustainable agriculture. *Critical Reviews in Plant Sciences* 34:2–3.

Ruhlen, M. 1991. *A Guide to the World's Languages: Classification*. Stanford, CA: Stanford University Press.

Ruhlen, M. 2001. Dene-Caucasian: A new linguistic family. *Pluriverso* 2:76–85.

Ruhlen, M. 1994. *The Origin of Language: Tracing the Evolution of the Mother Tongue*. New York: John Wiley & Sons.

Sagart, L. 2016. The wider connections of Austronesian. *Diachronica* 33:255–281.

Salo, D. 1999. *I Lauki: A Qenya Botany*. Ardalambion, Kjerrgarden. https://folk.uib.no/hnohf/ (accessed November 25, 2017).

Schaaffhausen, R. V. 1963. Dolichos lablab or hyacinth bean. *Economic Botany* 17:46–153.

Schaefer, H., P. Hechenleitner, A. Santos-Guerra et al. 2012. Systematics, biogeography, and character evolution of the legume tribe Fabeae with special focus on the middle-Atlantic island lineages. *BMC Evolutionary Biology* 12:250.

Sedivy, E. J., F. Wu, and Y. Hanzawa. 2017. Soybean domestication: The origin, genetic architecture and molecular bases. *New Phytologist* 214:539–553.

Shahwar, D., T. M. Bhat, M. Y. K. Ansari, S. Chaudhary, and R. Aslam. 2017. Health functional compounds of lentil (Lens culinaris Medik): A review. *International Journal of Food Properties* 20:1–15.

Shapiro, J. 1987. From Tupã to the land without evil: The christianization of Tupi-Guarani cosmology. *American Ethnologist* 1:126–139.

Shiraishi, H., M. Mizushima, and O. Kovan 2016. *Sound Materials of the Nivkh Language 13 (Schmidt Dialect)—Ol'ga Borisovna Kovan*. Sapporo, Japan: I-Word Printing.

Shiran, B., S. Kiani, D. Sehgal, A. Hafizi, M. Chaudhary, and S. N. Raina. 2014. Internal transcribed spacer sequences of nuclear ribosomal DNA resolving complex taxonomic history in the genus Vicia L. *Genetic Resources and Crop Evolution* 61:909–925.

Shurtleff, W., and A. Aoyagi. 2004. *The Soybean Plant: Botany, Nomenclature, Taxonomy, Domestication, and Dissemination*. Lafayette, CA: Soyinfo Center. http://www.soyinfocenter.com/ (accessed November 25, 2017).

Sicoli, M. A., and G. Holton. 2014. Linguistic phylogenies support back-migration from Beringia to Asia. *PLoS One* 9:e91722.

Sidwell, P., and R. Blench 2011. The Austroasiatic Urheimat: The Southeastern riverine hypothesis. In *Dynamics of Human Diversity*, N. J. Enfield (Ed.), pp. 317–345. Canberra, Australia: Pacific Linguistics.

Simon, M. V., A. M. Benko-Iseppon, L. V. Resende, P. Winter, and G. Kahl. 2007. Genetic diversity and phylogenetic relationships in Vigna Savi germplasm revealed by DNA amplification fingerprinting. *Genome* 50:538–547.

Singh, B. B. 2014. Botany and physiology. In *Cowpea: The Food Legume of the 21st Century*, B. B. Singh (Ed.), pp. 17–32. Madison, WI: Crop Science Society of America.

Sinjushin, A. A., and A. S. Belyakova. 2010. On intraspecific variation of Vavilovia formosa (Stev.) Fed. (= Pisum formosum (Stev.) Alef.: Fabeae). *Pisum Genetics* 42:31–34.

Sinjushin, A. A., and N. V. Demidenko. 2010. Vavilovia formosa (Stev.) Fed. (Fabeae, Fabaceae) on Meyen's "panel with a multitude of lamps." *Wulfenia* 17:45–57.

Smartt, J. 1985. Evolution of grain legumes. II. Old and new world pulses of lesser economic importance. *Experimental Agriculture* 21:1–18.

Smith, B. 2017. The last hunter-gatherers of China and Africa: A life amongst pastoralists and farmers. *Quaternary International*. doi:10.1016/j.quaint.2016.11.035.

Smýkal, P., C. J. Coyne, M. J. Ambrose et al. 2015. Legume crops phylogeny and genetic diversity for science and breeding. *Critical Reviews in Plant Sciences* 34:43–104.

Smýkal, P., Ž. Jovanović, N. Stanisavljević et al. 2014. A comparative study of ancient DNA isolated from charred pea (Pisum sativum L.) seeds from an Early Iron Age settlement in southeast Serbia: Inference for pea domestication. *Genetic Resources and Crop Evolution* 61:1533–1544.

Sonari. A, V. Manju, and K. Ashwin. 2015. Comparative study of Indian varieties of Lablab and field bean for phenotypic and nutritional traits. *Legume Genomics and Genetics* 6:1–7.

Sonnante, G., K. Hammer, and D. Pignone, 2009. From the cradle of agriculture a handful of lentils: History of domestication. *Rendiconti Lincei* 20:21–37.

Southworth, F. C. 2004. *Linguistic Archaeology of South Asia*. New York: Routledge.

Southworth, F. C. 2006. Proto-Dravidian agriculture. *Proceedings of the Pre-symposium of Rihn and 7th ESCA Harvard-Kyoto Roundtable*. Kyoto, Japan: Research Institute for Humanity and Nature.

Spataro, G., B. Tiranti, P. Arcaleni et al. 2011. Genetic diversity and structure of a worlwide collection of Phaseolus coccineus L. *Theoretical and Applied Genetics* 122:1281–1291.

Srithawong, S., M. Srikummool, P. Pittayaporn et al. 2015. Genetic and linguistic correlation of the Kra-Dai-speaking groups in Thailand. *Journal of Human Genetics* 60:371–380.

Starostin, G. S. 2006. *Dravidian Etymology*. The Tower of Babel, an International Etymological Database Project. http://starling.rinet.ru (accessed November 25, 2017).

Starostin, G. S. 2008. *Macro-Khoisan Etymology*. The Tower of Babel, an International Etymological Database Project. http://starling.rinet.ru (accessed November 25, 2017).

Starostin, G. S. 2009. On the origins of the three-way phonological distinction in Dravidian coronal consonants. *Aspects of Comparative Linguistics* 4:243–261.

Starostin, G. S. 2016. The Nilo-Saharan hypothesis tested through lexicostatistics: Current state of affairs. Draft 1.1, Academia.edu.

Starostin, G. S. 2017. Lexicostatistical Studies in East Sudanic I: On the genetic unity of Nubian-Nara-Tama. *Journal of Language Relationship* 15:87–113.

Starostin, G. *The Tower of Babel, Evolution of Human Language Project*. http://starling.rinet.ru.

Starostin, S. 2003a. *Abkhaz-Adyghe Etymology*. The Tower of Babel, an International Etymological Database Project. http://starling.rinet.ru (accessed November 25, 2017).

Starostin, S. 2003b. *Avar-Andian Etymology*. The Tower of Babel, an International Etymological Database Project. http://starling.rinet.ru (accessed November 25, 2017).

Starostin, S. 2005a. *Burushaski Etymology*. The Tower of Babel, an International Etymological Database Project. http://starling.rinet.ru (accessed November 25, 2017).

Starostin, S. 2005b. *Kartvelian Etymology*. The Tower of Babel, an International Etymological Database Project. http://starling.rinet.ru (accessed November 25, 2017).

Starostin, S. 2005c. *Kiranti Etymology*. The Tower of Babel, an International Etymological Database Project. http://starling.rinet.ru (accessed November 25, 2017).

Starostin, S. 2005d. *North Caucasian Etymology*. The Tower of Babel, an International Etymological Database Project. http://starling.rinet.ru (accessed November 25, 2017).

Starostin, S. 2005e. *Sino-Tibetan Etymology*. The Tower of Babel, an International Etymological Database Project. http://starling.rinet.ru (accessed November 25, 2017).

Starostin, S. 2005f. *Uralic Etymology*. The Tower of Babel, an International Etymological Database Project. http://starling.rinet.ru (accessed November 25, 2017).

Starostin, S. 2005g. *Yenisseian Etymology*. The Tower of Babel, an International Etymological Database Project. http://starling.rinet.ru (accessed November 25, 2017).

Starostin, S. 2006a. *Altaic Etymology*. The Tower of Babel, an International Etymological Database Project. http://starling.rinet.ru (accessed November 25, 2017).

Starostin, S. 2006b. *Japanese Etymology*. The Tower of Babel, an International Etymological Database Project. http://starling.rinet.ru (accessed November 25, 2017).

Starostin, S. 2006c. *Korean Etymology*. The Tower of Babel, an International Etymological Database Project. http://starling.rinet.ru (accessed November 25, 2017).

Starostin, S. 2006d. *Long-Range Etymologies*. The Tower of Babel, an International Etymological Database Project. http://starling.rinet.ru (accessed November 25, 2017).

Starostin, S. 2015. *Sino-Caucasian Etymology*. The Tower of Babel, an International Etymological Database Project. http://starling.rinet.ru (accessed November 25, 2017).

Starostin, S., Dybo, A., and O. Mudrak. 2003. *Etymological Dictionary of the Altaic Languages*. Brill, Leiden. http://starling.rinet.ru (accessed November 25, 2017).

Sterndale-Bennett, J. 2005. *Plant Names Explained: Botanic Names and their Meaning*. Newton Abbot, UK: David & Charles.

Stewart, J. M. 2002. The potential of Proto-Potou-Akanic-Bantu as a pilot Proto-Niger-Congo, and the reconstructions updated. *Journal of African Languages and Linguistics* 23:197–224.

Stoddard, F. L., S. Hovinen, M. Kontturi, K. Lindström, and A. Nykänen. 2009. Legumes in Finnish agriculture: History, present status and future prospects. *Agricultural and Food Science* 18(3–4):191–205.

Stolbova, O. 2006. *West Chadic Etymology*. The Tower of Babel, an International Etymological Database Project. http://starling.rinet.ru (accessed November 25, 2017).

Tanno, K. I., and G. Willcox. 2006. The origins of cultivation of Cicer arietinum L. and Vicia faba L.: Early finds from Tell el-Kerkh, north-west Syria, late 10th millennium BP. *Vegetation History and Archaeobotany* 15:197–204.

The Plant List. 2013. *The Plant List—A Working List of All Plant Species*. Royal Botanic Gardens, Kew/Missouri Botanical Garden, St. Louis. http://www.theplantlist.org (accessed November 25, 2017).

Tomooka, N., A. Kaga, T. Isemura, and D. Vaughanet. 2011. Vigna. In *Wild Crop Relatives: Genomic and Breeding Resources, Legume and Forages*, C. Kole (Ed.), pp. 291–311. Berlin, Germany: Springer.

Trask, L. 1997. *The History of Basque*. New York: Routledge.

Turner, N. C., G. C. Wright, and K. H. M. Siddique. 2001. Adaptation of grain legumes (pulses) to water-limited environments. *Advances in Agronomy* 71:193–231.

Turner, R. L. 1962–1966. *A Comparative Dictionary of Indo-Aryan Languages*. London, UK: Oxford University Press.

Ungureanu, D. 2014. Loanwords and substrate in romance languages. *Zeitschrift für Balkanologie* 51:127–144.

Van de Wouw, M., N. Maxted, K. Chabane, and B. V. Ford-Lloyd. 2001. Molecular taxonomy of Vicia ser. Vicia based on amplified fragment length polymorphisms. *Plant Systematics and Evolution* 229:91–105.

van der Maesen L. J. G. 1990. Pigeon pea: Origin, history, evolution, and taxonomy. In *The Pigeonpea*, Y. L. Nene, S. D. Hall, and V. K. Sheila (Eds.), pp. 15–46. Oxon, UK: CAB International.

van der Maesen L. J. G., N. Maxted, F. Javadi, S. Coles, and A. M. R. Davies. 2007. Taxonomy of the genus Cicer revisited. In *Chickpea Breeding and Management*, S. S. Yadav, R. J. Redden, W. Chen, and B. Sharma (Eds.), pp. 14–46. Wallingford, CT: CAB International.

van der Maesen, L. J. G., P. Remanandan, N. K. Rao, and R. P. S. Pundir. 1986. Occurrence of Cajaninae in the Indian subcontinent, Burma and Thailand. *Journal of Bombay Natural History Society* 82:489–500.

van der Maesen, L. J. G. 1998. *Revision of the genus Dunbaria Wight & Arn. (Leguminosae-Papilionoideae)*. Wageningen, the Netherlands: Wageningen Agricultural University.

van der Maesen, L. J. G. 2003. Cajaninae of Australia (Leguminosae—Papilionoideae). *Australian Systematic Botany* 16:219–227.

van der Maesen, L. J. G. 1995. Pigeonpea cajanus cajan. In *Evolution of Crop Plants*, J. Smartt and N. W. Simmonds (Eds.), pp. 251–255. Essex, UK: Longman.

van der Mey, J. A. 1996. Crop development of Lupinus species in Africa. *South African Journal of Science* 92:53–56.

van Driem, G. 1993. Language change, conjugational morphology and the Sino-Tibetan urheimat. *Acta Linguistica Hafniensia* 26:45–56.

van Driem, G. 2001. *The Languages of the Himalayas*. Leiden, the Netherlands: Brill.

Van Driem, G. 2005. *Limbu Dictionary*. The Tower of Babel, an International Etymological Database Project. http://starling.rinet.ru (accessed November 25, 2017).

Vasmer, M. 1953. *Russisches Etymologisches Wörterbuch*, 1 (A—K). Heidelberg, Germany: Carl Winters Universitätsverlag.

Vasmer, M. 1955. *Russisches Etymologisches Wörterbuch*, 2 (L—Ssuda). Heidelberg, Germany: Carl Winters Universitätsverlag.

Vasmer, M. 1958. *Russisches Etymologisches Wörterbuch*, 3 (Sta—Y). Heidelberg, Germany: Carl Winters Universitätsverlag.

Vaz Patto, M. C., R. Amarowicz, A. N. A. Aryee et al. 2015. Achievements and challenges in improving the nutritional quality of food legumes. *Critical Reviews in Plant Sciences* 34:105–143.

Vijaykumar, A., A. Saini, and N. Jawali. 2009. Phylogenetic analysis of subgenus Vigna species using nuclear ribosomal RNA ITS: Evidence of hybridization among Vigna unguiculata subspecies. *Journal of Heredity* 101:177–188.

Vishnyakova, M., M. Burlyaeva, J. Akopian, R. Murtazaliev, and A. Mikić. 2016. Reviewing and updating the detected locations of beautiful vavilovia (Vavilovia formosa) on the Caucasus sensu stricto. *Genetic Resources and Crop Evolution* 63:1085–1102.

Voisin, A. S., J. Guéguen, C. Huyghe et al. 2014. Legumes for feed, food, biomaterials and bioenergy in Europe: A review. *Agronomy for Sustainable Development* 34:361–380.

Vujaklija, M. 1980. *Leksikon stranih reči i izraza*. Belgrade, Serbia: Prosveta.

Waldman, K. B., D. L. Ortega, R. B. Richardson, and S. S. Snapp. 2017. Estimating demand for perennial pigeon pea in Malawi using choice experiments. *Ecological Economics* 131:222–230.

Wang, M. L., J. B. Morris, N. A. Barkley, R. E. Dean, T. M. Jenkins, and G. A. Pederson. 2007. Evaluation of genetic diversity of the USDA Lablab purpureus germplasm collection using simple sequence repeat markers. *The Journal of Horticultural Science and Biotechnology* 82:571–578.

Warkentin, T. 2014. A meeting with pulse beating. *Legume Perspectives* 7:4.

Welch, R. M., and R. D. Graham. 1999. A new paradigm for world agriculture: Meeting human needs: Productive, sustainable, nutritious. *Field Crops Research* 60:1–10.

Westengen, O. T., M. A. Okongo, L. Onek et al. 2014. Ethnolinguistic structuring of sorghum genetic diversity in Africa and the role of local seed systems. *Proceedings of the National Academy of Sciences of the USA* 111:14100–14105.

Westphal, E. 1975. The proposed retypification of Dolichos L.: A review. *Taxon* 24:189–192.

White, C. E. 2013. The emergence and intensification of cultivation practices at the Pre-pottery Neolithic site of el-Hemmeh, Jordan: An archaeobotanical study. *PhD. Graduate School of Arts and Sciences*, Boston, MA: Boston University.

Whitehouse, P., T. Usher, M. Ruhlen, and W. S. Y. Wang. 2004. Kusunda: An Indo-Pacific language in Nepal. *Proceedings of the National Academy of Sciences of the United States of America* 101:5692–5695.

Wiessner, P. W. 2014. Embers of society: Firelight talk among the Ju/'hoansi Bushmen. *Proceedings of the National Academy of Sciences of the USA* 111:14027–14035.

Wikipedia. 2017. *Wikipedia, the Free Encyclopedia*. Wikimedia Foundation. https://www.wikipedia.org/ (accessed November 25, 2017).

Wiktionary. 2017. Wiktionary, the free dictionary. https://www.wiktionary.org/ (accessed November 25, 2017).

Williamson, K., and R. Blench. 2000. Niger-Congo. In *African Languages: An Introduction*, B. Heine and D. Nurse (Eds.), pp. 11–42. Cambridge, UK: Cambridge University Press.

Winter, W. 1998. Tocharian. In *The Indo-European Languages*, A. Ramat Giacalone and P. Ramat (Eds.), pp. 154–168. London, UK: Routledge.

Winters, C. 2014. African and dravidian origins of the melanesians. *Indian Journal of Fundamental and Applied Life Sciences* 4:694–704.

Wisconsin State Herbarium. 2017. Lathyrus latifolius. Flora of Wisconsin. Madison/Steven's Point: University of Wisconsin. http://wisflora.herbarium.wisc.edu/ (accessed November 25, 2017).

Wright, R. P. 2009. *The Ancient Indus: Urbanism, Economy, and Society*. Cambridge, UK: Cambridge University Press.

Wuethrich, B. 2000. Peering into the past, with words. *Science* 288:1158.

Wurm, S. A. 1982. *The Papuan Languages of Oceania*. Tübingen, Germany: Gunter Narr.

Xie, P., X. Y. Jiang, Z. Bu, S. Y. Fu, S. Y. Zhang, and Q. P. Tang. 2016. Free choice feeding of whole grains in meat-type pigeons: 1. effect on performance, carcass traits and organ development. *British Poultry Science* 57:699–706.

Yeakel, J. D., B. D. Patterson, K. Fox-Dobbs et al. 2009. Cooperation and individuality among man-eating lions. *Proceedings of the National Academy of Sciences USA* 106:19040–19043.

Zander, R., F. Encke, G. Buchheim, and S. Seybold. 1993. *Handwörterbuch der Pflanzennamen*. Stuttgart, Germany: Ulmer Verlag.

Zepeda, O., and J. H. Hill. 1991. The condition of native American languages in the United States. In *Endangered Languages*, R. H. Robins and E. M. Uhlenbeck (Eds.), pp. 135–155. Oxford, UK: Berg.

Zeven, A. C., and P. M. Zhukovsky. 1975. *Dictionary of Cultivated Plants and their Centres of Diversity*. Wageningen, the Netherlands: Centre for Agricultural Publishing and Documentation.

Zhukovsky, P. M. 1929. A contribution to the knowledge of genus Lupinus Tourn. *Bulletin of Applied Botany, of Genetics and Plant Breeding* 11:16–294.

Zlatković, B., A. Mikić, and P. Smýkal. 2010. Distribution and new records of Pisum sativum subsp. elatius in Serbia. *Pisum Genetics* 42:15–17.

Zohary, D., M. Hopf, and E. Weiss. 2012. *Domestication of Plants in the Old World: The Origin and Spread of Domesticated plants in Southwest Asia, Europe, and the Mediterranean Basin*. Oxford, UK: Oxford University Press.

Zorić, L., A. Mikić, S. Antanasović, D. Karanović, B. Ćupina, and J. Luković. 2015. Stem anatomy of annual legume intercropping components: White lupin (Lupinus albus L.), narbonne (Vicia narbonensis L.) and common (Vicia sativa L.) vetches. *Agricultural and Food Science* 24:139–149.

Index

Note: Page numbers followed by f and t refer to figures and tables respectively.

D

E

F

N

O

P

S

T

U

V

W

Y

PGMO 06/29/2018